A. Teramoto, M. Kobayashi, T. Norisuye (Eds.)

Ordering in Macromolecular Systems

Proceedings of the OUMS'93
Toyonaka, Osaka, Japan, 3-6 June 1993

With 250 Figures and 20 Tables

Springer-Verlag
Berlin Heidelberg New York
London Paris Tokyo
Hong Kong Barcelona Budapest

Professor Akio Teramoto
Professor Masamichi Kobayashi
Professor Takashi Norisuye

Department of Macromolecular Science
Osaka University
Faculty of Science
Toyonaka, 560 Osaka, Japan

ISBN-13:978-3-642-78895-6   e-ISBN-13:978-3-642-78893-2
DOI: 10.1007/978-3-642-78893-2

Library of Congress Cataloging-in-Publication Data
OUMS'93 (1993: Osaka, Japan) Ordering in macromolecular systems: proceedings of the
OUMS' 93, Toyonaka, Osaka, Japan, 3 - 6 June 1993 / A. Teramoto, M. Kobayashi,
T. Norisuye (eds.)   p. cm.   Includes bibliographical references and index.
ISBN-13:978-3-642-78895-6
1. Polymers-Structure-Congresses. 2. Crystalline polymers-Congresses. 3. Polymer liquid
crystals-Congresses. I. Teramoto, A. 1933- . II. Kobayashi, M. (Masamichi), 1933- .
III. Norisuye, T. (Takashu), 1943- . IV. Osaka Daigaku. V. Title.
QD381.9.S87096 1993   547.7'0442--dc20   94-8717 CIP

© Springer-Verlag Berlin Heidelberg 1994
Softcover reprint of the hardcover 1st edition 1994

Typesetting: Camera ready by author
SPIN: 10127456   02/3020 - 5 4 3 2 1 0 - Printed on acid-free paper

Preface

This volume summarizes the papers presented at the First Osaka University Macromolecular Symposium OUMS'93 on "Ordering in Macromolecular Systems", which was held at Senri Life Science Center, Osaka, Japan, on June 3 through June 6, 1993. The symposium covered the three topics, (1) Crystallization and Phase Transitions, (2) Polymer Liquid Crystals and (3) Block Copolymers, Polymer Blends and Surfaces, and invited leading scientists in these fields. At present any of these topics is a hot issue in itself and frequently taken up separately in many occasions. It is noted however that all these topics are correlated with each other with the keyword "*Ordering*" and their combination provides a unique feature of the present symposium in reflecting the interactions among investigators working in these important fields with the common ground expressed by the keyword "Ordering". Nineteen invited lectures and 40 posters of both experiment and theory were presented at the symposium, and the eighteen lectures and ten poster presentations contribute to this volume.

In the first topic crystal structures and their transitions were discussed from kinetic as well as static points of view; attention was paid to give a molecular-level interpretation of the structure, phase transition and physical properties, using theories and simulations. The second topic was mainly concerned with static structures and thermodynamic properties of polymer liquid crystals including phase behaviours. It was shown that the structures and properties related to lyotropic liquid crystals are explained theoretically on the basis of simple molecular models; here were discussed not only nematic but also columnar and smectic phases. The final topic was concerned with various microscopic as well as macroscopic structures formed with block copolymers and polymer blends, which exhibit unique features in bulk and at surface. In all these topics both general insight and particular issues were included. The program of the symposium was arranged so that the three topics were properly mixed, encouraging discussions among these slightly different fields and with a particular emphasis on the interplay between experiment and theory. In some cases they are explained well theoretically, and in other cases either experiment or theory needs to be followed by the other. The papers contributed to this volume also should reflect the outcome from such interactions at the meeting. Thus we believe this volume will be useful for specialists working in these related fields on one hand, and serve as a reference book for those who wish to get familiar with these fields on the other hand.

The present symposium was sponsored by the Osaka University International Exchange Program in Macromolecular Science. Thus a brief account of this program may be pertinent here. In 1989 the Department of Macromolecular Science, Osaka University, celebrated the 30th anniversary and has founded in the Macromolecular Science Course, the Graduate School of Science, Osaka University, the Osaka University International Exchange Program in Macromolecular Science, with the generous support of many companies and individuals. The main purpose of the Program is to promote the international exchange in macromolecular science through international symposium and personnel exchange. The OUMS'93 symposium is the first of such symposiums. This symposium was supported in part by the Society of Polymer Science, Japan, to which acknowledgement is made.

January 1994

Akio Teramoto
Masamichi Kobayashi
Takashi Norisuye

Contents

A Unifying Scheme for Polymer Crystallization Based on Recent Experiments with Wider Implications for Phase Transformations

A. Keller, M. Hikosaka[†], S. Rastogi[††], A. Toda[†††] and P.J. Barham

H.H. Wills Physics Laboratory, University of Bristol
Tyndall Avenue, Bristol BS8 1TL

Abstract: This lecture is aimed to link the "main stream" subject of chain folded polymer crystallization to the "speciality stream" of extended chain crystallization, the latter as typified by the crystallization of polyethylene under pressure. This is achieved through a scheme based on some new experimental material comprising the recognition of thickening growth as a primary growth process of lamellae and of the prominence of metastable phases, specifically of the mobile hexagonal phase in polyethylene. The scheme relies on the consideration of crystal size as a stability determining factor, namely on melting point depression, which in general is different for different polymorphs. It is shown that under specificable conditions phase stabilities can invert with size, i.e. a phase which is metastable for infinite size can become the stable one when the phase is sufficiently small. When applying this condition to crystal growth it follows that a crystal in such a situation will appear and grow in a phase that is different from that in its state of ultimate stability, maintaining this state as a metastable one or transforming into the ultimate stable state during growth according to circumstances. The consequences of such deliberations, of potential significance to all phase transformations also beyond polymer crystallization, are being developed throughout the paper.

INTRODUCTION

The material of the present paper has emerged from varied and far ranging experimental works on polymer crystallization as initiated by one of us (MH) in Tokyo and currently pursued in Yamagata, subsequently extended in association with the Bristol Polymer Laboratory. While the individual works are being reported separately elsewhere (1-4) the experimental results have led to the

---

†   Prermanent address: Department of Materials Science, Yamagata University, Yonezawa, Japan

††   Permanent address: Department of Polymer Technology, Eindhoven Technical University, Eindhoven, The Netherlands

††† Permanent address: Department of Physics, Kyoto University, Kyoto, Japan

A. Teramoto. M. Kobayashi, T. Norisuje (Eds.)
Ordering in Macromolecular Systems

recognition of some new themes which promise to be of general relevance to polymer crystallization, to crystallization beyond the polymer field, and even to wider issues of phase transformation in general. It is these themes which we are trying to convey in the present paper with some references to the underlying works themselves

In broadest generality two factors have emerged in the course of the experimental works which have controlling influence on the crystallization process in those experiments. In the first instance metastable phases as the primary products of crystallisation, and secondly, the size dependence of phase stability. Neither of these are new in themselves yet their combination leads to new considerations which are the subject of the present paper. The role of metastability in phase transformations was recognised in the last century and is embodied in Ostwald's Rule of Stage (5). The role of phase sizes is of course familiar in the form of melting point, boiling point etc. depressions due to limited phase dimensions as expressed quantitatively by the Thomson-Gibbs equation. Such a phase of small dimension is metastable with respect to the one of infinite size. In the scheme to be developed in this paper we invoke a combination of the two kinds of metastability: that determined by phase type (e.g. different polymorphs of the same crystal phase) and that by the limited dimension of a given phase. As will be seen this will prove to be particularly helpful in envisaging the process of phase transformations in terms of phase type of competing stability.

In the case of polymer crystallization in particular it provides a unifying umbrella for two, so far largely separate areas of the subject: chain folded crystallization, the most widely studied form of crystallization here to be termed "main stream", and extended chain type crystallization, in the case of polyethylene (PE) - the model substance for such studies - arising at elevated pressure, here to be termed "speciality stream".

SOME BASIC FACTS

The factual material leading to the new perspectives to be presented fall into the two groups already stated: metastability and lamellar thickness. Taking lamellar thickness first it will be recalled that in the "main stream" subject of chain folded crystallization the crystals are thin platelets containing the chains in a folded conformation with the lamellar thickness corresponding (or closely related) to the fold length (6). The lamellae being uniformly thick crystal growth is envisaged as occurring only in the lateral direction with unaltered thickness, the corresponding growth process being the basis of present day theories of chain folded crystal growth. The lamellae can also increase their thickness either on heating, subsequent to the formation of the crystals, (annealing), or on storage at the crystallization temperature itself (isothermal thickening), in both cases corresponding to secondary processes of perfectioning of crystals already formed.

In the "speciality stream" of extended chain type crystallization (7, 8) the crystals can reach full chain extension which from the earliest days, was envisaged as arising from extension of initially chain folded crystals, which in turn can arise when the chains within the crystal are sufficiently mobile. The latter will be the case for the hexagonal phase of polyethylene, realisable under high hydrostatic pressure, hence the role of pressure in this type of experimentation (9) (for schematic PT phase diagram see Figure 5 further below). In past works, however, attainment of chain extension and the associated "thick" extended chain type lamellae has been envisaged, explicitly or implicitly, as the product of thickening subsequent to primary crystal growth.

It was thanks to latest developments that growth in the lamellar thickness direction was recognised as a primary growth process, i.e. growth as associated with formation of more crystal from the still uncrystallised matter, as opposed to the secondary process of perfectioning of crystals already present. Such "thickening growth" could be identified and assessed quantitatively from the cross-sectional view of isolated crystals such as in Figure 1. Here, in the knowledge of the linear growth rate (V), which can be assessed by in situ

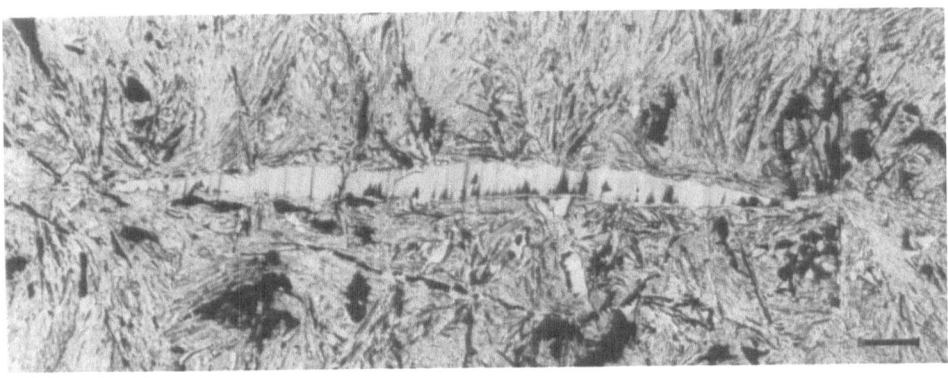

Fig. 1: Cross section of an isolated extended chain type lamellar crystal of polyethylene as revealed in replicas of exposed internal surfaces of pressure quenched samples by transmission electron microscopy. Note tapering cross section a result of unrestricted "thickening growth". This crystal was growing at P=4kbar and $(\Delta T)^h$=4°C. Scale bar = 3.5μm (ref. 2)

polarizing microscopy (1), the actual thickness (ℓ) can be mapped as a function of time (t) from the cross-sectional profile. This yields plots of ℓ v. t such as in Figure 3 giving the primary thickening growth rate U. directly. The distinction between (primary) "thickening growth" and the previously considered (secondary) "thickening" is basic. Morphologically, "thickening growth" corresponds to unrestricted growth in the thickness direction as the lamella increases its volume at the expense of the surrounding melt, as displayed by a layer in isolation such as the crystal in Figure 1, a process which ceases as the growing layers impinge along their basal planes (Fig. 2). At this latter stage lamellar "thickening", if and when it occurs, involves redistribution of matter within the overall crystallized volume leading to reduction of interface. The fundamentally different nature of the two processes will be apparent from Figure 4. As seen, not only is the dependence of the two thickening rates, U and W, on supercooling (ΔT) different functionally but is actually opposite in sign, U increasing and W decreasing with ΔT, with the underlying structural picture being represented schematically by the inserts in Figure 4. Nevertheless, there is one common feature on the molecular level: in both cases the molecules must be able to move along the chain direction within the lattice: in the case of "thickening growth" they must be able to thread

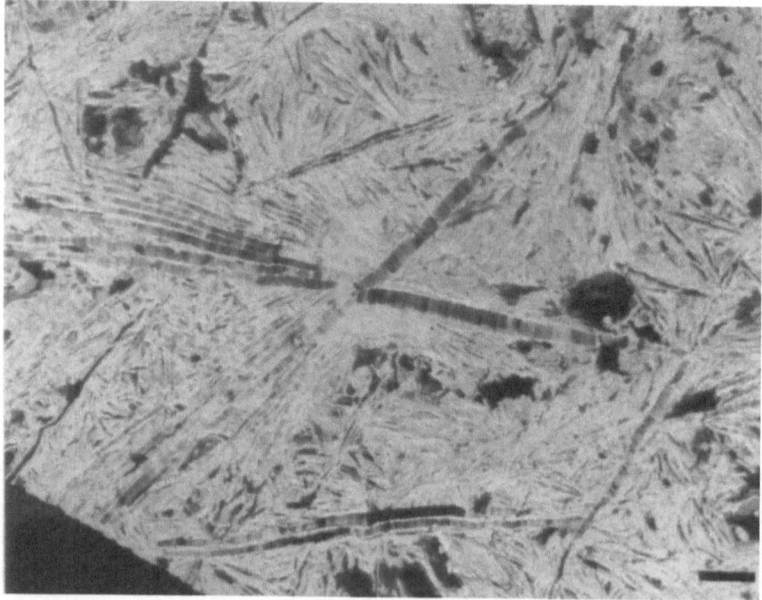

Fig. 2: Preparation similar to Fig. 1 with crystals grown at P-4Kbar and $(\Delta T)_h=7°C$. Note the pronounced multilayering (left side) where primary "thickening growth" has become arrested by face-on impingement of lamellae, from which stage on further "thickening" is by the secondary process of rearrangement (perfectioning). The photograph still contains two isolated lamellae (right side) displaying continuing thickening growth (ref. 2)

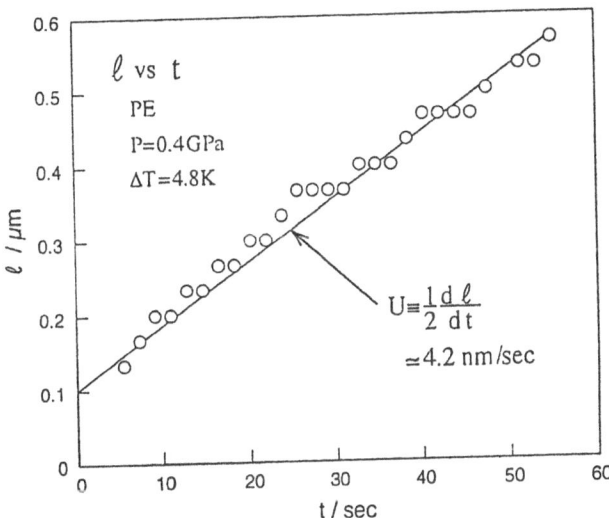

Fig. 3: Crystal thickness as a function of time representing primary "thickening growth" of an isolated crystal growing under the conditions stated in the figure, derived from a tapering cross-sectional crystal profile, allowing the determination of thickening growth rate (U).

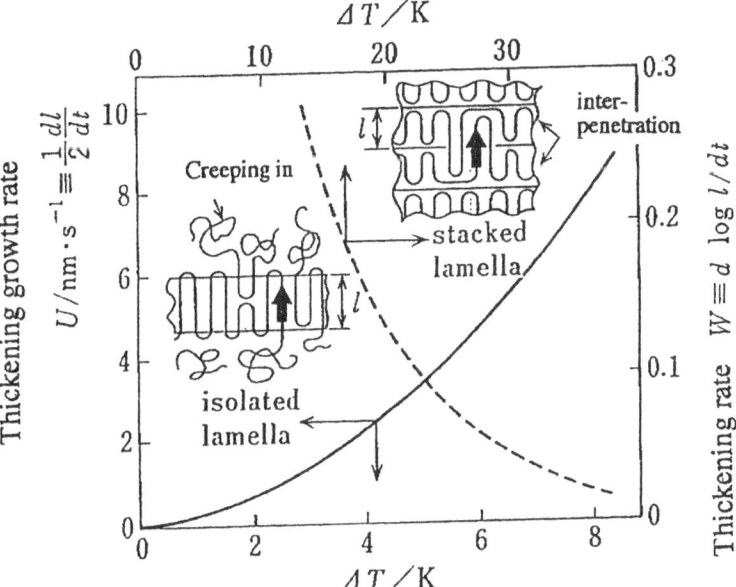

Fig. 4: Comparison of "thickening growth" (primary growth) and "thickening" (secondary, perfectioning). The curves show the opposing trends in their functional relation with temperature (supercooling ΔT); the inset diagrams are schematics of the underlying molecular mechanisms pertaining to isolated and stacked lamellae respectively.

into the growing crystals thus increasing its volume, while in that of "thickening" they thread *through* the lattice so as to reduce internal surfaces by refolding. The underlying physical process is sliding diffusion a mechanism incorporated into existing theories by one of us (M.H., 10, 11).

For the role of metastability we refer to previous works (9) by which it was recognised that crystallization can occur in the mobile hexagonal (h) phase even in the orthorhombic (o) phase stability regime, in which case the h phase is metastable. To this here we add that under the conditions of our own experiments not far below the triple point in the PT phase diagram (Figure 5) we find that crystallization occurs exclusively in the h phase, also while in the o stability regime, and further, that on transformation into the o phase, which occurs at some stage of the development of each crystal, crystal growth, both lateral (V) and thickening (U), stops, or at least slows down to an extent that it becomes unobservable on the time scale of our experiments.

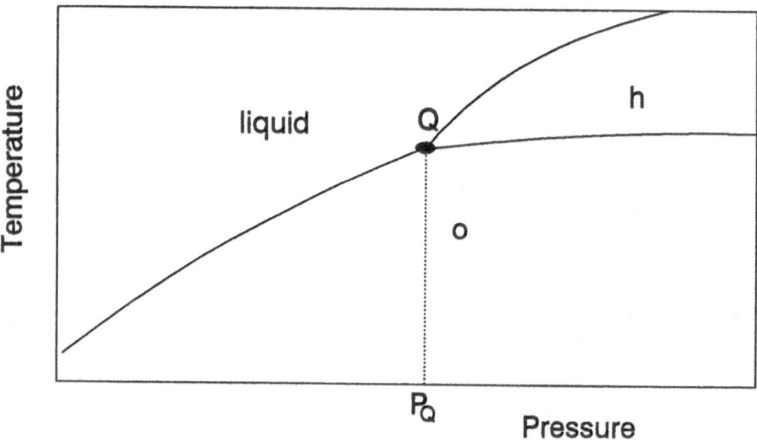

Fig. 5: P,T phase diagram for polyethylene displaying hexagonal (h), in addition to orthorhombic (o) phase regime and triple point (Q). Schematic based on refs. 8,9.

The scheme to follow is aimed at finding a connection between the above listed observations: metastability and thickening growth (and its stoppage on h → o transformation) by attempting to establish a link between the two types of stability/metastability: that determined by phase type in case of different phases and that by phase size in case of a given phase.

SIZE DEPENDANT PHASE STABILITY

The above facts focused attention on size determined phase stability for small phase sizes. While the theme to follow applies to all phases we shall confine the argument to lamellar crystals of polyethylene in its two modifications, orthorhombic (o) and hexagonal (h).

For finite size we invoke the Gibbs-Thompson equation and this as applied to lamellar crystals where the lamellar thickness ℓ is the stability determining factor. This leads to the familiar equation by Hoffman and Weeks,

$$T_m = \overset{\cdot}{T_m}\left(1 - \frac{2\sigma_e}{\ell\Delta H}\right) \qquad (1)$$

expressing melting point (T_m) depression as a function of ℓ. ($T^{\cdot}_m \equiv$ melting point for $\ell = \infty$, $\sigma_e \equiv$ basal surface free energy, ΔH heat of fusion).

Note that, with the appropriate $T^{\cdot}_m$, ΔH and σ_e eq. 1 applies both to the o and h phases (and also to the o → h transformation temperature T_{tr}), i.e. each of these will be depressed on lowering ℓ but to different extents pending on the parameters involved. More precisely, the appropriate T_m (or T_{tr}) will be a linear function with a negative slope in a plot of T_m (or T_{tr}) v.$1/\ell$. Here the slope is the appropriate $2\sigma_e/\Delta H$ (or the appropriate difference values for the o → h transformation) and the intercept along the T axis the appropriate $T^{\cdot}_m$ (or $T^{\cdot}_{tr}$), the value for $\ell = \infty$, i.e. the true equilibrium melting (or transformation) temperature. Now for PE at atmospheric P, where h is metastable $(T^{\cdot}_m)_h < (T^{\cdot}_m)_o < (T^{\cdot}_{tr})$. (Here $T^{\cdot}_{tr}$ is virtual, hence unrealisable). Thus the possibility arises that the $(T_m)_h$ v. $1/\ell$ and $(T_m)_o$ v. $1/\ell$ lines will intersect in which case the T_{tr} v. $1/\ell$ line will also intersect, the intersection of the three lines being at the same point (Q) in the T v $1/\ell$ plane. (Fig. 6). From eq. 1 the condition of intersection is clearly

$$\left(\frac{\sigma_e}{\Delta H}\right)_h < \left(\frac{\sigma_e}{\Delta H}\right)_o \qquad (2)$$

In Figure 6 the bold lines, including dashed and dotted, represent the demarcation lines corresponding to stable phases as a function of $1/\ell$, hence phase size. We shall denote such a representation "phase stability diagram" (reserving the notion "phase diagram" for infinite phase size as customary in equilibrium thermodynamics). In addition to the bold lines for stable phases we have their metastable and virtual (weak line) extensions with their customary significance in true equilibrium phase diagrams.

It follows from phase stability diagram such as Figure 6 that the stability of the h and o phases (or in general metastable and stable phases) can invert with size, i.e. a phase which is metastable for infinite size could become the stable one for sufficiently small size, which in our case of PE would mean that the h phase would be stable for lamellae which are sufficiently thin (even at atm. P.). The limit of stability is defined by the "triple" point Q, i.e. by the temperature T_Q and size ℓ_Q.

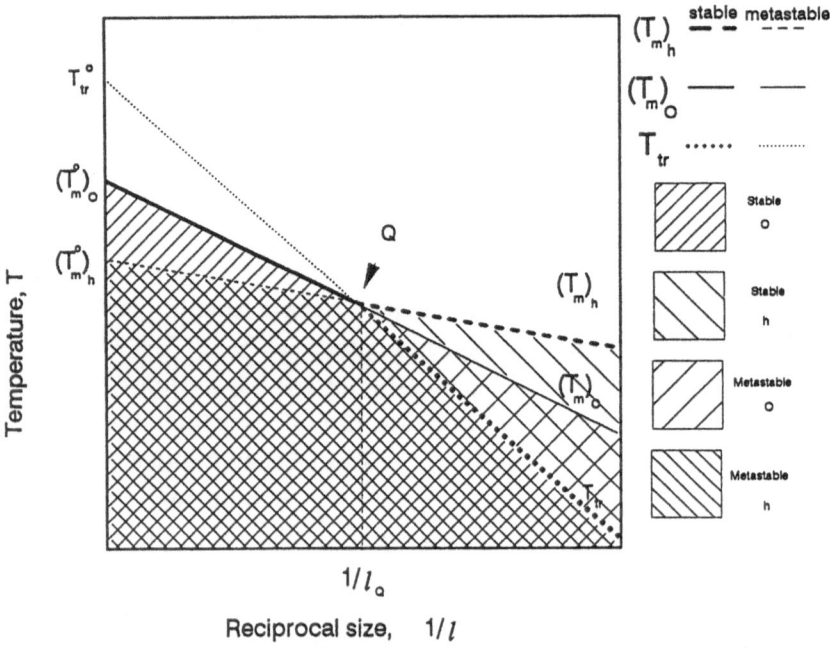

Fig. 6: Temperature (T) - reciprocal size (1/ℓ) "phase stability" diagram, displaying phase stability inversion with size. Subscripts refer to phases in polyethylene (o ≡ orthorhombic, h ≡hexagonal, tr ≡ h → o transition). Bold lines (continuous, dashed and dotted) delineate the "stable" phase regimes including the triple point (Q) For complete key to lines and shadings see insert. (ref. 12, 13).

Clearly, the above phase inversion relies on the condition that inequality 2 holds. Our analysis shows that this should in fact be so for the case of PE (12). More far ranging considerations imply that inequality 2 is in fact quite general, even if not a strict necessity, in fact it is implicit in Ostwald's Rule of Stages. It follows that size dependent phase stability inversions are expected to play a significant part in phase transitions in general and in polymer crystallisation in particular.

CRYSTAL GROWTH: NEW POSSIBILITIES

It is instructive to consider crystal growth (or in broadest generality phase growth) in terms of a phase stability diagram such as Figure 6, as represented by Fig 7a). Here isothermal growth is represented by horizontal arrows pointing towards $1/\ell = 0$, i.e. infinite size, chosen to lie in the two principal temperature regions, one above the other below T_Q, denoted by A and B respectively. In both cases, while in the liquid (L) stability region, the new phase entity is transient until the size $(1/\ell)$, corresponding to a phase line at a specified T, is reached. At this point the new phase (crystal) will become stable and capable of continuing growth. As seen, in region A it passes straight into the region of ultimate stability (which is the o phase in PE). There is a subdivision within A, according to whether the metastable phase (h in PE) can exist (A_2) or not (A_1), a point not to be enlarged upon at this place. In region B, the first crystal appears in a form which is only stable within a limited size range at and beyond its genesis, but is metastable for larger dimensions (which is the h phase in PE). It continues to grow in this phase until reaching the size $1/\ell^*_{tr}$) representing the boundary between the two phase regimes in T, $1/\ell$ space. After traversing this boundary two possibilities arise: i) the crystal stays in the form of its inception, hence becomes metastable and remains so in the final product, or ii) transforms into the phase of ultimate stability somewhere between ℓ^*_{tr} and $\ell = \infty$ in the course of its growth. In case (i) Ostwalds Rule of Stages will appear to be obeyed (by which phase transformations proceed through metastable states whenever such exist (5)), while in case ii) the previous history of phase development, i.e. that the crystals started life in a different phase, will be obliterated (except for our polymers where the texture at the stage of transformation remains preserved - see below).

In order to apply the above to chain folded crystallisation in polymers (here PE) we take ℓ, the phase stability determining dimension, as the lamellar thickness. Within region A the first crystal to appear will be in the o phase, and growth will proceed only up to a limited ℓ, (ℓ^*_g),from whereon further growth will only be in the lateral direction with this constant ℓ (Fig. 7b) i).) This mode of growth embraces the entire presently known and accepted body of material on chain folded polymer crystallisation. In region B growth will start in a different phase (h in PE) with a minimum stable lamellar thickness, which will increase in the course of continuing lamellar thickening growth (Fig. 7b) ii)), which it can do readily in view of the high chain mobility in this phase. Beyond a certain value of ℓ, within

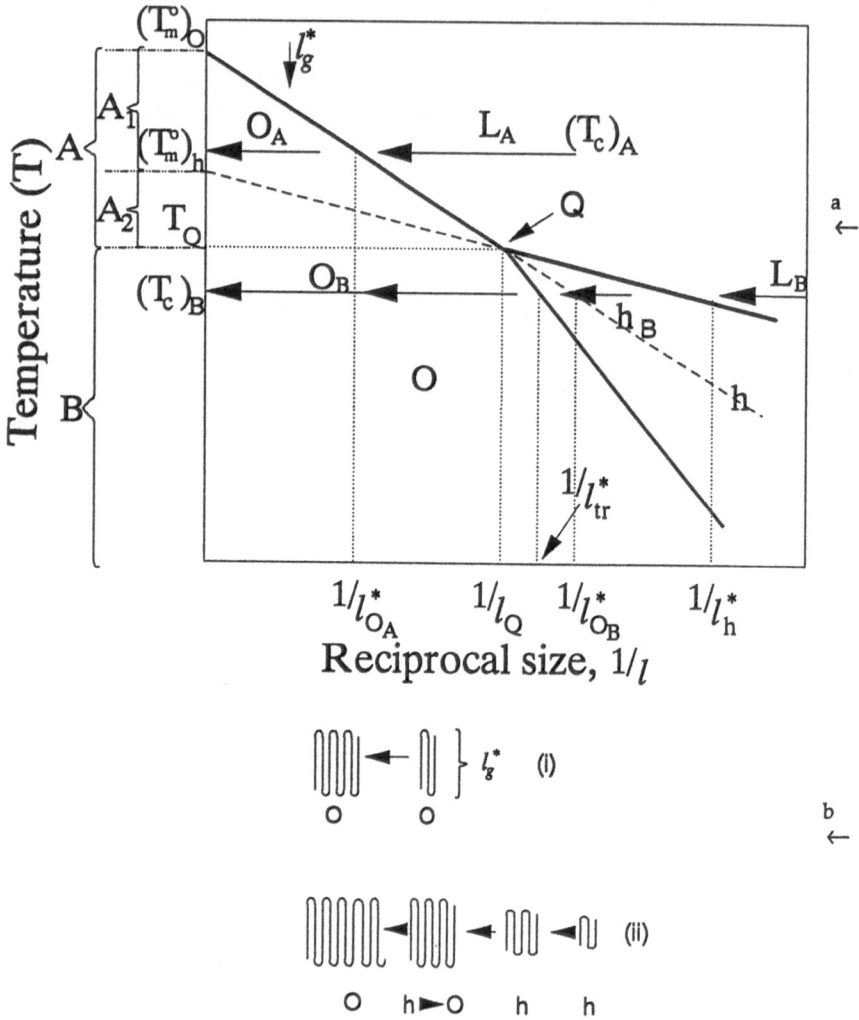

Fig. 7 (a) Phase growth in terms of a "phase stability" diagram as Fig. 6 with notation referring to crystal forms (h and o) in PE with L standing for the melt (liquid). The two sets of horizontal arrows, pointing towards $1/\ell = 0$, denote isothermal growth pathways at the two selected (crystallization) temperatures (T_c), here chosen to be in the two temperature regimes A and B. **(b)** Schematic representation of chain folded polymer crystal growth (for the example of PE) (i) Region A leading to lamellae of a specific restricted thickness (ℓ^*_g), in which they continue to grow laterally through direct growth in phase o, (ii) Region B where crystals arise in the h phase and develop by simultaneous lateral and thickening growth with the latter stopping (or slowing down) on h → o transformation and/or impingement (up to the stage of impingement crystals will be wedge shaped - not represented in sketch).

the newly attained o stability regime, h → o transformation will take place when, within the much less mobile o phase, the phase of ultimate stability, further growth will slow down or, particularly in the case of thickening growth, will practically come to a halt. The process just outlined is a new possibility not envisaged before. It has two salient consequences.

First, it would practically lock in the thickness at which h → o transformation has taken place thus imparting the characteristic thin lamellar feature to polymer crystals formed from flexible chains. Secondly, it would provide a new, alternative explanation for the commonly observed limited lamellar thickness, the principal characteristics of polymer crystallisation, a possibility of potentially far reaching implication.

THE P, T, 1/ℓ PHASE STABILITY DIAGRAM AND IT CONSEQUENCES

The T, 1/ℓ phase stability diagrams of Figs 6 and 7 apply to situations where the h phase (focusing on the case of PE) is metastable at infinite size, hence metastable in the sense of a usual equilibrium phase diagram, while the P,T phase diagram of Figure 5, applying to infinite size, reveals a truly stable h regime at elevated pressures. Connection between the two is revealed by a three dimensional P,T, 1/ℓ phase stability diagram such as Figure 8. Here, as seen, there is a continuous volume of stable h phase in P,T, 1/ℓ space which should be the complete description of the system involved, with a "triple line" (readily calculable (12)) forming the lower boundary of the h stability regime. Crystal growth can be represented in the same way as in Figure 7. For the most usual case of isobaric growth we need to take sections with P = const. and explore isothermal phase development, as done in the preceding section, following horizontal arrows along 1/ℓ. There will be a major distinction as to whether $P > P_Q$ or $P < P_Q$.

a. Take first section at $P > P_Q$. Here we shall have a stable h phase interval even for infinite ℓ. Thus for a range of crystallisation temperatures the crystals will appear in the h phase (as before) but now will stay in this phase, as a stable phase up to macroscopic size (ℓ). For our polymer this means that it starts as a chain folded crystal in the mobile h phase and will grow by thickening growth to increasing thicknesses up to full chain extension and beyond, until terminated by impingement on other growing crystals. This is the customary experience in high pressure crystallisation of polyethylene. (8)

b. Next consider sections at $P < P_Q$. Here the general form of the T, $1/\ell$ section will be as in Figures 6, 7. For P close to P_Q there will be a small A region which, however, will become larger for sections at increasingly lower P; correspondingly, T_Q will be shifting to increasingly lower temperatures as compared to T^*_m, and so will region B.

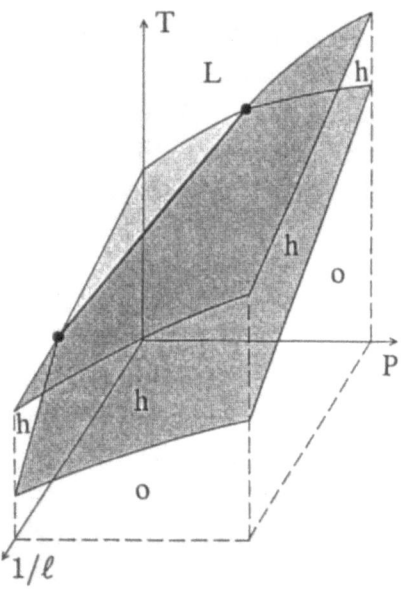

Fig. 8: Combined P,T, $1/\ell$ "phase stability" diagram. This, in case of PE, would create continuity between the pressure and (restricted) size generated mobile h phase. Isobaric crystal growth can here be represented as in Fig. 7a) for any T, $1/\ell$ section at a chosen const. P (including atmospheric P).

At this point we may profitably invoke some kinetic considerations. When region A is small we are close to $(T^*_m)_o$, throughout region A, hence at small supercoolings. Consequently, crystallisation will be slow, in fact for a sufficiently narrow A region it may become unrealisable in practise. In such a situation crystallisation, if it is to take place, will need to be in region B, hence in the case of PE, in the h form alone. On the other hand if region A is very wide, region B will lie far below $(T^*_m)_o$. In order to reach region B the system will need to be cooled through region A first, when depending on rate conditions, the system may crystallise in the phase of ultimate stability before it reaches T_Q. In this situation region B would become kinetically inaccessible. Referring to PE, in such a case the h phase will play no part in crystallisation.

APPLICABILITY TO POLYETHYLENE

The above scheme follows directly from basic principles of thermodynamics relying only on the validity of the Gibbs-Thompson equation. In that case if inequality 2 applies the rest follows. How far a given system conforms to the scheme depends on the input parameters, i.e. $(T^{\bullet}_m)$, ΔH and σ_e for the two phases involved which have to be assessed numerically in each case.

For polyethylene, the most widely studied model system, the two phases are o and h. Here the situation as under a), in the preceding section (i.e. for $P > P_Q$), is well established and has been the starting point for the present considerations. The main issue is whether considerations for case b) (i.e. for $P<P_Q$) also apply, in particular, whether there is any mode B crystallisation under the most widely studied circumstance at atmospheric pressure. If so, much of our present thinking would need to be reconsidered and past results re-evaluated.

The first question which arises is whether phase stability diagrams such as Figures 6, 7 apply at atmospheric P, i.e. whether a "triple point" temperature T_Q, hence region B, exists at all. If the answer is "yes" the second question is the value of T_Q itself.

To answer the above questions a computation was carried out where the input parameters were varied systematically. It emerged (12) that for all input values which were within the limit of plausibility, a T_Q , hence region B, does exist. Again, within limits of plausibility, T_Q was in the range , or just above the range, where PE crystallises at measurable rates from the melt at atmospheric P (122°C-130°C). This means that PE, as normally crystallised, could well do so in mode B, or at least mode B may compete with mode A, the issue requiring further critical attention. However, for crystallisation from solution (at atmospheric P) T_Q is much lower, at around room temperature. It follows that here, by the above argument, region B is kinetically inaccessible, hence crystallisation should proceed by mode A, as in fact has always been envisaged in the past. The origin of the difference between the melt and solution case is readily envisaged. Namely, it can be shown from simple free energy considerations, that not only is T_m lowered when going from melt to solution (melting becomes dissolution), but in addition, the intervals between the melting (dissolution) temperatures of polymorphic phases widens, thus widening region A with a consequent depressing of region B. (12)

CONNECTIONS WITH KINETICS

The first and most general connection with kinetics arises from the meaning of the phase stability diagrams themselves. Namely, the phase lines in the T, $1/\ell$ diagrams represent the size at which a particular phase can be stable at a given T, which in fact is the size of the critical nucleus (ℓ^*). Thus, when crossing a phase line along one of the arrows in Figure 7 we are in fact traversing the size of the corresponding ℓ^*, hence surpassing the principal activation barrier in the kinetics of crystal (or in general phase) growth. The connection between the purely thermodynamic considerations implicit in phase stability diagrams and that of the kinetics of the phase transformation will therefore be apparent. This connection will be unconditional as regards the nucleation of the phase (primary nucleation), but will also apply to the growth of the new phase on condition that the growth is nucleation (here secondary nucleation) controlled. The latter will be the case for chain folded polymer crystal growth which is being envisaged as a secondary nucleation controlled process.

For a further development of the above theme, namely the connection between thermodynamic consideration as arising from the preceding phase stability diagrams, and the kinetics of the phase transformation as arising from nucleation as the rate controlling step we need to refer to our more comprehensive publication (12) and also to a short preview (13). Here the conclusion only will be stated according to which in the case of competing phase variants (e.g. polymorphs in crystal growth) the phase which appears and develops first will be the one which is stable down to the smallest size (implicit in the phase stability diagrams), and that this same phase variant will, in addition, develop fastest. As out of two or more phase variants in general only one can be stable, with the rest being metastable (as referred to infinite size), it follows that the above assertion also embodies the kinetic competition between stable and metastable phases and, specifically, the long standing general experience that metastable phases usually evolve faster than the stable ones. It can be shown that this also leads to a combined kinetics - thermodynamics based justification of Ostwald's Rule of Stages within the appropriate, quantitatively definable boundaries where the Stage Rule can be expected to hold. (12)

CONCLUDING REMARKS

It will be apparent that the line of argumentation presented should have relevance to the wider sphere of phase transformations in general with potential consequences for the specific subject of polymer crystallization, the topic of our concern here. The wider issue is the possibility of phase stability inversion with size with all the far ranging implications for phase development, crystal growth in particular, laid out briefly above and more comprehensively in ref. 12. One consequence of the new conception is the possibility of transient phases during phase growth, specifically crystal growth, which may either persist as metastable phases in the final product, thus conforming to Ostwald's Rule of Stages, or may transform into the phase of ultimate stability during growth. In the latter case the past history of the phase growth would become obliterated, except in polymers where the transient phase is "mobile" allowing "thickening growth" during residence in this phase. It follows that in polymers such a transient phase should leave a mark on the final morphology, or conversely, some essential morphological features observed in the final product should be attributable to such a transient phase during growth. As shown, considerations along these lines can provide a unifying theme to so far largely disparate aspects of polymer crystallization.

References

1   Rastogi S, Hikosaka M, Kawabata H, Keller A (1991) Macromolecules 24:6384

2   Hikosaka M, Rastogi S, Keller A, Kawabata H (1992) J. Macromol. Sci. Phys. B. 31:87

3   Hikosaka M, Rastogi S, Keller A, Kawabata H, Toda A, in preparation

4   Hikosaka M, Amano K, Rastogi S, Keller A (1993) in Crystallization of Polymers Dosière M (ed) Nato ASI-C series in the press

5   Ostwald W, (1897) Z. Physik. Chem. 22:286

6   Keller A (1968) Reports on Progress in Physics 31:623

7   Wunderlich B, Melillo L, (1968) Makromol. Chem. 118:250

8   Bassett DC (1982) in Bassett DC (ed) Developments in crystalline Polymers Appl. Sci. Publ. London, New Jersey

9   Bassett DC, Turner B (1974) Philosophical Magazine 29:285 and 29:925

10   Hikosaka M Polymer (1987) 28:1257

11   Hikosaka M Polymer (1990) 31:458

12   Keller A Hikosaka M, Rastogi S, Barham PJ, Toda A, Goldbeck-Wood G (1993) J. Materials Sci. submitted

13   Keller A (1993) in Crystallization of Polymers Dosière M (ed) Nato ASI-C series in the press

Order-Disorder Transitions in Crystalline Polymers with Characteristic Mechanical and Electric Properties

Kohji Tashiro and Masamichi Kobayashi

Department of Macromolecular Science, Faculty of Science,
Osaka University, Toyonaka, Osaka 560, Japan

Abstract: Structural changes occurring in phase transitions of several polymers with characteristic physical properties have been investigated based on X-ray diffraction and infrared spectroscopic methods. (1) In the thermochromic phase transition of electrically conductive poly(3-alkylthiophene)s, the long alkyl side chains experience trans-gauche conformational change, which affects the planarity or the conjugation length of polythiophene skeletal chains, resulting in a drastic change of sample color. Alkyl chains were found to play an important role in determining the conjugation length and the cooperativity of phase transition. (2) A side-chain-type ferroelectric liquid crystalline polymer was found for the first time to show a variety of layer stacking structure or polytype phenomenon by changing a cooling rate from the isotropic state only slightly. A complicated transition scheme was clarified by measuring the temperature dependence of X-ray diffractions.

INTRODUCTION

Recently many novel polymers have been developed which possess a variety of characteristic mechanical and electric properties. Some typical examples are seen for ferroelectric fluorine polymers with excellent piezoelectric and pyroelectric effects [1], ferroelectric liquid crystalline polymers, highly electrically conductive polythiophene, and so on. These polymers are also unique in such a point that they exhibit a variety of phase transitions, which correlate intimately with the change in the physical properties. In this paper we will focus our attention onto two types of polymer; electrically conductive poly(3-alkylthiophene)s and ferroelectric liquid crystalline polymers. The structural changes occurring in the phase transitions will be investigated on the basis of X-ray diffraction and infrared spectroscopic techniques in association with their characteristic physical properties.

THERMOCHROMIC PHASE TRANSITION IN POLY(3-ALKYLTHIOPHENE)S

Poly(3-alkylthiophene)s have been developed so as to improve the processability of electrically conductive polythiophene by introducing alkyl chains as side groups [2]. These polymers are characterized by a phenomenon of thermochromic phase transition [3-5]. For example, in the case of

A. Teramoto, M. Kobayashi, T. Norisuje (Eds.)
Ordering in Macromolecular Systems
© Springer-Verlag Berlin Heidelberg 1994

poly(3-dodecylthiophene) (P3DT, n = 12), the thermochromic transition occurs around 50°C where the color of the sample changes from red to yellow. An organized combination of X-ray diffraction, infrared (IR) and visible-ultraviolet (VUV) spectroscopy has clarified what type of crystal structural change occurs in this thermochromic phase transition [6-8].

Temperature dependence of X-ray Diffraction, and IR and VUV spectra

In Figure 1 are shown the X-ray fiber diagrams taken at 20 and 110°C for uniaxially oriented P3DT sample . All the reflections observed at 20°C were indexed on the basis of an orthogonal unit cell with the parameters

$$a = 25.83 \text{ Å}, b = 7.75 \text{ Å}, \text{ and } c \text{ (fiber axis)} = 7.77 \text{ Å}.$$

The similar unit cell can be applied also for poly(3-hexyl thiophene) [P3HT, n=6].

$$a = 16.63 \text{ Å}, b = 7.75 \text{ Å}, \text{ and } c \text{ (fiber axis)} = 7.77 \text{ Å}.$$

The difference in unit cell parameters between these two polymers is detected only for the a-axial length, suggesting the alkyl chains jut out along the a axis direction approximately, as discussed in a later section. At high temperature the layer reflections become diffuse and a ringlike pattern is observed, which while not completely round shows some intensity distribution along the ring. The equatorial reflections, on the other hand, are still spotlike. Therefore we may consider that the molecular chains are packed with their chain axes still parallel along the draw direction, but the relative height between the neighboring chains is disordered, resulting in the diffuse layer-line reflections. In other words, the liquid crystalline like structure is obtained at high temperature while the coherence in relative height between the neighboring chains is diminished by the structural disordering caused by the thermal agitation.

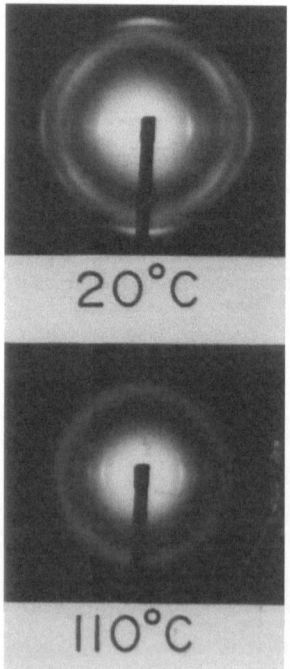
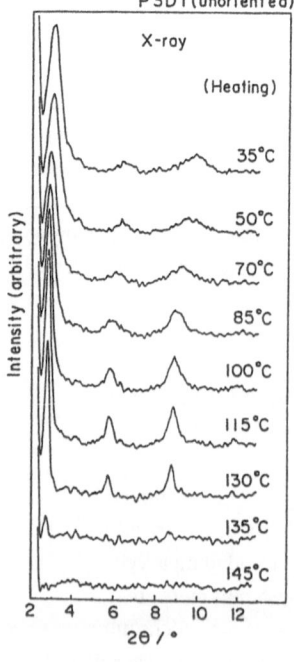

Figure 1 (left). X-ray fiber diagrams of uniaxially oriented P3DT sample taken at 20 and 110°C.

Figure 2. Temperature dependence of the h00 reflections measured for unoriented P3DT.

Figure 2 shows the temperature dependence of the X-ray diffraction profile of the *h00* reflections measured for the unoriented P3DT film. As the temperature increases, the reflections shift to the lower angles and the peaks become more intense and sharper. In the vicinity of the melting point, the peaks begin to decrease in intensity, and once above the melting temperature, they disappear completely. Figure 3 shows the temperature dependence of the integrated intensity and lattice spacing evaluated for the 100 reflection. The lattice spacing increases gradually as the temperature rises over a wide region of -100 - 0°C, and then relatively steeply between 0 and 80°C. Above this transition region the lattice spacing increases again but with a much lower slope. The integrated intensity increases gradually in the lower temperature region and then decreases in the transition region. As the temperature increases above 80°C, the intensity increases again. It begins to decrease as the temperature approaches the melting point. The half-width of the reflection shows a temperature dependence corresponding to those of the intensity and lattice spacing: in the transition region it decreases in particular.

The temperature dependence of the X-ray reflections mentioned above can be interpreted from the molecular structural level by measuring the infrared spectra. In Figure 4 are shown the polarized FTIR spectra of uniaxially oriented P3DT film measured at several temperatures. At low temperature a series of progression bands characteristic of the methylene trans segments can be detected clearly in the frequency region 700 - 1500 cm^{-1}. In the vicinity of the transition region these trans bands decrease in intensity and the methylene gauche bands can be observed. In particular the bands in the 1400 - 1300 cm^{-1} range can be assigned to such gauche conformations as GTG' (kink) at 1366 cm^{-1}, GG at 1353 cm^{-1}, GTT (end gauche) at 1341 cm^{-1},

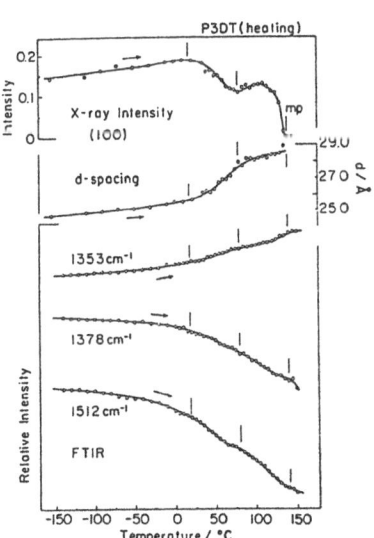

Figure 3. Temperature dependence of the integrated intensity and lattice spacing of X-ray 100 reflection and the relative intensity of infrared bands measured for P3DT sample.

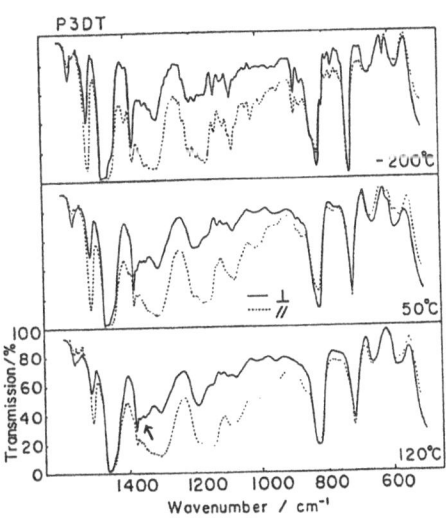

Figure 4. Polarized infrared spectra of oriented P3DT film measured at the various temperatures. The solid and broken lines represent the spectra taken with the electric vector of the incident infrared beam perpendicular and parallel to the drawing direction of the sample, respectively.

and GTG and GTG' at 1300 cm^{-1}, where T and G denote trans and gauche bonds, respectively [9]. These spectral changes indicate that the side alkyl chain experiences a large conformational disordering from the regular trans structure (in the low-temperature region) to the irregular structure of gauche and trans combinations (in the high temperature region). At the same time the bands characteristic of thiophene ring modes (1512, 840 cm^{-1},...) decrease also in intensity at the high temperature without losing their dichroism. As discussed also by Zerbi et al. [10], these bands are sensitive to the degree of electronic conjugation of polythiophene skeletal chain. In Figure 3 is made a comparison in temperature dependence of X-ray diffraction data and that of infrared data. As the temperature rises, the methyl and methylene trans bands and thiophene ring modes decrease in intensity and the gauche bands increases; the curves show inflection points in the temperature regions where the change in the X-ray diffraction pattern is observed. These structural changes correspond intimately to the change in electronic conjugation length as observed by VUV spectra shown in Figure 5 (a) and (b). In the low temperature the absorption peak is observed at about 530 nm. As the temperature increases in the transition region, a new peak begins to appear near 430 nm and the original peak decreases instead. As being plotted in Figure 5 (b) the 529 and 427 nm peaks exchange their intensities with some inflection points in the transition region, in agreement with the temperature behavior observed for X-ray and infrared data.

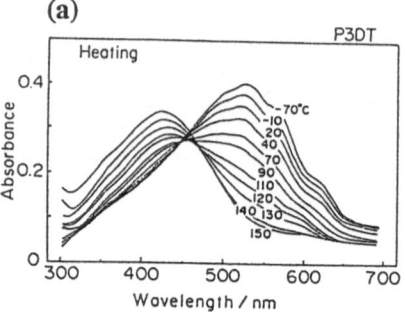

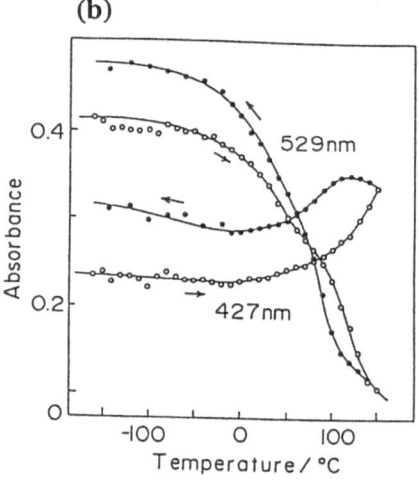

Figure 5. Temperature dependence of (a) the VUV spectra and (b) the relative absorbances of the bands at 529 and 427 nm obtained for the unoriented P3DT.

All the experimental data presented above lead us to describe the structural change occurring in the thermochromic phase transition of P3DT and P3HT as follows. At low temperature the polythiophene skeletal chain and the side group alkyl chains assume an almost fully extended trans conformation. As the temperature increases the alkyl chains begin to disorder through the conformational change between trans and gauche forms. As the temperature approaches the transition region (0 - 80°C for P3DT and 20-100°C for P3HT), the conformational disordering becomes more significant and thermal agitation within the alkyl chain segments affects the skeletal conformation and causes the structural disordering of the planar thiophene chain, resulting in the break down of the long π–π conjugation

system of thiophene rings as indicated by the temperature dependence of the VUV spectra
(Figure 5). One plausible mode of such a structural disordering in the planar conformation
of the thiophene ring sequences may be internal rotation about the ring-to-ring bonds. If such
an internal twisting of the thiophene main chain occurs actually, the scattering intensity of the
X-ray meridional reflection should be affected appreciably and the repeat period along the
chain axis should contract somewhat. In order to check these possibilities the temperature
dependence of the X-ray 002 reflection was measured for uniaxially oriented P3DT and
P3HT samples. Figure 6 shows the temperature dependence of 002 reflection profile
measured for P3DT and P3HT. At room temperature an intense peak is observed at $2\theta = 23^\circ$
which corresponds to the identity period of ca. 7.8 Å. The peak intensity decreases as the
temperature increases and a new peak is observed at about 24°, corresponding to the chain
contraction of ca. 5 %. This new peak becomes gradually more diffuse with increasing
temperature. These observations are consistent with the above predicted structural change of
polythiophene main chain induced through an internal torsional mode around the ring-ring
bonds.

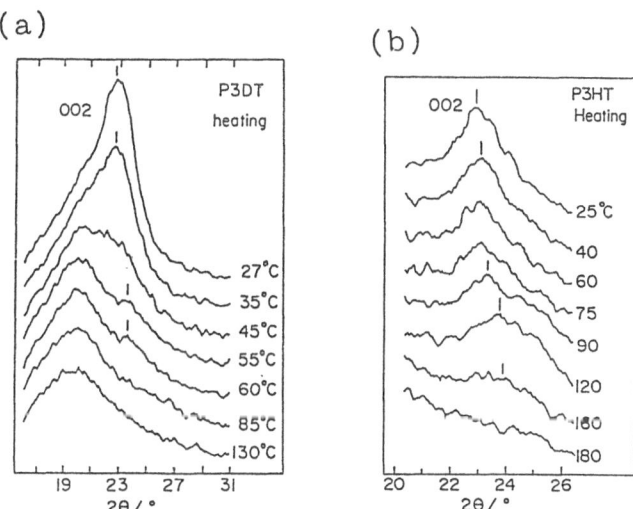

Figure 6. Temperature dependence of the X-ray 002 reflectional
pattern measured for the uniaxially oriented P3DT (a) and
P3HT (b) samples.

Crystal Structure at Room Temperature

Several authors proposed the crystal structure models of poly(3-alkylthiophene) at room
temperature [11,12]. But the structure has not yet been definitely established. We investigated
the crystal structure by an organized combination of the X-ray diffraction and infrared data and
the computer simulation technique [13]. Energy minimization was performed starting from some
reasonable structural models and four types of model have been found as energetically stable
structures. The calculation was made by using "Polygraf" (version 3.20, Molecular Simulation
Inc.). Figure 7 illustrates these structural models of P3DT. In models 1 and 2, the thiophene
rings of neighboring chains are packed closely along the b axis and the trans-zigzag alkyl side

P3DT

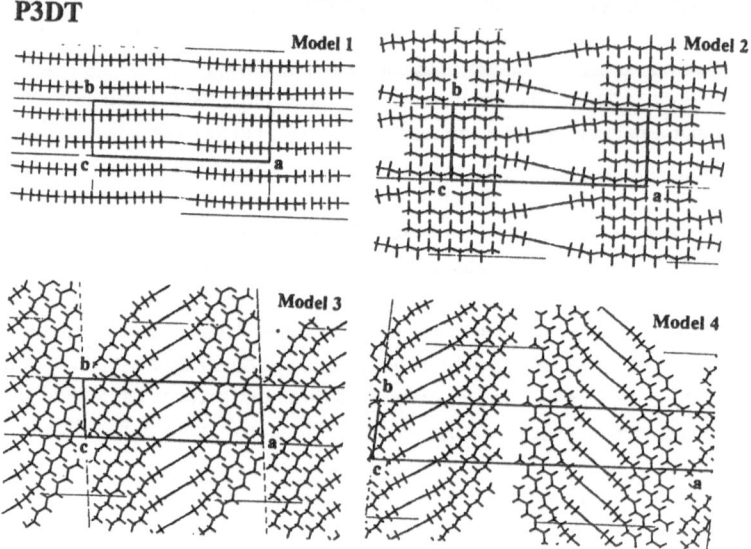

Figure 7. Energetically minimized crystal structural models of P3DT.

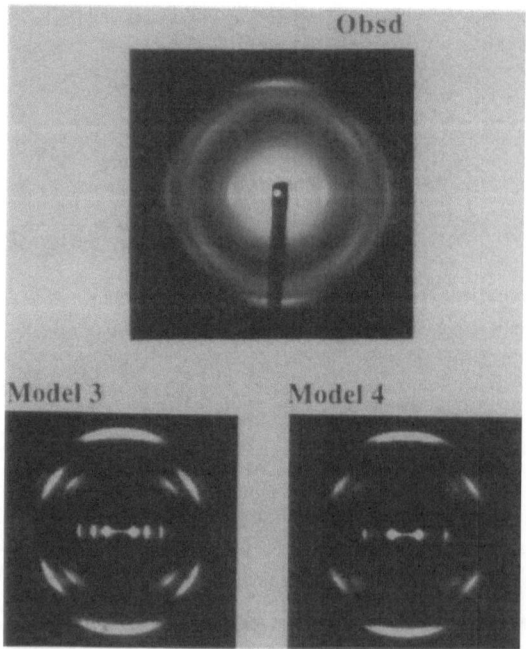

Figure 8. The observed and simulated X-ray fiber
diagrams of the uniaxially oriented P3DT.

chains are interdigitated partly along the a axis and form well packed subcell structure. The relative height of the neighboring chains is different by a quarter of the c axis length in model 1. In model 2 the thiophene rings are arranged side-by-side with a relative height difference of a half the c axis and the planar chains deflect as a whole slightly from the a axis direction. Models 3 and 4 consist of stacked layers, each of which is constructed by a parallel array of chains along the b axis. Mode 3 is of one layer type and model 4 is of double layer type. Although Winoker et al. proposed the structure similar to Model 4 [11], model 3 can not be also ignored. The X-ray fiber diagrams were calculated for these structural models by using a program "Cerius" (version 3.0; Molecular Simulation Inc.), where the regularity, size and degree of orientation of crystallites were taken into consideration. Through a comparison in relative intensities of equatorial and layer line reflections between the calculated and observed data, the models 3 and 4 were extracted as the most plausible candidates of the structure (see Figure 8). Even model 2, however, can reproduce the observed X-ray pattern reasonably, if some statistical disordering is taken into consideration in the calculation of the pattern. We are now trying to select out the best model which can reproduce the experimental data of X-ray fiber pattern and infrared spectra as quantitatively reasonably as possible.

Role of Alkyl Side Chains in Thermochromic Phase Transitions

(1) Conjugation Length at Low Temperature

Figure 9 shows the VUV spectra measured at room temperature for a series of poly(3-alkylthiophene) films. The peak position of the band shifts to longer wavelength side as the alkyl chain length increases from polythiophene (PT, n = 0) to P3DT (n = 12), indicating that the conjugation length increases with an increase of alkyl chain length. This consideration is supported by the infrared spectral data. The relative intensity of the thiophene ring mode at ca. $1500 \ cm^{-1}$ is sensitive to the conjugation length as discussed in the previous section. As shown in Figure 10, the relative intensity of this band of P3BT (n = 4) is only a half of that of P3DT even at liquid nitrogen temperature, indicating that the conjugation length of P3BT is appreciably short compared with that of P3DT. The half-width of the X-ray $h00$ reflections becomes sharper as the alkyl side chain is longer. Besides the rate of intensity decrement in a series of reflections becomes slower for the sample with longer alkyl chain: that is, the higher order reflections are observed for the sample with longer alkyl chains. These X-ray data indicate that the long-ranged structural regularity of the lattice cell increases with an increment in alkyl side chain length. As understood from the crystal structural models shown in Figure 7, a side-by-side overlap of the neighboring methylene segments may be more significant for the samples with longer alkyl chains . Such tight molecular packing may fasten the thiophene rings into more perfectly planar conformation, resulting in an increase of conjugation length as seen in Figures 9 and 10. This type of role of alkyl side chains is often called "zipper effect".

(2) Effect of Alkyl Chain on Phase Transitional Behavior

In Figure 11 is plotted the temperature dependence of relative intensity of VUV absorption peak near 500 nm measured for a series of poly(3-alkylthiophene)s. In every sample except for PT, the remarkable intensity exchange is observed between the two peaks near 500 and 430 nm as seen in Figure 5. But the steepness in intensity change is different among the samples: the intensity change or the change in conjugation length occurs more steeply in a lower temperature

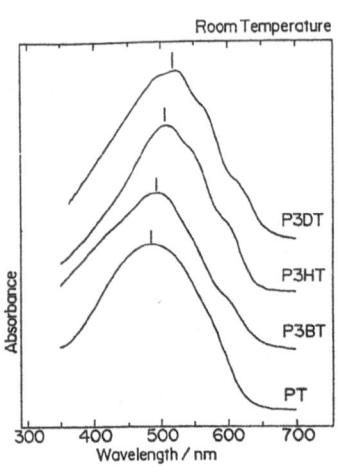

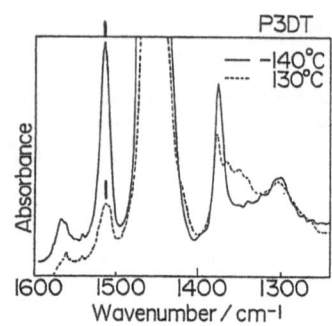

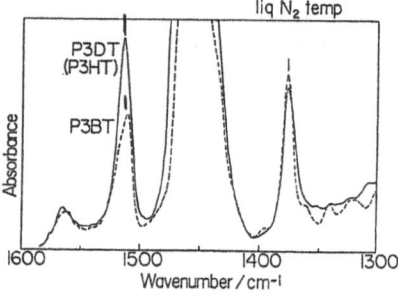

Figure 9. The VUV spectra measured at room temperature for a series of poly(3-alkylthiophene)s: PT (n=0), P3BT (n=4), P3HT (n=6), and P3DT (n=12).

Figure 10. (upper) Infrared spectra of P3DT measured at low and high temperatures and (lower) infrared spectra of P3DT and P3BT measured at liquid nitrogen temperature.

(b)

(a)

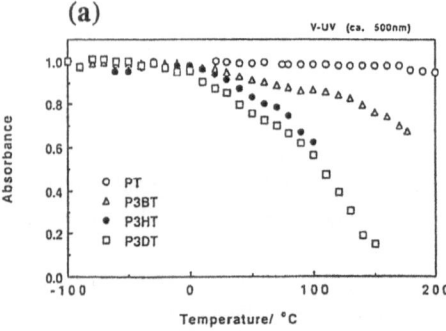

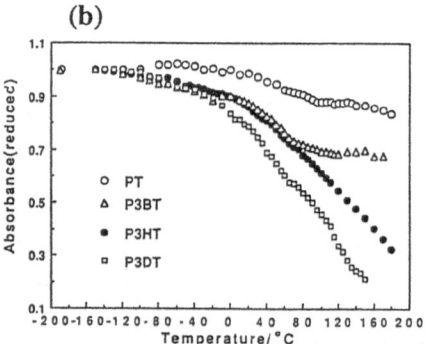

Figure 11. Temperature dependence of (a) VUV relative absorbance and (b) infrared absorbance measured for a series of poly(3-alkylthiophene).

region as the side chain length increases. Such a tendency in the VUV spectral change corresponds well to the structural change detected by infrared and X-ray diffraction measurements: the structural change becomes more diffusive and occurs in a higher temperature region as the alkyl side chain is shorter. These data indicate that the phase transition occurs in more cooperative fashion in a lower temperature region as the alkyl side chain length increases. As discussed above, for the sample with long alkyl chains, a zipper effect or an intermolecular overlap of trans-type side chains makes the conjugation length of the main chain more stable. Once the trans-gauche conformational change occurs in the alkyl chains and the side-by-side packing efficiency is reduced, then the twisting of the planar skeletal conformation is induced cooperatively and sensitively, resulting in an instantaneous shortening of the conjugation length. For the sample with short alkyl chain, such a zipper effect is not originally effective and the conjugation length of the skeletal chain approaches that of PT itself and the slight change in the short alkyl chain conformation does not affect the conjugation state so much.

In this way the longer side chains of poly(3-alkylthiophene)s have a potential to increase the conjugation length of polythiophene skeletal chain and induce more cooperative and sharper phase transition in lower temperature region. If the side chain is increased furthermore, a more enhanced effect may be expected in the conjugation length as well as in the phase transitional behavior. Then poly(3-docosylthiophene) (P3DCT, n = 22) was investigated as a case study. Figure 12 shows the VUV spectra of a series of Poly(3-alkylthiophene)s measured at low and high temperatures. Contrary to the above-mentioned expectation P3DCT shows the absorption peak at ca. 450 nm, appreciably shorter than those observed in the other members (n = 4 - 12). The position is rather close to that of the high-temperature phase of shorter conjugation length. That is to say, P3DCT has only short conjugation length even at low temperature. This is supported also by the infrared spectral data: the thiophene ring bands are low in intensity (Figure 13). The methylene sequences are considered to take trans form preferentially but with small portion of gauche parts, as judged from the infrared pattern (Figure 13). That is to say, P3DCT is considered to take the structure of the slightly twisted polythiophene main chain with long trans methylene side groups and some gauche bonds, although the position of such gauche part is not known definitely at present.

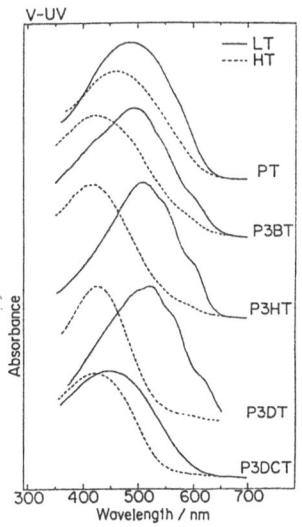

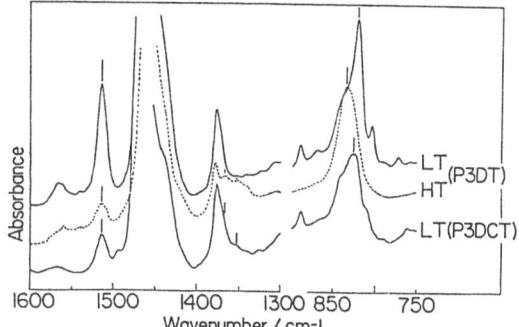

Figure 12 (left). VUV spectra of poly(3-alkyl thiophene)s at low (LT) and high (HT) temperatures.
Figure 13 (above). Infrared spectra of P3DT and P3DCT at low (LT) and high (HT) temperatures.

Measurements of VUV spectra at high temperature revealed that the conjugation length of P3DCT increases slightly in the temperature region immediately below the melting point of ca. 75°C: shoulders are detected in this temperature region as seen in Figure 14. This change is related with the structural change occurring in this temperature region. For example, as shown in Figure 15, the trans methylene bands decrease in intensity and the gauche bands increase and at the same time the intermolecular correlation splitting observed for the CH_2 rocking band (730 and 720 cm^{-1}) disappears also into a singlet. That is to say, the trans-gauche conformational change is attendant by the packing structure change in methylene side chains from orthorhombic to hexagonal type, as illustrated in Figure 16. The measurement of X-ray diffraction at high temperature supports this consideration. For P3DT and P3HT with long conjugation structure, the methylene rocking bands are observed as singlet even at low temperature, suggesting a hexagonal (or monoclinic) side chain packing. The change of side-chain subcell structure into hexagonal type, therefore, might cause the better side chain packing, which results in the similar situation observed for P3DT and P3HT, i.e., the longer conjugation length. Of course such an effect is not so high because of the existence of gauche bonds and also because the thermal agitation might disturb the regular arrangement of thiophene rings.

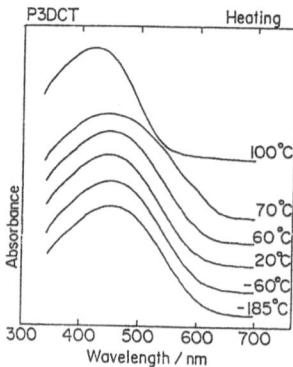

Figure 14. Temperature dependence of VUV spectra of P3DCT.

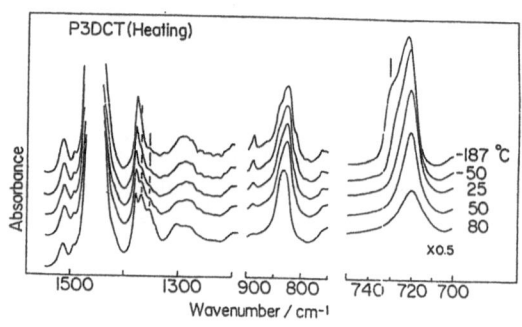

Figure 15. Temperature dependence of infrared spectra of P3DCT.

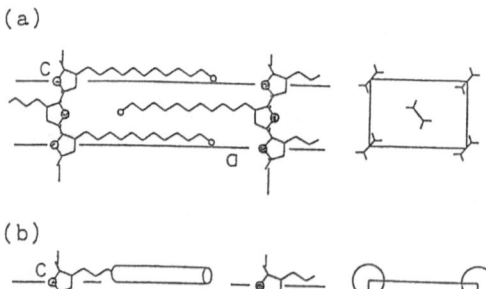

Figure 16. Illustrated structural change in P3DCT. (a) low temperature and (b) high temperature.

POLYTYPE PHENOMENA IN FERROELECTRIC
LIQUID CRYSTALLINE POLYMERS

Because of their characteristic electric properties, ferroelectric liquid crystalline materials have been widely investigated from both scientific and industrial points of view. In particular, ferroelectric liquid-crystalline polymers (FLCP) containing chiral mesogenic groups in the skeletal or side chains are attracting many attentions because of their potentials in controlling the rate of electric dipole inversion under external electric field, in producing mechanically tough materials, in developing easily accessible materials, and so on. In spite of these practical importance, however, molecular-dimensional structural studies have been quite limited in number.

In a series of papers concerning ferroelectric liquid-crystalline compounds, we have clarified a high sensitivity of layer stacking structure to a slight change in sample preparation conditions such as cooling rate from the isotropic phase, type of solvent, etc [14,15]. High sensitivity of layer stacking structure is not limited only to the low-molecular-weight liquid crystalline substances but is expected to see also for the above-mentioned FLCPs. In fact we have found that a side-chain-type FLCP experiences a drastic change in layer stacking structure when the cooling rate from the isotropic state is varied only slightly from 0.2°C/min to 0.1°C/min, for example [16]. In the latter half part of this paper, we will clarify a variety of stacking structure and a complicated phase transitional behavior of a side-chain-type FLCP sample on the basis of X-ray diffraction and infrared spectroscopic measurements.

The samples used in this study were supplied from Canon Co. Ltd; FLCP (P1) with the following chemical structure and its monomer (M1). The M1 was used here in order to compare the transitional behavior with that of P1.

M1

$$CH=CH_2$$
$$|$$
$$C=O$$
$$|$$
$$O-(CH_2)_{11}-O- \bighexagon -COO \bighexagon \bighexagon -O-CH_2-CH^*(CH_3)-O-C_6H_{13}$$

P1

$$-CH-CH_2-$$
$$|$$
$$C=O$$
$$|$$
$$O-(CH_2)_{11}-O \bighexagon -COO- \bighexagon \bighexagon -O-CH_2-CH^*(CH_3)-O-C_6H_{13}$$

Polytype Phenomenon in P1

Figure 17 shows the DSC thermograms taken for M1 and P1 samples. The X-ray and optical microscopic observations revealed the transitions of M1 as indicated in the thermograms of this figure. The DSC thermogram of P1 is apparently simple but a quite complicated change occurs actually as clarified here in this paper. Figure 18 shows the X-ray diffraction patterns measured at room temperature for the P1 samples prepared by cooling from the Iso state at the various rates

28

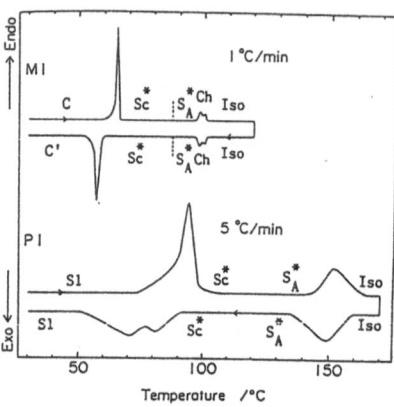

Figure 17. DSC thermograms of M1 and P1 samples.

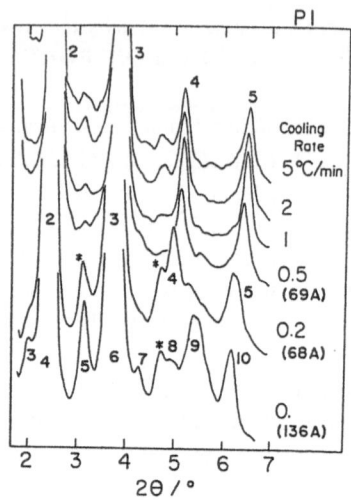

Figure 18. X-ray diffraction patterns of the P1 samples obtained at the cooling rate of 5 - 0.1°C/min.

of 5 to 0.1°C/min. A series of intense X-ray reflections can be observed in the small angle region. For the samples obtained at the rate of 5 - 0.5°C/min, the X-ray pattern is essentially the same with each other. The peaks are indexed as 002, 003, 004, 005, and so on, where the c axis is assumed to be perpendicular to the layer plane with the repeating period of ca. 69 Å. The sample obtained at the cooling rate of 0.2°C/min shows the reflections at the positions slightly different from those of 0.5°C/min but the essential pattern is almost the same with the latter. The layer repeating period is estimated as 68.4 Å. Remarkable change is observed in the X-ray diffraction pattern when the cooling rate is changed only by ca. 0.1 from 0.2 to 0.1°C/min. That is, the sample cooled at 0.1°C/min shows the X-ray reflections not only at almost the same positions with those of the sample cooled at 0.2°C/min but also at the positions intermediate between these common reflections. If the commonly observed reflections are indexed as $l = 2$, 3, 4, etc., the additional reflections should be indexed as $l = 2.5, 3.5, 4.5$, and so on. By doubling the repeating period as 136.8 Å, all the reflections can be indexed using integers, i.e., $00l$ with $l = 3, 4, 5, 6$ and so on. Figure 19 shows the 2-dimensional X-ray patterns measured for the 0.1°C/min sample with the incident X-ray beam parallel and perpendicular to the sample plane. In case of parallel incidence a highly oriented pattern is obtained, while the perpendicular incidence gives Debye-Sherrer rings, indicating that the layers are highly uniaxially oriented around the axis perpendicular to the sample plane. This high degree of orientation is reduced gradually with an increment of the cooling rate as shown in Figure 20, suggesting that the orientational ordering of the layer stacking proceeds quite slowly: a rapid cooling results in a frozen-in of disordered layer stacking. In Figure 21 shows the X-ray diffraction pattern taken in a wide scattering angle range. Compared with the remarkable difference of the pattern in the small angle region, the reflectional positions in the wide angle region are almost the same between the two samples cooled at 0.2 and 0.1°C/min although the relative intensities are different from each other because of the difference in the degree of orientation (refer to Figures

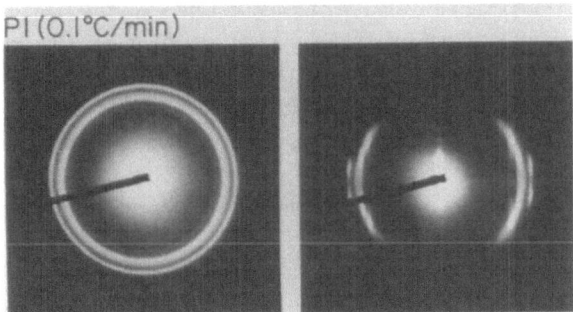

Figure 19. X-ray diagrams taken for the P1 sample obtained at cooling rate of 0.1°C/min. The incident X-ray beam is (a) perpendicular and (b) parallel to the sample plane.

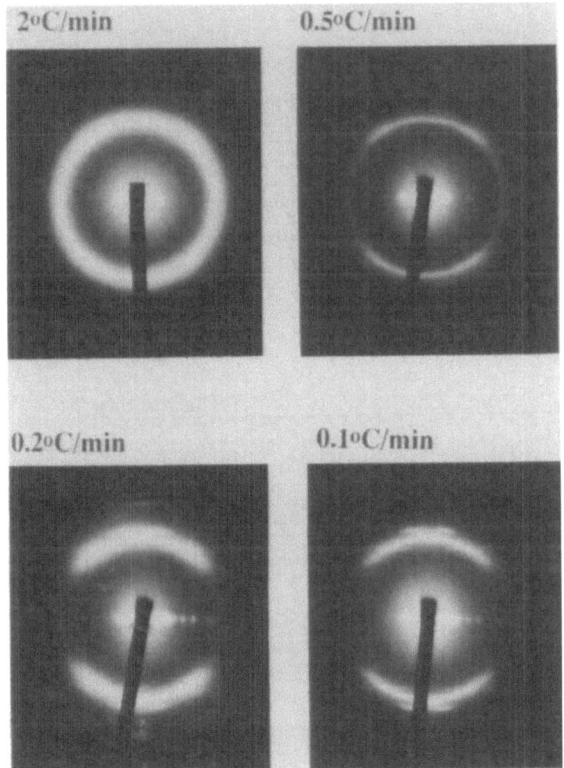

Figure 20. X-ray diagrams taken for the P1 samples obtained at cooling rate of 2 - 0.1°C/min.

19 and 20). Therefore, from Figures 18 and 21, we may conclude that the two samples discussed here are different in the layer repeating period, 68.4 Å and 136.8 Å, but the layer structure itself is essentially the same with each other. This type of phenomenon, i.e., a modification of the layer stacking structure is called polytype, which is frequently observed for the layer-type substances such as mica, silicon carbide, n-paraffin, and n-fatty acids [17,18]. The present report may be the first example to succeed in finding out such a polytype phenomenon in the side-chain-type ferroelectric liquid crystalline polymers. The two types of the liquid crystalline modification found at room temperature are called S1 and S2. The M1 was treated in the similar way with P1 but the X-ray patterns were almost the same Debye-Scherrer rings, although the intralayer structure might be changed somewhat depending on the cooling rate, as seen from slight difference in the wide angle X-ray pattern. In this way the polytype phenomenon is considered to be characteristic of the polymer system P1. The reason might be correlated with such a situation of polymer system that the spatial motions of mesogenic groups are confined by being linked covalently to the skeletal chain and a slight change in a mesogenic group results in a cooperative structural change of the whole space, as clarified later in an observation of a complicated phase transitional behavior.

Stacking Structure of Layers

As judged from the similarity in infrared spectra (Figure22), the conformation of the mesogen groups of the P1 at room temperature and high temperatures is considered to be similar to that of the liquid crystalline state of the M1. The length of fully extended planar-zigzag chain conformation of M1 is calculated as ca. 44 Å. The layer spacing ca. 41 Å observed for the S$_A$ phase of the M1 is close to this calculated value: the slight difference may come from the conformationally disordered structure of molecules at high temperatures. Then the side chain of P1 may be assumed basically to have a value close to 41 Å. The SA phase of P1 takes the interlayer spacing of 68.8 A. How does it correlate with the value 41 A ? If the mesogenic

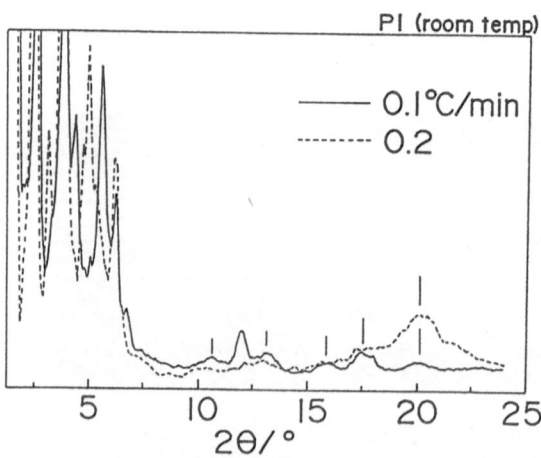

Figure 21. A comparison of X-ray diffraction patterns of the P1 samples cooled at the rate of 0.1 and 0.2°C/min.

groups stand vertically in the layer sheets as expected from the conventional structure of general SA phase, the side chain length should be 68.8/2 = 34.4 Å. The value is too short and contradicts to the above-mentioned 41 Å. Rather we might assume that the layers are constructed by an interdigitation of two side chain groups by ca. 7 A, as illustrated in Figure 23. The interlayer spacing observed at room temperature is 68.4 Å, close to that in the SA phase. If the degree of interdigitation is almost equal in both the SA and S1 (S2) phases, then the side chain is considered to tilt by ca. $\cos^{-1}(68.4/68.8) = 6°$ from the normal to the layer plane (The wide-angle X-ray reflections should be interpreted reasonably based on this tilt angle). The side groups jut out of the main chain in an *atactic* fashion and the orientation of side chains is considered in two ways as illustrated in Figure 24. By taking all these factors into consideration, models of S1 and S2 are deduced as shown in Figure 25 (a). Although the models are speculative at present, we may have such a possibility that S1 is ferroelectric and S2 is anti-ferroelectric if the electric dipole of the layer is the same in both the phases.

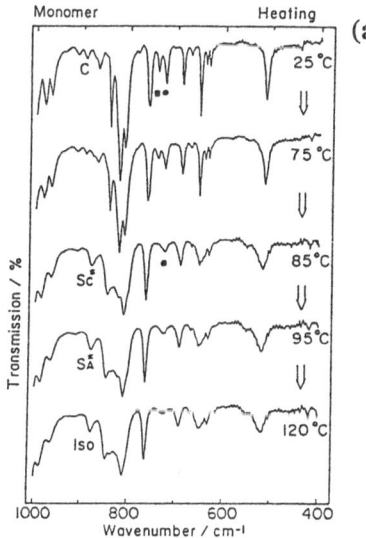

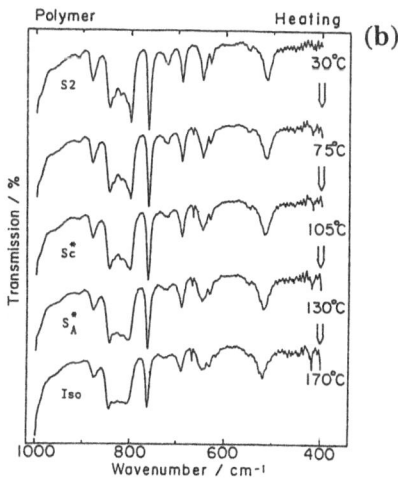

Figure 22. Temperature dependence of infrared spectra taken for (a) M1 and (b) P1 samples.

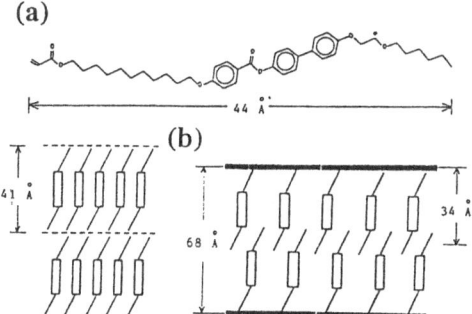

Figure 23. (a) fully extended M1 molecule,
(b) layer models of SA phase of M1 (left) and P1.

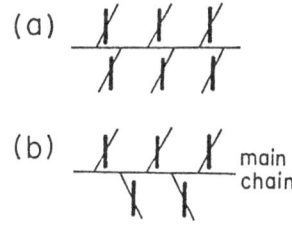

Figure 24. Possible orientation of side chain groups linked to the main chain.

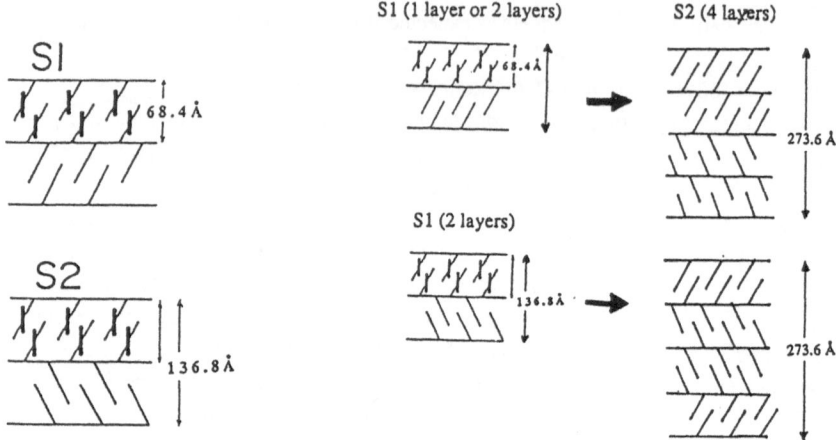

Figure 25. (a) A possible packing structure of S1 and S2. (b) Another models of S1 and S2 with the layer repeating distance 2 times longer than that of model (a). The upper and lower models are two possibilities of S1 and S2 structure. (The two models of S2 are equivalent to each other.)

As shown in Figure 18, the X-ray 00l reflections with odd l are detected in the sample of S2. If the layer stacking is repeated regularly and the structural factor of the layer is constant, then the odd-numbered reflections should disappear due to the requirement of space symmetry. An observation of odd-numbered reflections might come from the following situations. (1) As shown in Figure 25 (b), the repeating periods of the S1 and S2 might be doubled, 136.8 and 273.6 Å, respectively. Then the originally odd reflections are reindexed as even-numbered ones. (2) The structure factor of the two types of layer might be different from each other, and some irregular replacement of layer might occur in the S2 phase. For example, the S1 layer is introduced in an irregular fashion into the main S2 layer structure, and then the structure factor should become

For $l = 2n$ $F(00l) = F_2 + [F_2\omega + F_1(1-\omega)] = (1+\omega)F_2 + (1-\omega)F_1 \neq 0$

For $l = 2n + 1$ $F(00l) = F_2 - [F_2\omega + F_1(1-\omega)] = (1-\omega)(F_2-F_1)$

where F_1 and F_2 are structure factors of S1 and S2 layers and ω is the probability of the S1 layer within the S2 layers. Therefore if $\omega \neq 1$ and $F_1 \neq F_2$, then we may observed both the odd and even reflections. Difference between F_1 and F_2 may be possible if the contribution from the main chain structure is taken into account in addition to the side chain structure, but it is not considered so large. Therefore the odd-numbered reflections should be weak compared with those of even-numbered reflections. The actually observed tendency is consistent with such a speculation.

Phase Transitions

Figure 26 shows an example of temperature dependence of X-ray reflectional patterns starting from the S1 phase. Analysis of these X-ray data in the heating and cooling processes revealed the transitional scheme shown in Figure 27. The S1 and S2 phases change into Sc*, SA, and cholesteric phases via new phases having different layer repeating periods (S3 with L = 218 Å and S4 with L = 62.8 Å). In the cooling process from the Sc* phase, the S3 appears at first and then transfers into S1 or S2, depending on the cooling rate. The S3 phase is obtained frequently at room temperature when the cooling rate is not well controlled: the reason may be obtained from the above-mentioned phase transitional scheme.   In Figure 28 is shown a series of 2-dimensional X-ray patterns taken during heating of the S3 sample as an example. These X-ray pattern changes are consistent with the above-mentioned transition scheme; the S3 transfers into Sc*, SA, and Cholesteric and Iso phases. The orientation of layers is kept unchanged during the transition into the liquid crystalline phases and disappears finally in the Iso state.

As understood from Figures 26 - 28, the P1 sample experiences a variety of phase transition and the layer stacking structure changes quite sensitively by a slight change in the temperature during heating or cooling of the sample. As already discussed, the side chains are connected side by side to the main chain by covalent bondings through flexible alkyl groups. Therefore a slight change in one mesogenic group may induce a cooperative motion of the neighboring side chains and result into a drastically large structural change in the whole layer structure, i.e., a variety of polytype phenomenon and complicated phase transitional behavior as clarified in the above sections.

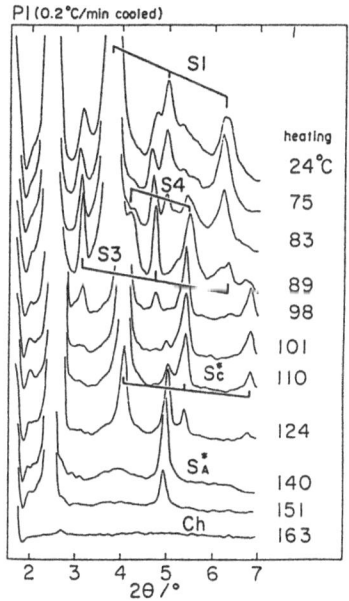

Figure 26. Temperature dependence of X-ray diffraction pattern of P1 starting from the S1 state.

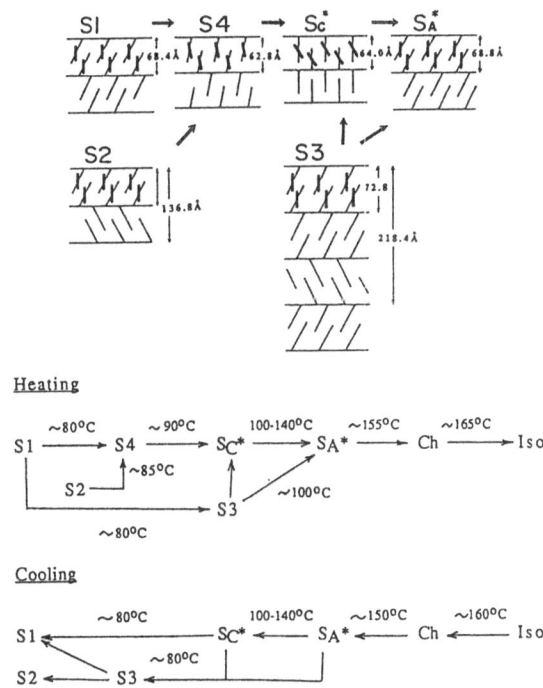

Figure 27. (upper) Illustrated phase transitional scheme of the P1 in the heating process. (lower) The summarized phase transitional schemes of P1 in the heating and cooling processes.

34

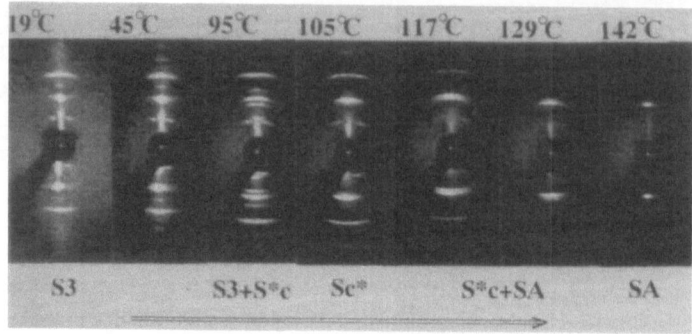

Figure 28. Temperature dependence of X-ray imaging plate diagrams of the oriented P1 sample starting from the S3 state.

ACKNOWLEDGMENTS

The authors wish to thank professor Katumi Yoshino, Dr. Tsuyoshi Kawai, and Dr. Shigenori Morita of Faculty of Engineering, Osaka University (Suita, Osaka 565, Japan) for their fruitful discussion on poly(3-alkylthiophene)s. They are also deeply indebted to Canon Co. Ltd., Japan for their kind supply of the P1 and M1 samples. These studies are supported in part by Grant-in-Aid for Scientific Research from the Ministry of Education, Science, and Culture, Japan (No. 03555189 and No. 04403020).

REFERENCES

1 Tashiro K, Kobayashi M (1989) Phase Transitions 18:213
2 Sugimoto R, Takeda S, Gu HB, Yoshino K (1986) Chem Express 1:635
3 Yoshino K, Nakajima S, Park DH, Sugimoto R (1988) Jpn J Appl Phys 27:L716
4 Yoshino K, Nakajima S, Onoda M, Sugimoto R (1989) Synth Metals 28:C349
5 Themans B, Salaneck WR, Bredas JL (1989) Synth Metals 28:C359
6 Tashiro K, Ono K, Minagawa Y, Kobayashi M, Kawai T, Yoshino K (1991) Synth Metals 41-43:571
7 Tashiro K, Ono K, Minagawa Y, Kobayashi M, Kawai T, Yoshino K (1991) J Polym Sci:Part B:Polym Phys 29:1233
8 Tashiro K, Minagawa Y, Kobayashi M, Morita S, Kawai T, Yoshino K (1993) Synth Metals 56: in press
9 Hagemann H, Strauss HL, Snyder RG (1987) Macromolecules 20:2810
10 Zerbi G, Chierichetti B, Ingänas O (1991) J Chem Phys 94: 4646
11 Prosa TJ, Winokur MJ, Moulton J, Smith P, Heeger AJ (1992) Macromolecules 25:4364
12 Mårdalen J, Samuelsen EJ, Gautun OR, Carlsen PH (1992) Synth Metals 48:363
13 Tashiro K, Kobayashi M, Morita S, Kawai T, Yoshino K (1993) Polym Prepr Jpn 42:1442
14 Tashiro K, Hou J, Kobayashi M (1992) J Phys Chem 96:2729
15 Hou J, Tashiro K, Kobayashi M (1991) Mol Cryst Liq Cryst 200:145
16 Tashiro K, Hou J, Kobayashi M (1992) J Phys Chem 96:2729
17 Yeomans J (1988) Solid State Phys 41:151
18 Kobayashi M, Kobayashi T, Itoh Y, Sato K (1986) Bull Mineral 109:171

Computer Simulation of Macromolecular Crystals and Their Defects

Bernhard Wunderlich, Bobby G. Sumpter, Donald W. Noid, and Guanghe L. Liang

Chemistry Division, Oak Ridge National Laboratory, Oak Ridge, TN 37831-6182, USA, and Department of Chemistry, The University of Tennessee, Knoxville, TN 37996-1600, USA

Abstract: Computational results on the dynamics of polyethylene, derived by our research group during the last five years are reviewed and connected to experimentally known facts. It could be demonstrated that conformational defects can be created at temperatures as much as 100 K below the melting point and that the concentration continues to increase exponentially with temperature, ultimately leading to disordered crystals along the polymer chains (high temperature CONDIS crystals). Although the rate of formation of these defects is high, approximately 1×10^{10} s^{-1} at 350 K, the defects do not, by themselves, lead at low temperature to macroscopic motion that could give rise to lamellar thickening or deformation. The diffusion mechanisms involve coupling of large-amplitude torsional motion with transverse and longitudinal vibrations of the crystal, which lead to the formation of disclinations, dispirations, and twists. Such defects can, under proper conditions, move towards the end of the crystal, thereby causing a chain diffusion process that is probably at the root of lamellar thickening and deformation processes. Collective twisting of the chains without major influence of conformational defects leads to hexagonal or pseudo-hexagonal structures of the asymmetric motifs, involving dynamic multidomain arrangements of the chains. The especially efficient molecular dynamics simulation-code developed for this research produces reasonable agreement with experimental data on density, defect concentration, heat capacity, vibrational spectra (including stress-induced frequency shifts), melting temperature, and speed of sound. The simulations produced data on crystals of up to 30,000 atoms for times of up to 100 picoseconds (10^{-10} s). The total simulation efforts needed a massive effort of approximately 8,000 hours of supercomputer CPU time between 1988 and 1993.

INTRODUCTION

Much of the understanding of modern science is based on the microscopic description of matter. The length scale in question is the nanometer (nm) or ångstrom (Å), distant from our macroscopic senses by at least a factor of 10^4. Although it is well known, that the small-amplitude molecular vibrations, measurable by infrared spectroscopy, have a frequency of up to 10^{14} Hz, it has only recently become possible to follow the large-amplitude motion by realistic molecular dynamics simulation. These simulations are done with supercomputers and enable the extremely slow motion presentation that is needed to follow the fate of atoms and molecules on their movement in the solid state. The surprising result was that such large-amplitude motion can have an atomic time-scale in the picosecond range (10^{-12} s), slower by only a factor 1/100 than the fastest vibrations, but faster by a factor of 10^9 than direct human observation (assumed to be milliseconds). It is thus a much more difficult task to mentally bridge the microscopic, atomic time-scale to the macroscopic, human time-scale than to bridge the corresponding length scales. Experimental verification of such fast large-amplitude motion is seen presently in the interpretation of solid state NMR results, as well as quasi-elastic neutron scattering.

A. Teramoto, M. Kobayashi, T. Norisuje (Eds.)
Ordering in Macromolecular Systems
© Springer-Verlag Berlin Heidelberg 1994

In this paper the first results from a five-year effort of molecular dynamics simulation of polyethylene crystals is discussed, with a special effort to establish a connection to the available experimental results. These simulations have led to the identification of the mechanics of formation of conformational defects, twists, disclinations, and dispirations of chains, and crystal domain structures. Overall, a new view of the defect solid state of polymers has become possible. A more detailed discussion with close to 150 references has been prepared recently [1]. The reader is directed there, and at the limited key references repeated in this paper for an entry to the fast-growing literature in this field.

MOLECULAR DYNAMICS SIMULATION OF CRYSTALS

The molecular dynamics method (MD) involves the numerical solution of Hamilton's equations of motion, starting from some chosen initial positions and velocities of the atoms or particles of the system. In our highly efficient MD simulations, the integrations of the equations of motion are carried out in Cartesian coordinates, thus giving an exact definition of the kinetic energy and coupling [2]. Integrations are performed using an up to 12th order predictor/corrector routine with variable time steps. This method (ODE) allows the integrations to be performed to a high accuracy, permitting conservation of all constants of motion to as many as four digits without the need for any *ad hoc* scaling of the momenta (total energy, angular and linear momenta).

The Hamiltonian is composed of the kinetic and potential energy for the system:

$$
H = \sum_{n=1}^{m} \left[\sum_{i=100(n-1)+1}^{100n} \frac{p_{x_i}^2 + p_{y_i}^2 + p_{z_i}^2}{2M} + \sum_{i=100(n-1)+1}^{100n-1} V_{\text{bond}}(r_{i,i+1}) \right.
$$

$$
\left. + \sum_{i=100(n-1)+1}^{100n-2} V_{\text{bend}}(\theta_{i,i+1,i+2}) + \sum_{i=100(n-1)+1}^{i=100n-3} V_{\text{introt}}(\tau_{i,i+1,i+2,i+3}) \right]
$$

$$
+ \sum_{i=1}^{100m} \left[\sum_{j} V_{\text{nonbonded}}(R_{i,j}) + \sum_{k=1}^{100l} V_{\text{nonbonded}}(R_{i,k}) \right]
$$

(1)

Equation (1) refers to a crystal of m dynamic chains of 100 CH_2-groups each. The second brackets represent the nonbonded interactions among all dynamic molecules i with j, where the latter are the dynamic atoms not considered in Eqs. (2, 4, and 5) for the given i. The last term, summing over the interactions of the atoms i with k, is only needed if the crystal is enclosed with l chains of rigid (nondynamic) atoms, fixed to keep the volume constant. The potential functions for the adjacent, bonded atoms (i and $i+1$) in kJ/mol are:

$$
V_{\text{bond}}(r) = 334.72(1 - e^{-19.9(r-0.153)})^2
$$

(2)

where 0.153 nm represents the equilibrium bond length (Morse equation). The interaction between any two nonbonded atoms is represented by the Lennard-Jones expression (note that along the chain the atoms must be separated by a sequence of three atoms or more):

$$V_{nonbonded}(R) = 1.9748\left[\left(\frac{0.4335}{R}\right)^{12} - \left(\frac{0.4335}{R}\right)^{6}\right]$$ (3)

where R represents the appropriate distance between i and j or k. For the bending of a three-atom, bonded sequence the potential is:

$$V_{bend}(\theta) = 65.061(\cos 113° - \cos\theta)^2$$ (4)

with 113° representing the equilibrium bond angle. The torsional coordinate τ between four successive atoms is linked to the internal rotational potential:

$$V_{introt}(\tau) = 8.3704 - 18.4096\cos\tau + 26.78\cos^3\tau$$ (5)

The parameters of Eq.(5) have been fitted to give a barrier height of 16.7 kJ/mol and a *gauche* − *trans* energy difference of 2.5 kJ/mol. These potential energy functions have been demonstrated to yield good spectroscopic, thermodynamic, and kinetic data, as well as to provide the atomistic details of temperature-dependent phase transitions for crystalline polyethylene and correspond to a collapse of the CH_2-groups into a single particle of mass 14 amu [1].

Models with explicit hydrogens have also been simulated, with appropriately modified Eqs. (1−5). Since such full simulation increases the number of atoms (and the computation time) considerably, only selected simulations have been done with explicit hydrogens [1]. Furthermore, as will be discussed in the next section, it is not possible to reproduce the ideal crystal structure with the present hydrogen force fields. Many other properties, in turn, can be represented as well, or better, and certainly faster, by the united atom model.

For all simulations, the initial conditions involve imparting a randomly chosen momentum, subject of a zero center-of-mass velocity constraint. Due to the large number of atoms in the system, and thus the high density of vibrational states, thermal equilibrium, as described by a Boltzmann distribution, is rapidly achieved. An effective temperature can then be determined from the average kinetic energy $<KE>$ by:

$$<KE> = \frac{3}{2}NkT = \sum_{i-1}^{100m} \frac{p_{x_i}^2 + p_{y_i}^2 + p_{z_i}^2}{2M_i}$$ (6)

where k is Boltzmann's constant, T is the temperature, p the Cartesian momentum and M the mass of the atoms i. In general, analyses of the simulations were started after allowing the kinetic energy to redistribute.

The simulations produced data on crystals of up to 30,000 atoms for times of up to 100 picoseconds (10^{-10} s). The total simulation efforts needed, thus, a massive effort of approximately 8,000 hours of supercomputer CPU time between 1988 and 1993 [1−3]. Further progress with the next generation of parallel computers, already installed in our institutions, awaits the development of advanced molecular dynamics simulation codes. Considerable time is expected to be needed to accomplish this task and to find the necessary financial support for this effort.

IDEAL CRYSTALS

Helped by the high external symmetry, the understanding of crystals has always been at the forefront of the natural sciences [4]. Kepler, for example, explained the multiplicity of the shapes of snow flakes by the large number of possible regular packings of spheres some 200 years before the accepted proof of the atomic nature of matter by Dalton in 1808. Figure 1 reproduces a modern-looking crystal, drawn in 1690 by Huygens. About 250 years later, the knowledge of crystals had advanced from the general principle, displayed in Fig. 1, to the detailed, ideal structures determined by X-ray diffraction. The structure of orthorhombic polyethylene (Pnam) is shown in Fig. 2. The projection along the c-axis places the zig-zag chains at the corners of the unit cell and in the center. Following the packing rules developed by

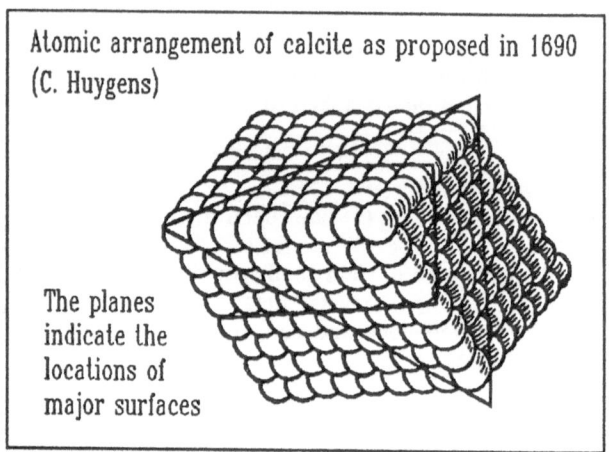

Atomic arrangement of calcite as proposed in 1690 (C. Huygens)

The planes indicate the locations of major surfaces

Figure 1

Kitaigorodskii [5], and typical van der Waals radii, one can derive that Pnam is, indeed, the densest possible crystal with a coordination number of six for the chains, and a packing fraction k of 0.70 (k = van der Waals volume / unit cell volume). Four of the neighboring chains show strong triple-contacts (marked by arrows in Fig. 2), the other two have only single contacts. The structure is thus critically dependent on packing the CH_2-groups of the center chain into the appropriate depressions of the surrounding chains.

During such analysis, of packing in polyethylene,[4] it was discovered that, in order to agree with experimental lattice parameters, an artificially large, lateral van der

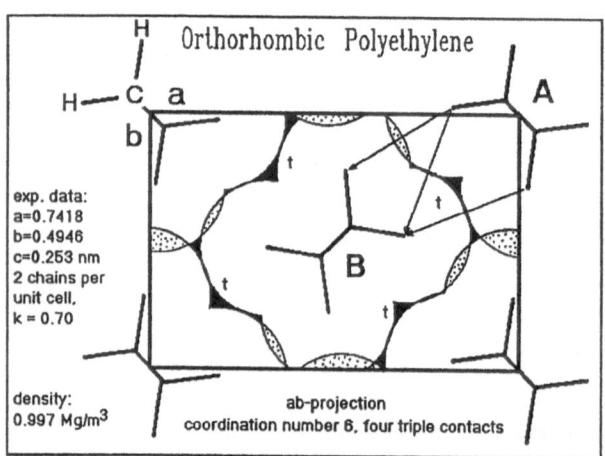

Orthorhombic Polyethylene

exp. data:
a=0.7418
b=0.4946
c=0.253 nm
2 chains per
unit cell,
k = 0.70

density:
0.997 Mg/m³

ab-projection
coordination number 6, four triple contacts

Figure 2

Waals radius of 0.135 nm had to be used (instead of the commonly accepted value of 0.117 nm, derived from contacts in the bond direction). Without this correction, the packing fraction turned out to be 0.85 [4]. The large variation in van der Waals radius with direction is unique with hydrogen and has been attributed by Bunn [6] to a nonspherical shape of the covalently bound hydrogen. Naturally, the packing of hydrogen in polymer crystals is of central importance and has not yet been resolved.

Our first MD simulations with explicit hydrogens used a spherical van der Waals interaction, employing an expression similar to Eq. (3) [1]. As should perhaps have been expected, the resulting crystal structure had an excessively high density (and preferred a somewhat different structure). The higher density resulted in reduced conformational defect concentrations. At present, one must thus be cautious in the interpretation of all MD simulations involving nonbonded interactions with explicit hydrogen atoms. By clearly identifying the problem of the nonspherical hydrogen shape, it is hoped that progress in solving the hydrogen shape will be forthcoming soon, although a more complicated Eq. (3) for the nonbonded interaction will, again, increase the computation time.

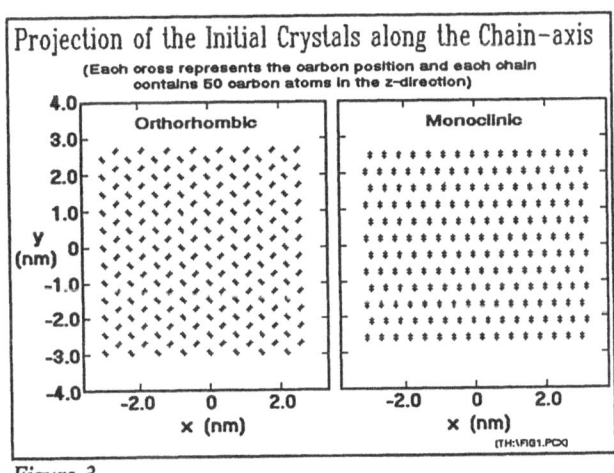

Figure 3

The united atom force fields of Eqs. (2−5), in turn, are well chosen and produce crystals with close to experimental densities. Since the united atom lacks the detailed mm2 point symmetry of the CH_2-group, indicated in Fig. 2, the resulting crystal structure is hexagonal and can serve as a model of the high-temperature (and -pressure) form of polyethylene. The orthorhombic structure, can, however, also be simulated by enclosing the dynamic chains in a ring of orthorhombically placed rigid chains (constant volume simulation). Figure 3 shows two of the initial ideal crystal structures chosen for simulation (ORTH = orthorhombic, as in Fig. 2, and MONO = monoclinic). A third one, included a random placement of the zig-zag chains, to simulate a hexagonal structure (HEX). The size of each crystal was initially about $6.0 \times 6.0 \times 6.3 = 227$ nm^3. This is sufficiently large to serve as a model for polymer crystals, which are typically between 2 and 20 nm thick. The simulations were conducted in a temperature range from 55 to 410 K for 10 to 100 ps with the nonbonding potential represented by Eq. (3), being

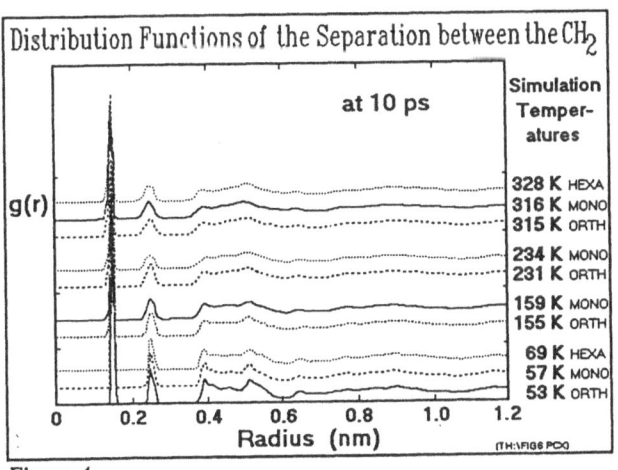

Figure 4

cut-off at 1.0 nm to conserve computation time. The crystals quickly reached a steady-state, enabling us to compare dynamic and structural behavior with experimental results. After about 4 ps, the radial distribution functions ceased to be dependent on time and became indistinguishable among the different initial structures, which means that all three starting structures (ORTH, MONO, and HEXA) had been transformed to the same "steady-state", as can be seen from the radial distribution curves of Fig. 4.

Once the crystal structure is established, its properties can be assessed. Figure 5 shows a density plot. Oscillations occur about the average, which is close to the experimental density of polyethylene. At about the same temperature, different initial structures have slightly different density fluctuations in direction and amplitude, but the same frequency ($\approx 3 \times 10^{11}$ Hz). Assuming that one sees the first fundamental vibration of the crystal with a wavelength of about 12 nm (twice the crystal size), one can estimate the speed of the wave to be about 4 km/s, which, in turn, is typical for measurements of the speed of sound in polymers (1.4 – 5.9 km/s).

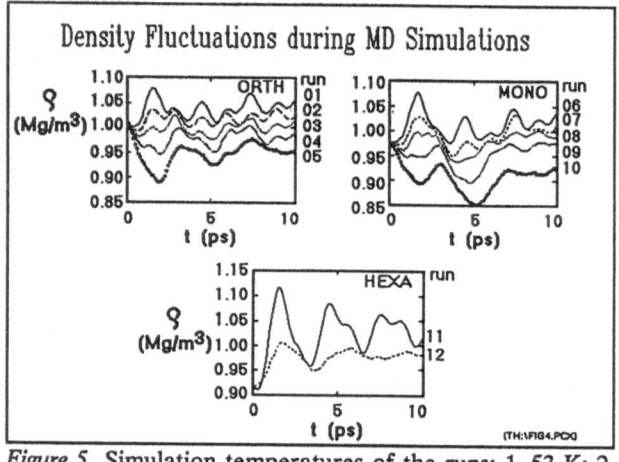

Figure 5 Simulation temperatures of the runs: 1, 53 K; 2, 155 K; 3, 231 K; 4, 315 K; 5, 410 K; 6, 57 K; 7, 159 K; 8, 243 K; 9 316 K, 10, 403 K; 11, 69 K; 12, 328 K.

Once a crystal structure has been established, it is thus possible to establish its properties. A number of these, such as the just illustrated density and speed of sound, as well as modulus and heat content, agree well with the ideal crystal structure. These properties are called *structure insensitive*, to contrast with the *structure sensitive* properties (such as the ultimate strength, stress required for plastic deformation, chemical reactivity, and diffusivity within a crystal).

The thermal properties of macromolecules can, next, be derived from the vibrational behavior. The motion of a polyethylene crystal, simulated at constant volume and at low temperature, is displayed in Fig. 6. It is easy to see the three basic,

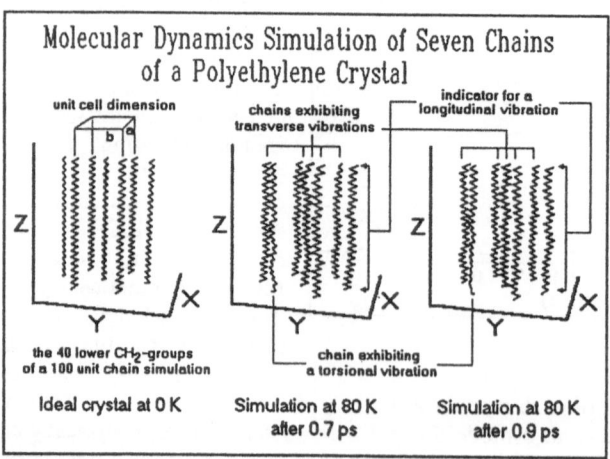

Figure 6

skeletal vibrations that are coupled over several CH_2-groups along the chain: transverse, torsional, and longitudinal vibrations.

The quantitative analysis of simulations of this type is able to reproduce the full frequency spectrum of a polyethylene crystal [1]. The vibrational spectrum, in turn, is easily linked to the heat capacity by summing the contribution of each vibrational frequency to the heat capacity using an Einstein function [7]:

$$C_v/NR - E(\Theta_E/T) - \frac{(\Theta_E/T)^2 \exp(\Theta_E/T)}{[\exp(\Theta_E/T) - 1]^2} \tag{7}$$

where N represents the multiplicity of the vibration and Θ_E is the Einstein temperature (frequency), expressed in $h\nu/k$. Once the heat capacity is known from absolute zero of temperature, the enthalpy, entropy, and Gibbs energy can easily be calculated. Our *ATHAS* data bank [8] has been used over the last twenty years to accomplish this correlation for more than 100 polymers. Figure 7 shows the example of polyethylene. The skeletal vibrations, indicated in Fig. 6, and the additional group vibrations, consisting mainly of C–H motion that can be simulated with explicit hydrogen models, make up the heat capacity at constant volume C_v. After conversion to heat capacity at constant pressure, C_p, the comparison with experiment shows agreement from 0 to 300 K.

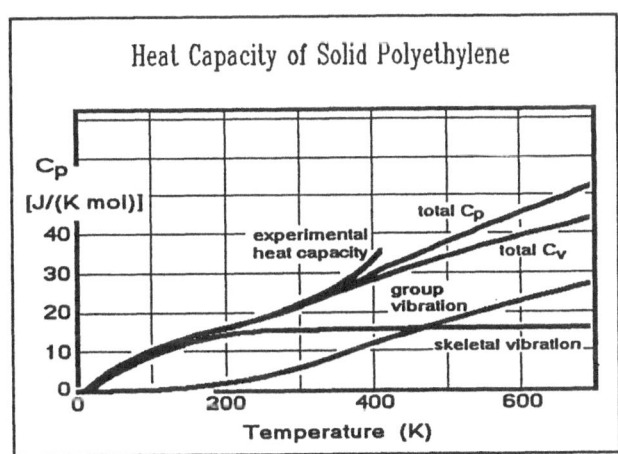

Figure 7

Not only can the heat capacity, and with it also all integral thermodynamic functions, be established with an approximate frequency spectrum, it is also possible to predict heat capacities of polymers with similar chemical structures and establish heat capacities for copolymers in the solid state. Similar results are expected for all other structure-insensitive properties.

At about 300 K the experimental heat capacity starts to deviate significantly from the calculation that was carried out under the assumption that only vibrations govern the thermal energy. This is a first indication of defect contribution to the thermal properties and will be discussed in the remaining part of this paper.

GENERAL DESCRIPTION OF DEFECTS IN POLYMER CRYSTALS

To describe semicrystalline polymers, a two-phase model has been commonly used over the last 50 to 60 years. One of the two phases being represented by perfect crystals, the other by a fully amorphous phase. Structure-insensitive properties vary approximately linearly with the weight or volume fraction of this crystallinity. Since such a two-phase structure in a one-component system can, according to the phase rule, not be in equilibrium, there must be constraints hindering further crystallization of the amorphous phase, *i.e.* there must be a connection between the two phases (microphases). Because of this link between crystal and amorphous phase, the latter has also been called a *three-dimensional, "amorphous defect"* [4]. Its structure and mobility is still not well understood, but may in the future be amenable to MD simulations of the type presented here.

Characteristic for polymeric materials are also *two-dimensional defects*, represented by surfaces. External surfaces are always the prime defect of a single crystal. For polymer crystals they are particularly important because of the usually small crystal size. The highly regular surfaces of solution-grown crystals could be analyzed in detail by optical and electron microscopy, and more recently also by atomic force microscopy. These investigations gave much information about chain folding, tie molecules, fold-sector boundaries, stacking faults, twin boundaries, kink bands, *etc.*

Figure 8 illustrates MD simulations of the hexagonal crystal. One can see domains of different orientation of the CH_2-zig-zag planes (setting angle). Even at 57 K (Fig. 8a) the crystal structure deviates from perfect order (compare to Fig. 3). Not only are the projections of the chains broadened by vibrations, but one can also see a break-up into at least two domains, separated by boundaries of two to three layers of chains. At 234 K

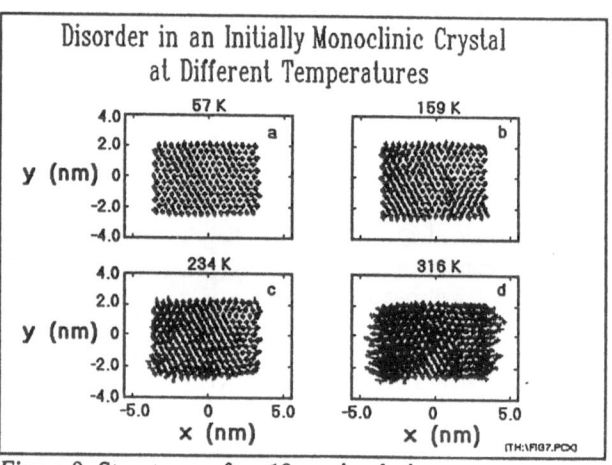

Figure 8 Structures after 10 ps simulations.

the crystal becomes more disordered and develops easily recognizable domains (Fig. 4c). The disorder increases with temperature and becomes significant at 316 K where, in addition, a "fuzzy" surface structure is evident (Fig. 4d). This occurs at a temperature that is still about 50 K below the experimental melting point. The scale of disordering is not uniform throughout the entire crystal, but seems to vary from one domain to another. It will be shown, below, that these nanometer-size domains are dynamic and change orientation and boundaries on a picosecond time scale, providing the average needed to fit an overall hexagonal symmetry.

Decreasing the dimensionality of the defects leads one to *dislocations (one-dimensional defects)*. The classical edge- and screw-dislocations of crystals of small molecules could also be documented in crystals of macromolecules. Particularly obvious are growth spirals with a Burgers vector of the size of the lamellar thickness, *i.e.* 5 – 50 nm, which are produced during crystal-growth [4]. Dislocations with a Burgers vector of unit-cell dimensions were seen by electron microscopy with the help of moiré patterns and were linked to chain ends in the crystal. A connection to deformation processes, as was found in crystals of small molecules was not possible because of the sessile nature of many dislocations and the absence of a source for new defects during deformation.

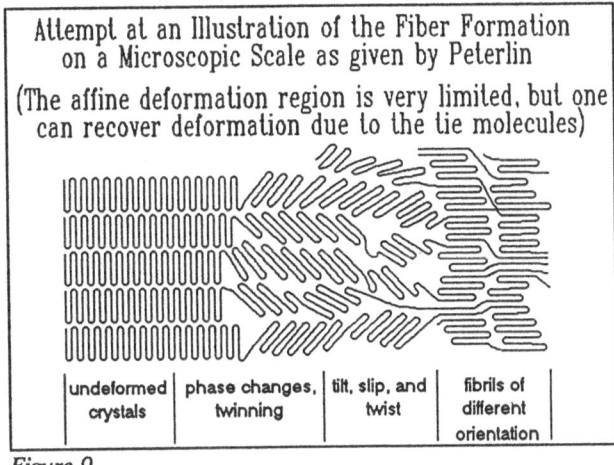

Attempt at an Illustration of the Fiber Formation on a Microscopic Scale as given by Peterlin

(The affine deformation region is very limited, but one can recover deformation due to the tie molecules)

| undeformed crystals | phase changes, twinning | tilt, slip, and twist | fibrils of different orientation |

Figure 9

The most important deformation process of polymer crystals is that occurring on drawing, cold-extrusion, and rolling. These processes involve a quite chaotic process, as is illustrated in Fig. 9. Based on the MD simulations the defects involved in these processes involve the conformational motion coupled to skeletal vibrations, and fall into the class of *point defects*. They are much different for polymer crystals when compared to small-molecule crystals and provide the ultimate element in the understanding of structure-sensitive properties. The differences arise from the connectiveness of the permanent covalent bonds of the backbone chain. Vacancies can only arise together with other defects such as kinks, chain ends, or wrong substituents in the chain (copolymers). Interstitial atoms must be linked with chain disorder, copolymerization, or monomeric impurities. Similarly, substitutional positions must be caused by prior copolymerization or by cocrystallization with small molecules (rare) [4]. Chain disorder, such as kinks, folds, jogs, torsions (twists) are the basic point defects in macromolecular crystals [4].

A starting point of defect evaluation can be the rotational isomer model. It makes use of the three-fold minimum in rotational energy [Eq (5)]. To produce a defect in an all-*trans* sequence of polyethylene, one can, for example, represent a defect conformation as a sequence of *gauche* (g$^+$, g$^-$; $\tau = 300°$ and $60°$, respectively) and *trans* (t; $\tau = 180°$) conformations. A simple defect in polyethylene is a "2g1 kink," represented by the sequence of g$^{\pm}$tg$^{\mp}$. The symbol 2g indicates the number of *gauche* conformations, and the second numeral, the amount of shortening of the chain in multiples of $c/2$ (c is the unit cell dimension along the chain axis). A 2g1 kink displaces the chain in addition by about 0.2 nm in the directions at right angles to the zig-zag plane. The latter displacements were proposed to be removed by distortion of the chain. In case the displacement are bigger, as in 2g2, they reach a neighboring zig-zag plane and are called jogs. The intermolecular energy of a 2g1 kink was estimated from the increased volume multiplied with the cohesive energy density. Overall, it was estimated that an equilibrium concentration of 1 kink per 250 carbon atoms should be possible at 400 K [9].

Defects that do not involve rotational isomers, but gradual twists, compressions or expansions of the chain have also been proposed earlier. Successive small rotations about the backbone bonds and bond angle deformations are involved in such defects. A 360° rotation about the chain axis leads to proper register with the neighboring chains in the crystal above and below the defect. Defects of this type are called disclinations. Finally, a rotation of 180° can be combined with a translational defect of $c/2$ to achieve crystalline register above and below the defect. Such combination of disclination and dislocation is called a dispiration [10].

Before it became possible to simulate a sufficiently large crystal by MD, the detailed behavior and energetics of point defects was not well understood. The often applied molecular mechanics calculations to establish the lowest energy conformations are not able to map out the actual reaction path and often lead, when considering intermolecular interactions, to much too high energies and activation energies.

SIMULATIONS OF POINT DEFECTS

The immediate observation on MD simulation, even for the case of a single chain surrounded by fixed chains, was that more than 100 K below the expected melting temperature a large number of conformational defects could be seen. Based on molecular mechanics calculations, it was deduced that such defects should be rare. Rotating a single bond from *trans* to *gauche* should completely remove the ability to fit into the array of parallel chains of the crystal. Figure 10 illustrates how such disorder

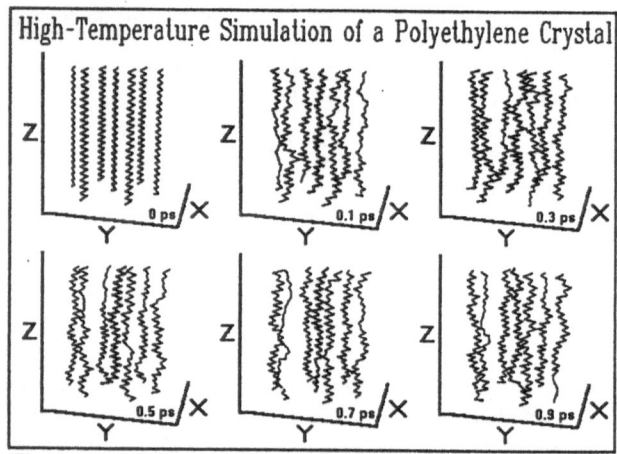

Figure 10

is spread over larger areas of the crystal with torsional and bending vibrations being involved in the existence of the conformational defects (compare to Fig. 6). Figure 10 represents results from simulations above actual melting. Due to the presence of a static ring of 12 additional chains that maintain a constant volume, melting is avoided. The short-time simulation can in this way show several defects.

The rate of formation of *gauche*-defects (counted as soon as the rotation angle τ exceeded $\pm 90°$ from the *trans* conformation) as a function of the size of the model crystal and temperature is displayed in Fig. 11. As the number of surrounding chains increases, the restriction due to the rigid

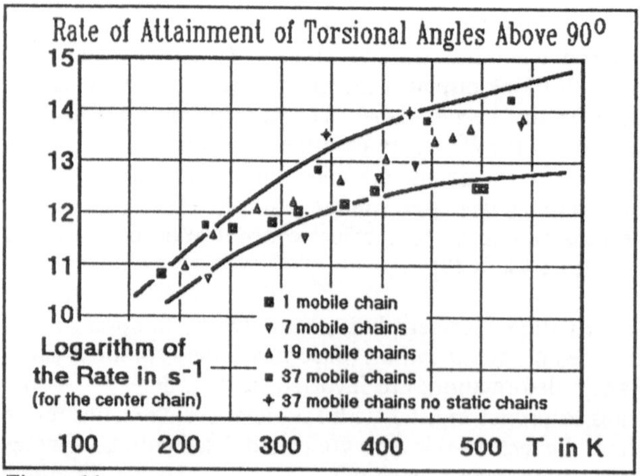

Figure 11

crystal decreases, and the rate of formation of *gauche* bonds increases by 0.5 to 2.0 orders of magnitude. The upper curve represents a limit of no enclosure of rigid chains (constant pressure simulation). At 350 K there are on the order of 10^{12} large, internal rotations per second about the backbone bonds, or about 10^{10} per bond. Such an enormously active source of defects can naturally drive many defect-linked processes in polymer crystals. From rate data, as shown in Fig. 11, the activation energies for the formation of *gauche* bonds could be derived using the transition state theory. It was found to be close to the potential energy barrier introduced in Eq. (5), varying from 13 to 25 kJ/mol, with 16.7 kJ/mol most common for the less restrictive models. The relatively low steady state concentration of *gauche* conformations suggests an extremely short life-time of a *gauche* defect. So short, in fact, that experiments that average over long time periods do not observe the deviations.

The population of rotated bonds for the constant-volume, 19 mobile chains surrounded by 18 static chains, with each chain consisting of 100 CH_2 groups, is shown in Fig. 12 as a function of temperature. Adding the shallow peak about the *gauche* angle at 322 K, for example, leads to a total concentration of only about 0.5%. Note that for small deformation angles the (vibration) amplitudes are truncated heavily, with the actual percentages written at the top for each standard interval. Note also, that the crystal environment has changed the *gauche* maximum from cos $\tau = 0.5$ to cos $\tau \approx 0.4$ (rotation from 180° to 66° or 294° instead of 60° or 300°). The increased potential energy due to the conformational defects was closely duplicated by the observed increase in heat capacity in the same temperature range, as shown in Fig. 7 [3].

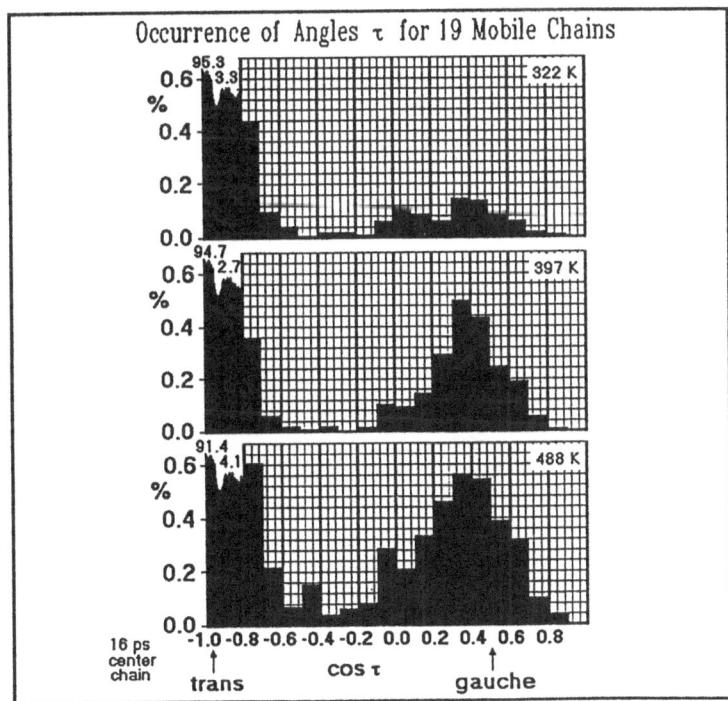

Figure 12

Examination of the dynamics depicted, for example, in Fig. 10 enables the search for possible point defects and a microscopic mechanism for structure-sensitive macroscopic processes. Figure 13 shows a typical plot of the life-time of *gauche* defects. The lifetime of the defects is, as expected in the picosecond range, and shows predominant sequences that can be characterized as $g^{\pm}tg^{\mp}$ kink defects and cause a shortening of the chain by $c/2$, in addition to some displacement in the a and b crystallographic directions. From the measured average chain-length at 440 K there is a shortening of about 1 nm, or an equivalent of four 2g1 defects, in approximate agreement with Fig. 13.

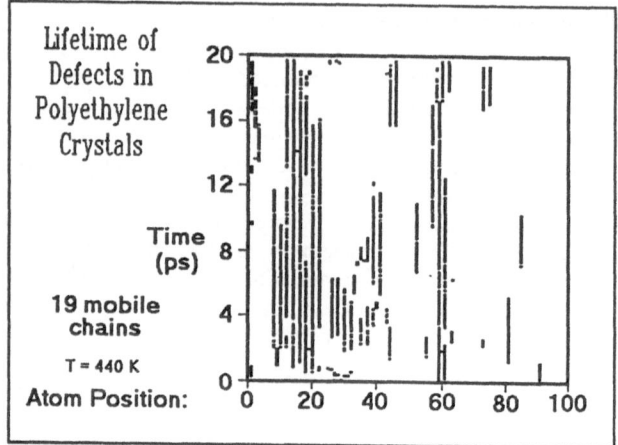

Figure 13

Mechanistic details for the formation of a 2g1 kink can be derived from Fig. 14

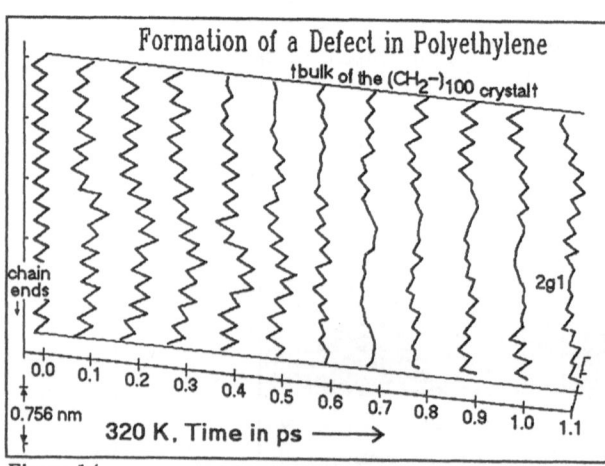

Figure 14

which depicts the conformation of a single chain as a function of time. The first 1.1 ps of a simulation at 320 K are displayed for the bottom portion of a chain in a crystal. One can easily recognize the process that leads in the end to the defect formation, a transverse vibration initiated at time 0.1 ps, a compression of the chain at 0.7 to 1.1 ps, and a torsional vibration moving into the field of view at time 0.5 ps. Their unique collision between 0.7 and 1.0 ps causes the 180° turn of the chain end at 0.8 ps and the indicated localized 2g1 kink at 1.1 ps. The kink caused the remaining end of the chain to be in register 1/2 unit cell length removed upwards. In this particular event, registered in Fig. 14, the life-time of the 2g1 defect was about 2 ps [3].

Although the majority of bonds are at any one time close to *trans* "conformers", each of them is found to frequently deviate more than expected from normal vibrations (see Fig. 12, ±40°). The zig-zag chains may then twist by sequential deviations in the same rotation direction. The formation of such twists is illustrated in Fig. 15. A series of time snap-shots at 316 K is given for the center-chain of a simulated crystal of the type shown in Fig. 3. The chain undergoes rapid twisting motion that results in a change of the orientation of the zig-zag plane of the molecule. After about 1.5 ps the chain is rotating *via* segmental motion, starting at the top of the crystal. Within about 0.5 ps the chain has turned by about 90°. Similarly, a new side-on view of the zig-zag develops between 7 and 8 ps, indicating another overall 90° rotation, turning roughly

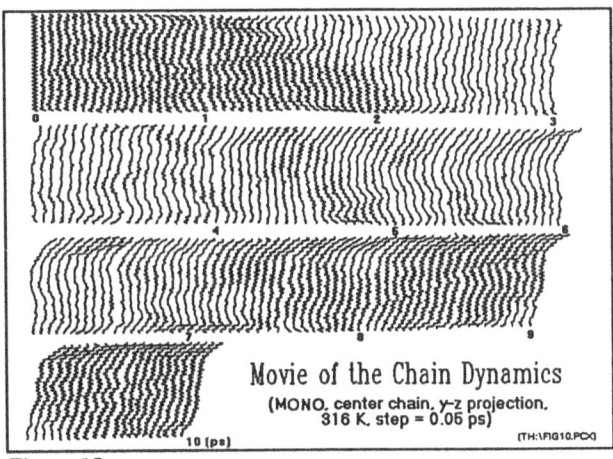

Movie of the Chain Dynamics
(MONO, center chain, y-z projection.
316 K, step = 0.05 ps)

Figure 15

back to the initial orientation. That the rotation was not a full 180° turn can be deduced from the direction of the terminal group of the zig-zag chain in Fig. 15. As temperature increases, the frequency and magnitude of this "rotational" motion increases. In addition, the accordion mode of vibration causes diffusion of the whole chain in the crystallographic *c*-direction, and is followed by tilting of the chain-heads.

The next step of analysis was the introduction of an external force to the chain. Adding a free energy gradient on the thermal motion of the crystal permits a diffusion of the chains in the direction of the gradient. To study the mechanism in more detail, the defects were noted in Fig. 16, simultaneously with the positions of the upper and lower chain ends in Fig. 17, and the end-to-end distances in Fig. 18. The illustrated crystal has a chain length of 100 CH_2-groups. The immediate observation is that the *gauche* defects do not move through the crystal, *i.e.* they are *not* involved in the chain diffusion, although without their presence, the transport appears to stop. The actual chain transport must thus be based on gradual twists or accordion-like chain-vibrations, initiated in the gradient direction at the *gauche* defects. From Fig. 17 it can be seen that the chain moved about 2.5 nm out of the crystal. The six nearest neighbor positions are marked by the horizontals and have, outside the longitudinal vibration, no translational motion. A move in the chain direction, making use of the amplitude of the longitudinal vibration (longitudinal acoustic vibration, LAM), means full unit cell motion (≈0.25 nm) would be neces-

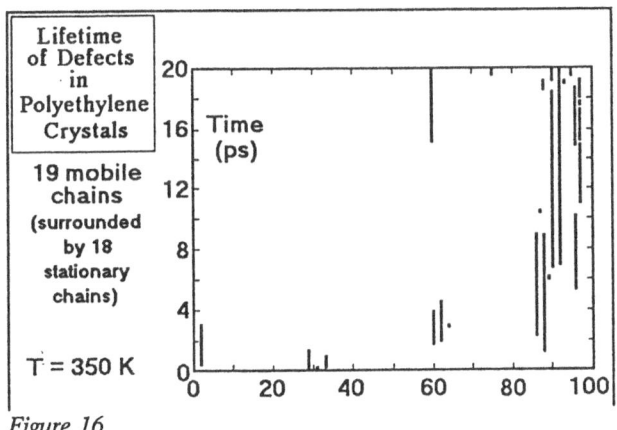

Lifetime of Defects in Polyethylene Crystals

19 mobile chains (surrounded by 18 stationary chains)

T = 350 K

Figure 16

48

sary to maintain crystallographic register with the neighboring chains. Involving the lower frequency torsional oscillations permits steps of half the unit cell, coupled with a 180° rotation of the chains. Inspection of Figs. 17 and 18 seems to indicate steps of c/2 superimposed on the LAM, *i.e.* during the compression phase, the chain end is pulled into the crystal and during the expansion, it is expelled, coupling the LAM with the torsional vibration and involving

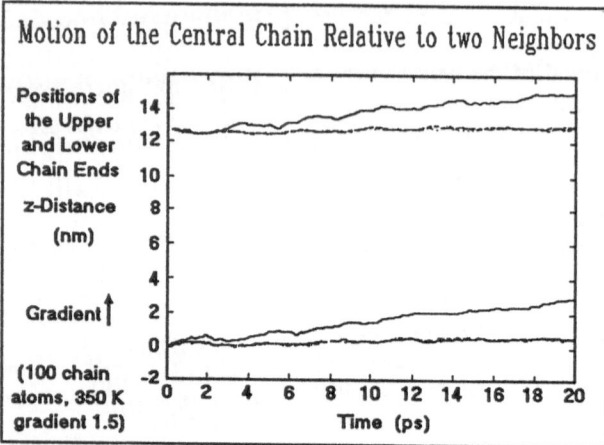

Figure 17

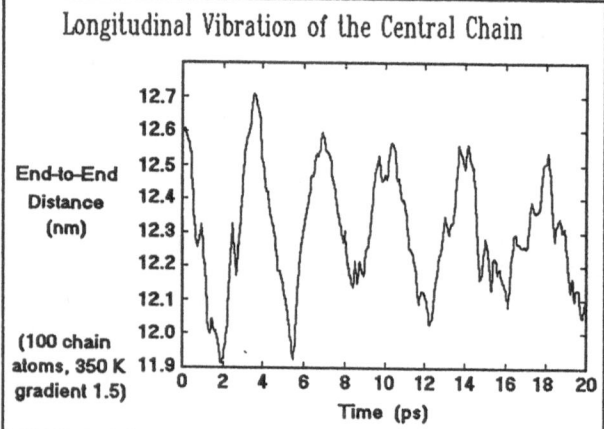

Figure 18

the conformational defect to initiate the motion. During, the approximately 5 vibrations, about 20 c/2-steps occurred, as can be counted from the small irregularities superimposed on the LAM.

Comparison with experimental data shows that the LAM frequency is of the order of magnitude expected, and its change with chain length has the required $1/\sqrt{n}$ functional relationship. Figure 19 illustrates further, that an increase in thickness of

the lamellar crystal, as derived from chain diffusion, is that found for annealing of crystals. Similarly, the slowdown of motion with lamellar thickness, as shown in Fig. 20, is what is expected from experiment.

CONCLUSIONS

Many experimental properties of defect crystals are caused by picosecond, large-amplitude motion that can already be simulated by MD

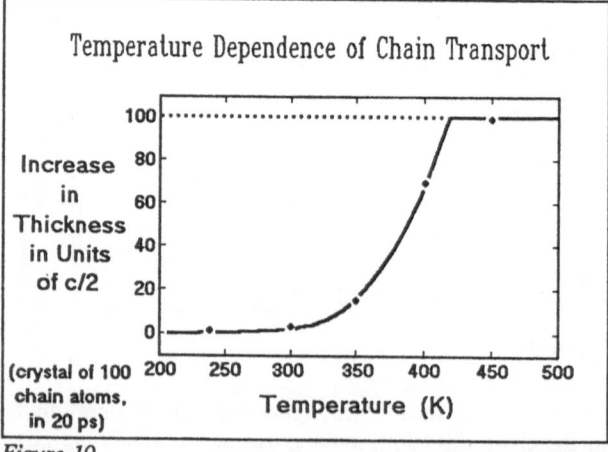

Figure 19

49

on sufficiently large crystals to give a realistic mechanism. A more definitive picture of the defect solid state of the macromolecules is emerging from this work. The new generation of supercomputers should also permit the solution of problems arising from the inclusion of hydrogen atoms in the simulation, let us extend the simulations into the nanosecond range, and solve the questions of structure and dynamics of the crystal interface.

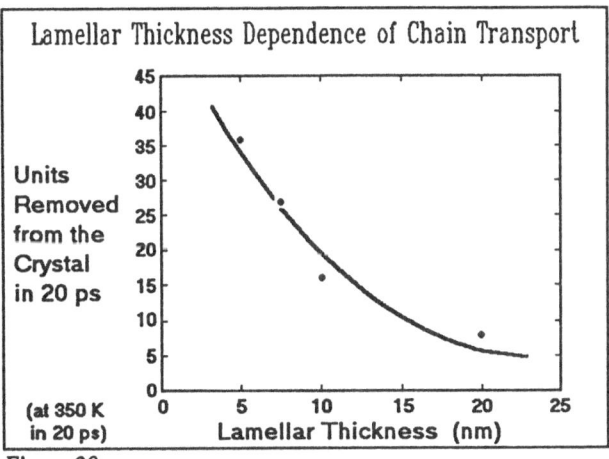

Figure 20

Acknowledgements: This work was supported by the Division of Material Sciences, Office of Basic Energy Sciences, U.S. Department of Energy, under Contract No. DE-AC05-84OR21400 with Martin Marietta Energy Systems, Inc., and by the Polymer Program of the National Science Foundation, present Grant No. DMR-92-00520. Computations were performed on the UTCC computers at the University of Tennessee, Knoxville, the Cray X-MP at Oak Ridge National Laboratory, the Cray Y-MP at the National Center for Supercomputing Applications, and the Cray Y-MP at the Pittsburgh Supercomputer Center.

References:

1 Sumpter BG, Noid DW, Liang L, Wunderlich, B (1993/4) Atomistic dynamics of macromolecular crystals, *Adv Polymer Sci.* to be published

2 Noid DW, Sumpter BG, Wunderlich B, Pfeffer GA (1990) Molecular dynamics simulations of polymers: methods for optimal fortran programming, *J Comp Chem*, 11:236; Noid DW, Sumpter BG, Cox RL (1991) Computational strategies for molecular dynamics simulations of polymer processes: numerical integration schemes, *J Comp Polym Sci* 1:161

3 Sumpter BG, Noid DW, Wunderlich, B (1990) Computer experiments on the internal dynamics of crystalline polyethylene, mechanistic details of conformational disorder, *J Chem Phys* 93:6876; (1992) Computational experiments on the motion and generation of defects in polymer crystals, *Macromolecules* 25:7247

4 Wunderlich B (1973, 1976, 1980) Macromolecular physics; vol I, crystal structure, morphology, defects; vol II, crystal nucleation, growth, annealing; vol III, crystal melting, Academic Press, New York

5 Kitaigorodskii AI (1955) Organicheskaya Kristallokhimiya, Press of the Akad Sci USSR, Moscow

6 Bunn CW (1939) The crystal structure of long chain normal paraffin hydrocarbons; the "shape" of the $>CH_2$-group, *Trans Farad Soc* 35:483

7 Wunderlich B (1970) Heat capacities of linear high polymers, *Adv Polymer Sci* 7:151

8 For an up-to-date copy write to the author

9 Pechhold W (1970) Theorie der Phasenumwandlung in Polymeren, *Ber Bunsenges* 74:784

10 Reneker DH, Mazur J (1988) Small defects in crystalline polyethylene, *Polymer*, 29:3

AFM Observation of Surface Morphology
for Oriented PET Films

Y. Sasaki, X.J.Shao, T. Suzuki, H. Ishihara

Toyobo Research Center, Toyobo Co.,Ltd
1-1, Katata 2-chome, Ohtsu, Shiga, 520-02 JAPAN

ABSTRACT: The atomic force microscopy (AFM) was used to examine surface morphologies for biaxially oriented PET films, and the results were discussed from the view point of strain induced structure formation. For the most of oriented films, the nodular microstructure was observed. Its height and diameter were several nm and 20-50nm, respectively. Although the microstructure was observed also on the surface of as-cast film, its height was less than 1nm. The plot of surface roughness measured by AFM as a function of the draw ratio showed that the microstructure had developed gradually with drawing until a certain draw ratio (λ c), but the development had declined after λ c. The AFM images and the behavior of roughness change also showed that a higher order structure like a spherullite was easily formed by annealing for the films undrawn or drawn less than λ c, while the films drawn more than λ c were little influenced by the annealing. The strain induced crystallization were examined by wide angle X-ray scattering (WAXS), density and average refractive index measurements. The experimental results suggested that the development of surface microstructure was closely related to the strain induced crystallization, and the strain induced crystallization in the bulk slightly delayed compared with the development of surface microstructure.

INTRODUCTION

Surface properties of biaxially oriented polyethyleneterephthalate (PET) films are quite important for its various industrial or consumer use as well as bulk properties. The requirements for surface properties include tribological properties such as coefficient of friction, abrasion resistance, adhesiveness, while the requirements for bulk properties include thermal and mechanical properties such as Young's modulus, tensile strength, dimensional stability and so on. These requirements for the surface must closely relate to the *top surface structure*, which may be different from the bulk structure.

There may be two approaches to clear the *top surface structures* of oriented polymer films. One is a structural analysis of near surfaces, *e.g.*, attenuated total

A. Teramoto, M. Kobayashi, T. Norisuje (Eds.)
Ordering in Macromolecular Systems
© Springer-Verlag Berlin Heidelberg 1994

reflection IR spectroscopy (ATR–IR) [1–9], fluorescence polarization method for vacuum–deposited dye molecules on film surfaces [10] and polarized reflection spectroscopy [11]. Some of these studies reported higher degree of molecular orientation in the surface region than that in the bulk [1,10]. But some others reported no significant difference between the surface and the bulk [2,11]. Although we can presume the *top surface structures* of oriented polymer films from these analysis of near surfaces, it still not be fully cleared.

Another approach to clear the *top surface structure* may be a direct observation of surface microstructures. Geil *et al.* [12–14] observed surface morphologies of oriented and unoriented PET films using transmission electron microscopy (TEM). They found two kinds of microstructures. One is a smaller structure (75–100 Å) observed in unoriented films. The other is a larger structure (200–500 Å) observed in oriented films. They proposed that these structural units were originally present in the amorphous PET and did not change by deformation. These structural units were called "nodule" for smaller units, and "supernodule" for larger units. They assumed that the "nodule" was a paracrystalline structure, and the "supernodule" was a unit of deformation, which was composed from an aggregate of smaller nodules. Although the existence of such units could not be confirmed by other methods, it might be a fact that some kinds of microstructures were present on the surfaces of PET films. We believe that the detailed study of these microstructures may contribute to clear the *top surface structures* of oriented PET films.

In this work, the microstructures on the surfaces of unoriented and oriented PET films with different draw ratios were directly observed using an atomic force microscopy (AFM). The AFM is a very powerful tool for the observation of surface morphologies of polymer solids [18–21], since it is possible to observe the surfaces as they are in the air or liquids without any pretreatments like metallizing, and makes it possible to get quite high lateral resolution and vertical accuracy down to atomic scale, in principle. Using these characteristics of AFM, we investigated the development process of *top surface* microstructure. Furthermore, the development process of the microstructure was compared with the development process of strain induced crystallization in the bulk, which is well known structure formation by drawing [12,15–17]. We will report the results herein.

EXPERIMENTAL

Film specimens were prepared of polyethyleneterephalate (PET). PET was polymerized in TOYOBO Co., Ltd. with a number–average molecular weight of 20000. Unoriented, as–cast, amorphous PET sheet of 250 μ m thickness were prepared by melt–quenching method through an extruder and T–dies.

Biaxially drawn films with different draw ratios were prepared by using T.M.Long biaxial stretcher, employing blown–air heating. The draw temperature was 90 °C, and the nominal strain rate was 5000%/min. The draw ratio was varied from MD × TD = 1.7 × 1.7 to 4.0 × 4.0. After drawing, some of the specimens were heat treated (annealed) at 220 °C for 30sec, under fixed dimensions. In order to get reliable data in following measurements, ten films were prepared for each condition at random.

AFM observations were carried out in the air for oriented or annealed films and in KCl water solution for as-cast films, using a SFA300 scanning probe system (SEIKO Instruments Inc., Tokyo, Japan) . The surfaces of the samples were scanned by a tiny pyramidal Si_3N_4 tip (Radius = 25 μ m) which is attached to a microfabricated cantilever (200 μ m triangular base). Deflections of the cantilever were registered via deflection of a laser beam. The scan area was 1 μ m and 5 μ m square. In order to verify the observed images, the images were taken repeatedly at least ten times with different scanning directions and different specimens.

WAXS was carried out by use of a rotating-anode generator (RU-200, Rigaku Corporation, Japan) with the maximum power of 4KW. Density was measured at 25 °C in a density gradient column made of CCl_4 / n-hepthane mixed solvent. Average refractive index was calculated from the refractive indices measured in the three principal directions by use of an Abbe refractometer.

RESULTS

Figure 1 shows three-dimensional images of 1 μ m square AFM scans for the films undrawn as-cast (a) and drawn to the ratio of MD × TD = 4.0 × 4.0 (b). On the surface of the drawn film, nodular microstructure having about 30nm of lateral diameter was observed. The surface appearance of the drawn film was similar to the TEM image reported by Geil and co-workers [13,14], so called "supernodule". The surface of the drawn film was rougher than that of the undrawn film. The average surface roughness of the drawn film was about 1nm, while that of the undrawn film was only about 0.1 nm. The roughness increased ten times by drawing. Top views of 5 μ m square AFM scans for the films annealed at 220 °C for 30 sec were shown in figure 2, comparing undrawn and drawn films. Annealing effected a remarkable change on the surface morphology of undrawn film, that is, the surface roughness was markedly increased. This increment of surface roughness may be due to a formation of higher order crystalline structure by annealing. While, the annealing effect on drawn films was rather small, namely, the surface microstructure caused by drawing did not change by annealing.

In order to investigate the growth of surface microstructure by drawing, the average surface roughness obtained from 1 μ m square AFM scans were plotted as a function of draw ratio in figure 3. It was found, from figure 3, that the average roughness of unannealed films (solid line) increased with increasing draw ratio, and then it became steady after reaching a certain roughness. The draw ratio at which the average roughness reached steady was about MD × TD = 2 × 2 , that

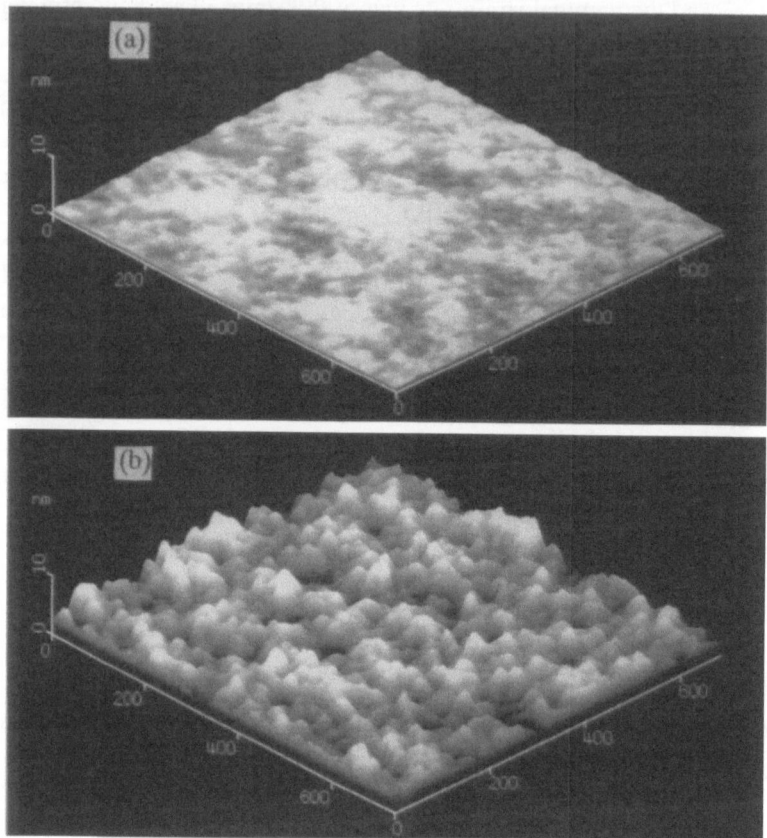

Fig.1 Three-dimensional images of 1 μm square AFM scans for the films undrawn as-cast (a) and drawn to the ratio of MD x TD = 4 x 4 (b)

was the ratio at which the surface microstructure occurred in AFM observation. In the region of constant surface roughness, the appearance of surface microstructure did not change. These observations suggest that the average roughness can be applied as an indication of the growth of surface microstructure. The surface microstructure grew with increasing draw ratio, and once it had been formed, it did not change by further drawing. The average roughness may also be applied as an indication of annealing effect on the surface microstructure (dashed line in figure 3). For the films with smaller draw ratios (up to MD $\times$ TD = 2 $\times$ 2), annealing effected a remarkable change on the surface roughness, suggesting a formation of higher order crystalline structure. While, in the region of larger draw ratio, the annealing effect was rather small, that is, the surface roughness and appearance of AFM images were almost the same as that of unannealed films. These results suggested that the formation of surface microstructure inhibited the formation of higher order crystalline structure like a spherullite by annealing.

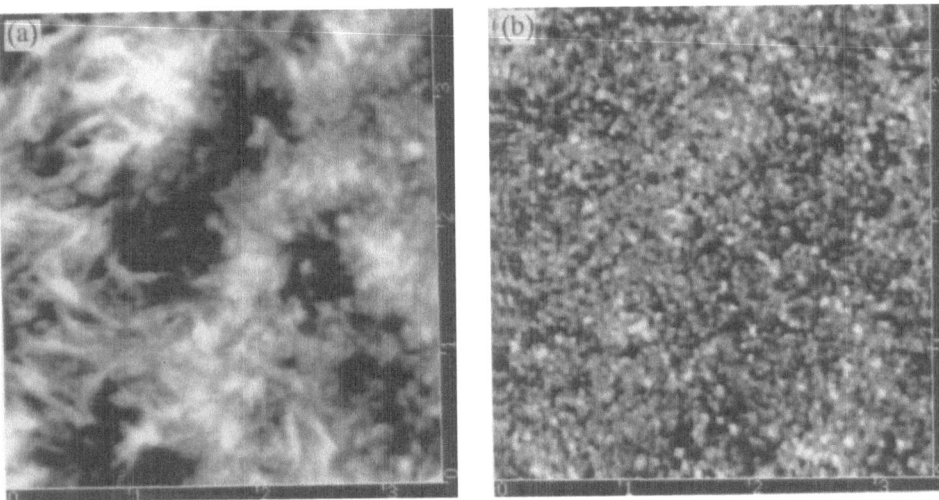

Fig.2 Top views of 5 μm square AFM scans for annealed films;
undrawn and annealed at 220℃ for 30 sec (a), drawn to the
ratio of MD x TD = 4 x 4 and annealed at 220℃ for 30 sec (b)

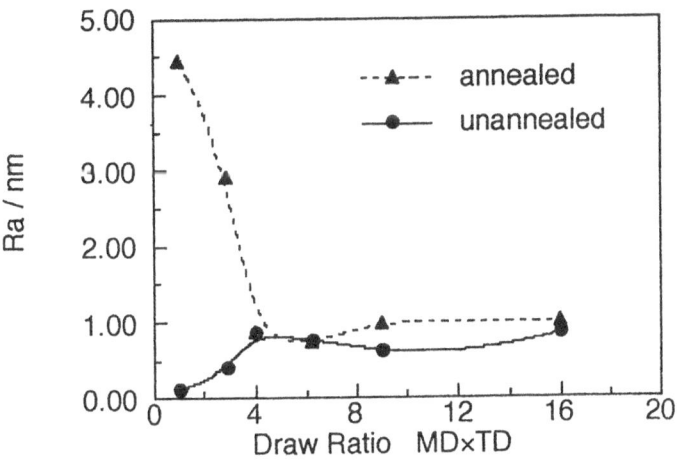

Fig.3 The average roughness of biaxially oriented PET films
before and after annealing plotted against draw ratio

In order to compare the growth of microstructure with the formation of strain induced crystallization in the bulk, WAXS, density and refractive index measurements were carried out. Figure 4 shows the WAXS patterns for undrawn film and biaxially drawn PET films for scattering angles 2θ between $15°$ and $30°$. For the films with smaller draw ratios up to $MD \times TD = 2 \times 2$, diffuse halo patterns indicated that the films were not crystallized. The peak resulted by crystallization, which was identified to the plane of (100), could be observed for the film with a draw ratio of $MD \times TD = 2.5 \times 2.5$, and the intensity increased with increasing draw ratio.

Figure 5 shows the draw ratio dependence of density for unannealed PET films. It was found that the density increased steeply between the draw ratios of $MD \times TD = 2 \times 2$ and 3×3, and then the increment declined gradually. Figure 6 shows the draw ratio dependence of average refractive index for unannealed PET films. The behavior was similar to that of the density. The results of WAXS, density and refractive index measurements showed that the strain induced crystallization in the bulk developed with increasing draw ratio, and declined after a certain draw ratio. Comparing the surface morphological behavior with the bulk behaviors, it was suggested that the formation of nodular microstructure on the surface closely related to the strain induced crystallization in the bulk.

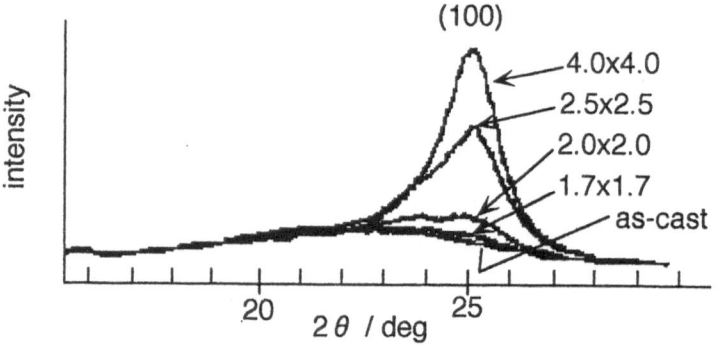

Fig.4 WAXS of biaxially oriented PET films

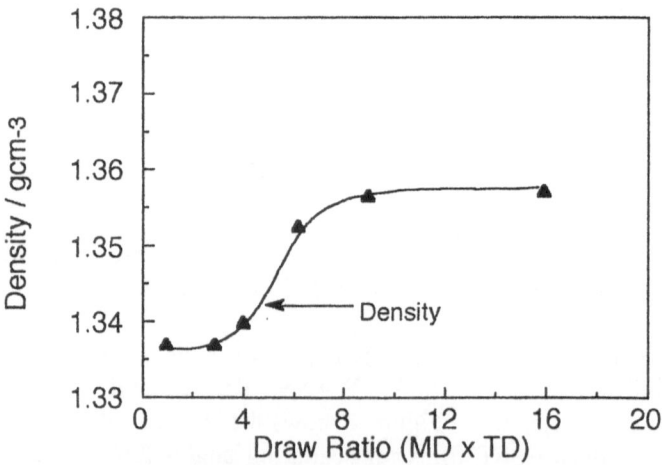

Fig.5 The density of biaxially oriented PET films plotted against draw ratio

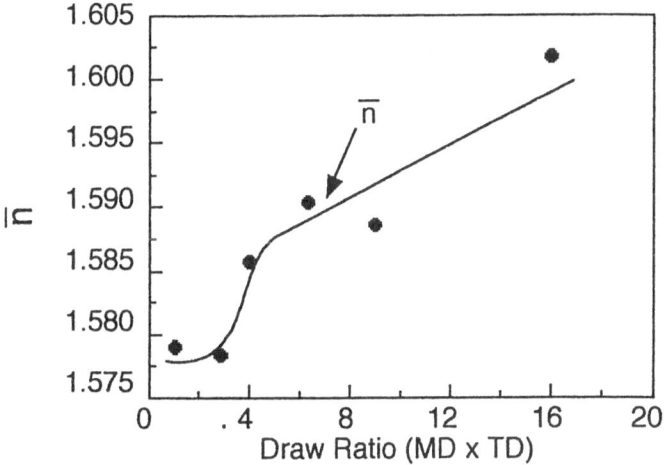

Fig.6 The average refractive index of biaxially oriented PET films plotted gainst draw ratio

DISCUSSION

Geil *et al.* noted that the nodular microstructure so called "supernodule" did not change by drawing, and concluded that the nodular microstructure was a unit of deformation. However, their observation lacked dimensional accuracy in vertical direction. The AFM observation revealed that the average roughness increased markedly by drawing. This behavior coincided with that of strain induced crystallization in the bulk. These results may be explained as follows. The strain induced crystallization resulted in the fluctuation of the density or the rigidity near the surface. This fluctuation appeared as a surface morphological microstructure, *i.e.*, surface roughness. Namely, it may be said that the surface microstructure is an indication of the strain induced crystallization in the *top surface*.

Comparing the development process of surface microstructure with that of strain induced crystallization (figure 3 to figure 6), the surface microstructure developed at a smaller draw ratio than the strain induced crystallization in the bulk. That is, the AFM images may reflect the *top surface structure*, while the density and WAXS represent the degree of crystallization in the bulk, and the average refractive index may represent the crystallinity in the near surface owing to its principle of total reflection at the surface. If we compare the dependences of the results by these four measurements on the draw ratio, the surface microstructure developed at first, then the refractive index increased, and finally the density and the intensity of WAXS peak increased. These results may suggest that the structure development in the *top surface* is much activated than that in the bulk or that in the near surface.

REFERENCES:

1 Walls DJ, Coburn JC (1992) J Polym Sci Polym Phys 30:887
2 Sung CSP (1981) Macromolecules 14:591
3 Yuan P, Sung CSP (1991) Macromolecules 24:6095
4 Ito M, Pereira JRC, Porter RS (1982) J Polym Sci Polym Lett 20:61
5 Xue G (1985) Makromol Chem Rapid Commun 6:811
6 Zerbi G, Gallino G, Fanti ND, Baini L (1989) Polymer 30:2324
7 Tshmel AE, Vettegren VI, Zolotarev VM (1982) J Macromol Sci Phys B21:243
8 Mirabella FM (1984) J Polym Sci Polym Phys 22:1293
9 Wang LH, Porter RS (1983) J Appl Sci 28:1439
10 Ohmori S, Ito S, Onogi Y, Nishijima Y (1987) Polym J 19:1269
11 Kaito A, Nakayama K, Kanetsuna H (1988) J Polym Sci Polym Phys 26:1439
12 Yeh GSY, Geil PH (1967) J Macromol Sci Phys B1:251
13 Klement JJ, Geil PH (1971) J Macromol Sci Phys B5:505
14 Klement JJ, Geil PH (1971) J Macromol Sci Phys B5:535
15 Heffelfinger CJ, Schmidt PG (1965) J Appl Polym Sci 9:2661
16 Vries AJD, Bonnebat C, Beautemps J (1977) J Polym Sci Polym Symp, 58:109
17 Spruiell JE, McCord DE, Beuerlein RA (1972) Tran Soc Rheol 16:535
18 Snetivy D, Vancso GJ (1992) Polymer 33:432
19 Occhiello E, Marra G, Garbassi F (1989) Polym News 14:198
20 Stocker W, Bar G, Kunz M, Moller M, Magonov SN, Cantow HJ (1991) Polym Bull 26:215
21 Yang ACM, Terris BD, Kunz M (1991) Macromolecules 24:6800

Characterization of Polymer Interface by Micro Raman Spectroscopy

S.Hosoda, S.Hoshi, K.Kojima, A.Uemura, H.Yamada and *M.Kobayashi

Sumitomo Chemical Co.Ltd., Chiba Research Laboratory,
Ichihara, Chiba 299-01, JAPAN

*Department of Macromolecular Science, Faculty of Science,
Osaka University, Toyonaka, Osaka 560, JAPAN

Abstract: Micro Raman spectroscopy was applied to characterize the interface region of multi-layered film specimens in which ethylene-propylene random copolymer(EPR) was sandwiched with polyethylene(PE) or polypropylene(PP) sheets. Micro Raman line scan of the interface region for PE/EPR pair indicated that PE molecules migrated into the EPR phase and crystallized there, forming a mixed phase where the PE crystallites were embedded in EPR matrix. Apparent diffusion constant(D) evaluated from the thickening rate of the mixed phase strongly depended on the chemical composition of EPR, i.e., D increased with increasing the ethylene content of EPR. On the contrary, the mixed phase thickness of PP/EPR pair showed the opposite dependence on the composition of EPR.

1. Introduction

Many kinds of immiscible polymer systems have been increasingly used as industrial materials such as multi-layered extrusion films and polymer alloys. In such systems, the interfacial structure and the morphology strongly influence their properties. Then stabilization and quantitative evaluation of the interface region are essentially important subjects for achieving excellent properties.

Polyolefin blends composed of PE, PP and EPR are one of the most important multi-component systems that are industrially used in various fields. In the blends it is well known that the mechanical properties, for instance, are drastically changed by the structural factors of each constituent polymers. Then we investigated the chemical composition, the morphology and the orientation in the interface region of the layered sheets of PE/EPR and PP/EPR by using micro Raman spectroscopy combined with other techniques. This technique is useful to investigate the structure of micrometer scale taking advantage of its superior space resolution (about 1 μm), and has been used in the field of polymer materials[1], blends[2] and composites.[3-5]

2. Experimental

Materials: Fig.1 illustrates a model of three-layered specimens used in this study. The specimens in which EPR was sandwiched with PE or PP sheets were prepared by hot compression mold at 180 °C. PE and PP used were high density polyethylene and homo-polypropylene, respectively. Their melt flow rate(MFR) were 0.94 and 8.00 g/10min, respectively. The propylene content and Mooney viscosity of EPR used in these experiments are listed in Table 1. These specimens were cut into cross-sections with a microtome and were subjected to these measurements. The PE/EPR blends used for the observation of the morphology were melt-blended in a melt extruder for 10 min at 170 °C.

Measurements: Micro Raman spectra were recorded at 4 cm^{-1} resolution with a Dilor multichannel Raman spectrometer using the excitation beam of 514.5 nm focused through a 50 x objective(space resolution 2 μm). The morphology of the melt blends of PE and EPR was observed with a transmission electron microscope(TEM) by staining the micro-tomed specimen with RuO$_4$. Wide-angle X-ray diffraction patterns were obtained with a micro-diffractometer(collimator 10 μmϕ) and imaging plates, using a Ni-filtered Cu Kα-radiation generated by a Rigaku rotating anode x-ray generator, RU-200, operated at 50 kV and 150 mA.

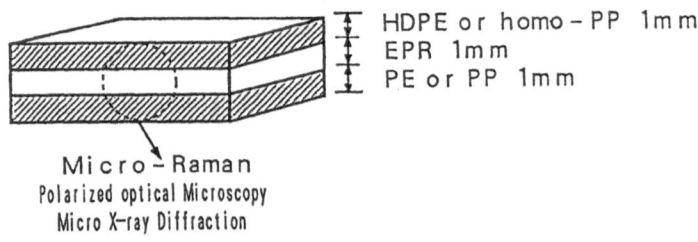

Fig.1 Schematic of a three-layerd specimen.

Table 1 Chemical composition and Mooney viscosity of EPR.

C$_3$' (wt%)	ML (121 °C)
22	26
27	33
54	29
67	25

3.Results and Discussion

Fig.2 shows micro Raman spectral profiles scattered from the interfacial region between the PE and EPR(C_3'=27%) phases. One sample(a) was kept in a hot press at 180 °C for 5 hours and another(b) was not heated. The spectrum was obtained for each 5 μm step. The 21 spectra were measured along the direction perpendicular to the interface. In the spectra of PE region, the crystallite band at 1416 cm^{-1} due to the CH_2 bending mode in the orthorhombic lattice is observed for both samples. It is very interesting that this crystallite band can be clearly seen in the spectra of EPR side too, for the heat treated sample. On the contrary, for the non-treated sample, this band disappears in the spectra of the EPR side. The depth profiles was obtained by using the intensity of the crystallite band to the total integrated intensity of the entire CH_2 bending region(Fig.3). For the heat-treated sample, the 1416 cm^{-1} band in the EPR side decreases in intensity rather gradually with increasing distance from the interface, and is still detectable even at a point 200 μm apart from the interface. This phenomenon suggests two possibilities; one is the crystallization of the PE molecules which migrate into the EPR phase, and the other is the crystallization of the crystallizable components in the EPR molecules. In order to confirm this point, we measured the same kind of micro Raman spectrum on the deuterated PE(d-PE) / protonated EPR pair. As a result, in the heat treated sample the 1416 cm^{-1} band was not observed in the both sides of the interface. Therefore, the 1416 cm^{-1} band observed in the interface region of the PE/EPR blend is considered to be due to PE molecules migrated into the EPR phase. We define this diffusion distance of PE into EPR as "the mixed phase thickness".

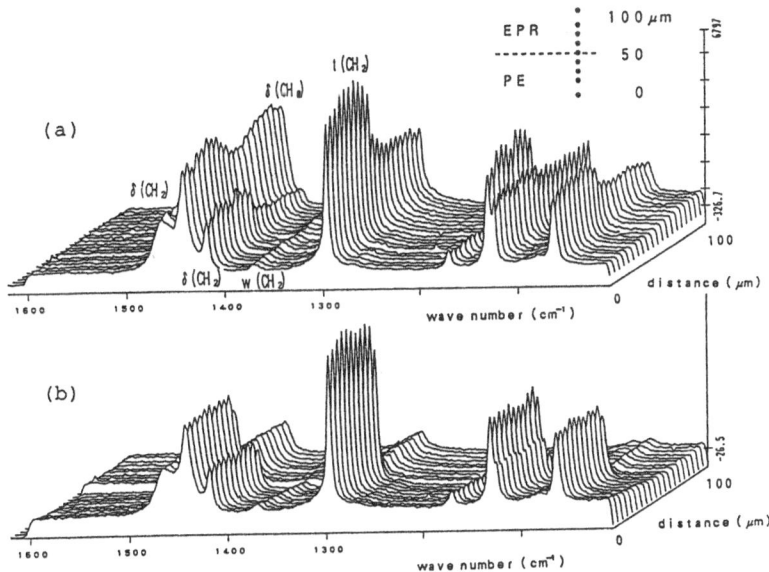

Fig.2 Micro Raman spectral profiles obtained for the interface region between PE and EPR(C_3'=27%) pair; (a)heat treated at 180 °C for 5 hours, (b)non-treated.

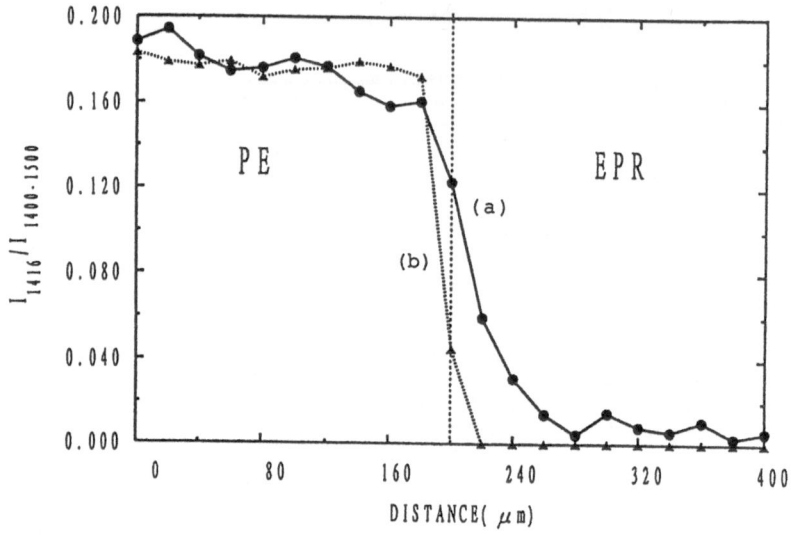

Fig.3　The depth profiles obtained from micro Raman spectra of PE/EPR(C_3'=27%) pair, (a)heat treated at 180 °C for 5 hours, (b)non-treated.

Fig.4 shows the time dependence of the mixed phase thickness for the PE/EPR(C_3'=27%) pair. The mixed phase thickness increases with time and the diffusion distance is evaluated as a few hundred micrometer in the case of the sample heated for 10 hours. The mixed phase thickness was found to follow the simple diffusion equation and the diffusion coefficient was obtained by the least-squares fitting to be 4.2×10^{-8} cm^2/s as shown in Fig.5. This value seems to be in agreement with the diffusion coefficients for the additives in polymers[6].

Fig.6 shows polarized optical micrographs taken on the interface region of the PE/EPR(C_3'=27%) pair, for different periods of heat treatment. The well-organized spherulites formed in the PE phase and the distinct boundary between two phases are observed. Furthermore, the crystals are observed in the EPR phase, too. These crystals seem to orient near the boundary, and both the total amount of the crystals in the EPR phase and the thickness of the oriented crystal at the boundary increase with heating time. These results correspond well with those of Raman spectroscopy described above.

The orientation of the crystalline phase near the interface was investigated by X-ray micro-diffraction at several points along the direction perpendicular to the interface of the PE/EPR(C_3'=27%) pair heated for 10 hours. Fig.7 shows the schematic diffraction patterns, and the 110/200 peak intensity ratios obtained from the intensity profiles in the vertical and the lateral directions of these diffraction patterns. In the PE side and at a point in the EPR 500 μm apart from the boundary, the 110/200 peak intensity ratios agreed with that of the unoriented PE. On the other hand, in the EPR side near the boundary, the intensity ratios are different from that of the unoriented PE. The result indicates the presence of oriented crystallites near the boundary in the EPR phase as expected by the result of polarized optical microscope.

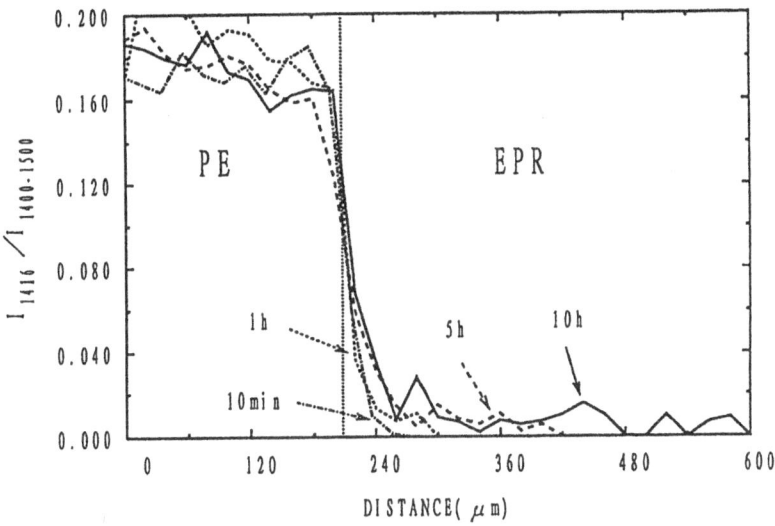

Fig.4 The depth intensity profiles at the interface region of PE/EPR(C_3'=27%) pairs. Number denotes the time of heating at 180 °C.

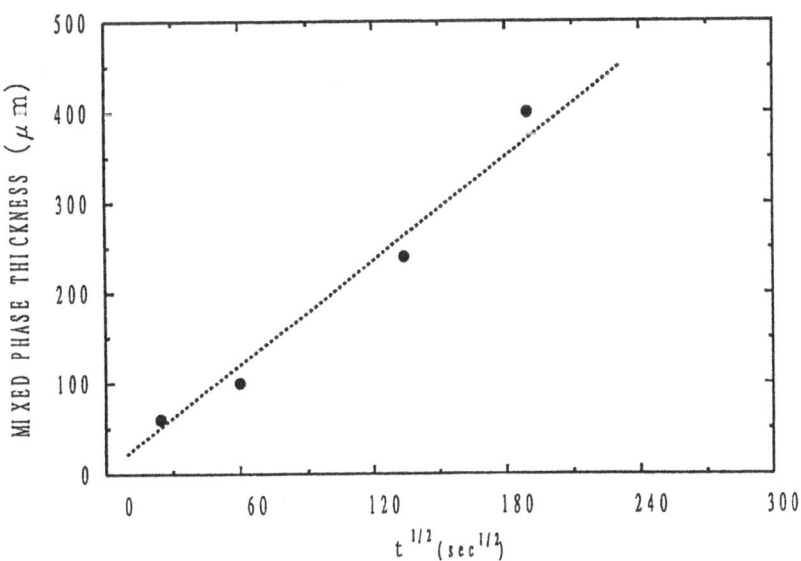

Fig.5 Diffusion coefficient obtained from the mixed phase thickness of PE/EPR(C_3'=27%) pair. The calculated diffusion coefficient was 4.2×10^{-8} cm^2 s^{-1}.

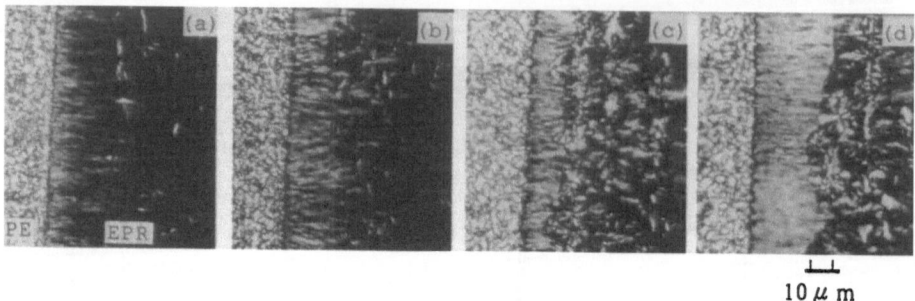

Fig.6 Polarized optical micrographs of the interface region between PE/EPR(C$_3$=27%) pair heated at 180 °C in different periods of heating; (a)10 min, (b)1 hour, (c)5 hours, (d)10 hours.

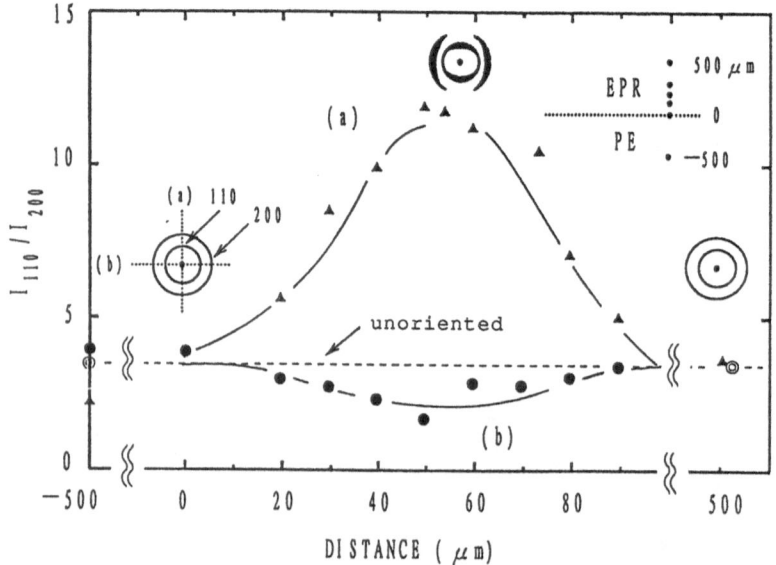

Fig.7 Intensity ratios of the 110 and 200 reflections obtained from X-ray micro-diffraction intensity profiles of PE/EPR(C$_3$'=27%) pair, (a)in the vertical direction, (b)in the lateral direction of the PE/EPR model figure. Diffraction patterns are also shown. ◎ ; unoriented PE.

Fig.8 shows the dependence of the mixed phase thickness and the diffusion coefficient(D_{app}) on the chemical composition of EPR. Both are strongly dependent on the chemical composition of EPR, i.e., the higher the propylene content, the smaller the D_{app} and the thickness. On the contrary, for PP/EPR pair the mixed phase thickness was rather thin compared with that of PE/EPR pair, and increase with increasing propylene content of EPR. The results seem quite reasonable, because the higher content of the common monomer(C_2' or C_3') results in the longer sequence length of ethylene unit or propylene unit in an EPR molecule, which gives higher affinity with respective homopolymer.

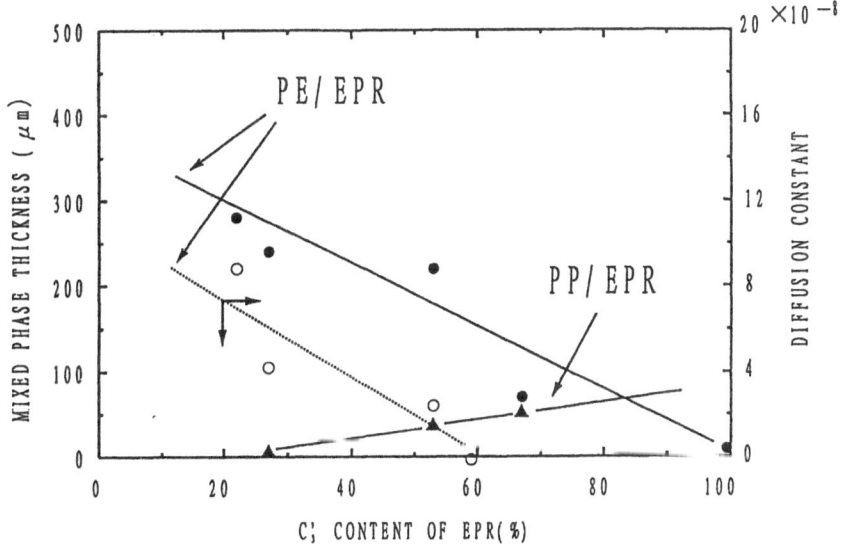

Fig.8   Dependencies of the mixed phase thickness and the diffusion coefficient on the chemical composition of EPR. The mixed phase thickness were determined from the profiles obtained by micro Raman spectra of PE/EPR and PP/EPR pairs heated at 180 °C for 5 hours.

Finally, the results of the TEM observation for PE/EPR melt blends are shown in Fig.9. The domain-matrix morphology was observed for each kind of blends, but the domain size of EPR drastically changed with the chemical composition of EPR, i.e., the higher the propylene content, the larger the domain size. In the case of the EPR of 27 % propylene content, which gives the smallest domain size, the boundary between domain and matrix is obscure, and many PE lamellae interpenetrating the EPR domain are observed. These results suggest the high affinity between PE and EPR. On the other hand, the blend makes a clear boundary and the lamellae stop at the boundary in the case of EPR with high propylene content.

66

The domain size of the melt blend is considered to be governed mainly by the chemical affinity between PE and EPR, because the melt viscosity of EPR used in this study is in the same range. It is interesting that the morphology of the boundary region and the EPR domain size of the melt blends are interpreted well in terms of with the affinity between PE and EPR estimated from the results obtained by Raman spectroscopy. The presence of the mixed phase and its thickness are able to be measured by the micro Raman technique. Characterization of the micro-phase formed at the interface is very important for understanding the compatibility of polymer blends.

（1）

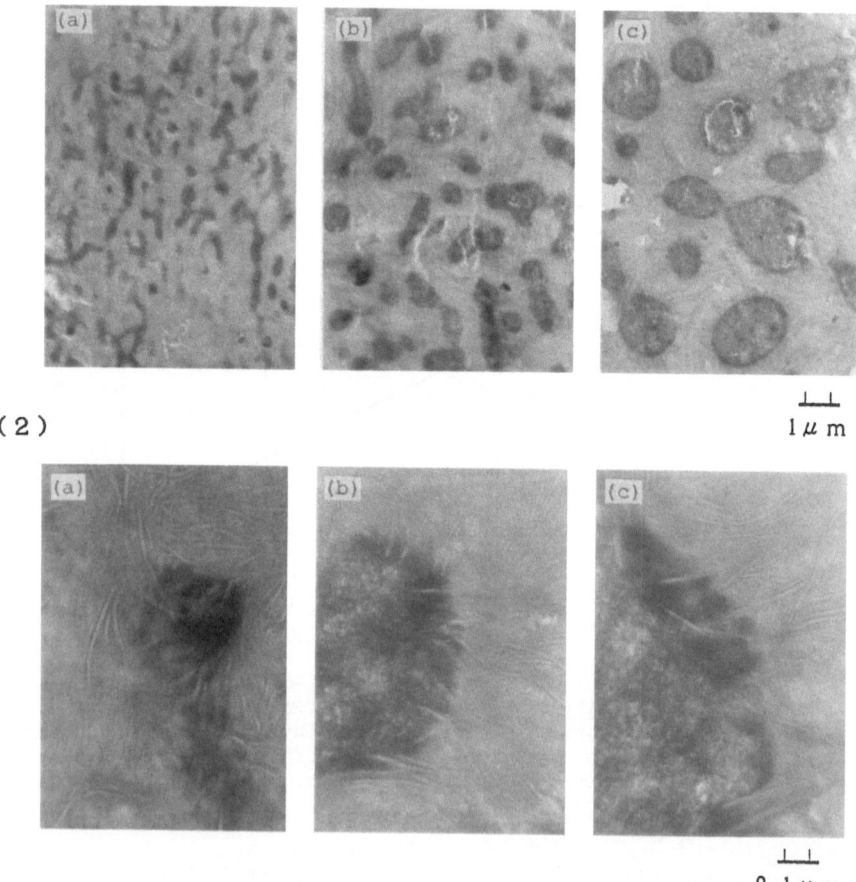

（2）

1 μ m

0.1 μ m

Fig.9 Transmission electron micrographs of PE/EPR blends(PE/EPR=70/30) taken at a magnification of (1) × 10000, (2) × 100000 (staining agent: RuO_4). (a)PE/EPR(C_3'=27%), (b)PE/EPR(C_3'=53%), (c)PE/EPR(C_3'=67%) and the average particle diameters were (a)0.4 μm, (b)0.9 μm, 2.4 μm, respectively.

References:

1 Kobayashi M, Ishioka T (1989) Raman Spectroscopy: Sixty Years On Vibrational spectra and structure, Vol.17B 369-390 Elsevier, Amsterdam.

2 Tanaka T, Nishi T, Ikeda T (1984) Polymer Preprints, Japan 33:73

3 Young RJ, Lu D, Day RJ (1991) Polymer J 24:71

4 Satoh N, Kurauchi N (1992) Koubunshi 41:334

5 Boogh LCN, Meier RJ, Kausch HH, Kip BJ (1992) J Polym Sci B30:361

6 Hsu SC, Lin-Vien D, French RN (1992) Appl Spectrosc 46:225

Supramolecular Assemblies of Cyclodextrins with Polymers and Preparation of Polyrotaxanes

A. Harada, J. Li, and M. Kamachi

Department of Macromolecular Science, Faculty of Science, Osaka University, Toyonaka, Osaka, 560 Japan

Abstract: Cyclodextrins (CDs) have been found to form inclusion complexes with various polymers with high selectivities to give stoichiometric compounds in crystalline states. α-CD formed complexes with poly(ethylene glycol)(PEG) of molecular weight higher than 300, although β-CD did not form complexes with PEG of any molecular weight. However, β-CD formed complexes with poly(propylene glycol)(PPG), although α-CD did not form complexes with PPG of any molecular weight. γ-CD formed complexes with poly(methyl vinyl ether)(PMVE), though α- and β-CD did not form complexes with PMVE. Polyrotaxanes in which many CDs are threaded on a single PEG chain were prepared by capping the chain ends of the complex with bulky groups. The neighboring CDs in a polyrotaxane are crosslinked by epichlorohydrin and then stoppers at the both ends and the polymer chain of the crosslinked polyrotaxane were removed by strong base to give a tubular polymer, a molecular tube.

INTRODUCTION

Biological systems are highly organized assemblies of macromolecules, which express unique properties and functions. In the synthetic polymer systems, new materials with novel properties and functions could be designed and constructed by combining specific intermolecular interactions. We take inclusion compounds of cyclodextrins with polymers, for example, and describe molecular assemblies [1].

There are some cases in which unusual functions, which cannot be achieved only by an individual molecule, can be observed by arranging molecules properly and make them assembled. Even in an assembly of the same molecules, such as crystals and bilayer membranes, depending on the way in which they are arranged, the properties and functions are different. There appears so many combination and structures in the assemblies of different molecules that they show high diversity and infinite possibility.

In the polymer systems, there are some ways to achieve multi-phase materials using different polymers: blends, alloys, and composites. Although some new structures and properties have been found in these systems, their functions have some statistical nature.

A. Teramoto, M. Kobayashi, T. Norisuje (Eds.)
Ordering in Macromolecular Systems
© Springer-Verlag Berlin Heidelberg 1994

In biological systems, macromolecules form assemblies through specific interactions to give organs, tissues, and cells. The living creatures are composed of such a stepwise structure and maintain their lives. Although the textures of synthetic polymers are quite different from those of living systems, if synthetic macromolecules can be organized in molecular levels, as can be seen in the biopolymers, synthetic polymers will show a lot of potentials.

In this proceedings, we describe the formation of inclusion complexes of some polymers with cyclodextrins, synthesis of polyrotaxanes, and their application to the synthesis of a molecular tube.

FORMATION OF INCLUSION COMPLEXES OF CYCLODEXTRINS WITH POLYMERS

Cyclodextrins, which are cyclic oligomers consisting of glucose linked by α-1,4-linkages (**Fig. 1.**)(**Table 1**), are known to form inclusion complexes with a variety of low molecular weight compounds, ranging from nonpolar organic molecules to highly polar ions and even to rare gases [2]. However, the guest molecules have been limited to small molecules and simple ions. There were no reports on the inclusion complexes of cyclodextrins with polymers when we started our project.

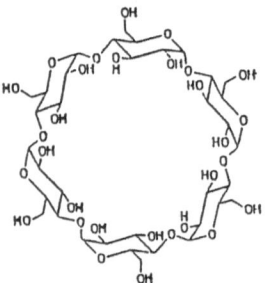

Fig. 1. α-Cyclodextrin

Table 1. Molecular dimensions of cyclodextrins

CD	No. of Glucose Units	Cavity Diameter (Å)
α	6	4.7~5.2
β	7	6.0~6.4
γ	8	7.5~8.3

First, we tested whether cyclodextrins would form complexes with some water-soluble polymers. **Table 2** shows the results of the formation of the complexes of cyclodextrins with some non-ionic polymers. We found that poly(vinyl alcohol), polyacrylamide, and poly(N-vinyl pyrrolidone) did not form complexes with cyclodextrins. However, we found that α-cyclodextrin(α-CD) formed complexes with poly(ethylene glycol)(PEG) of various molecular weights to give crystalline complexes in high yields [3]. When aqueous solutions of PEG (or PEG as it is) were added to a saturated aqueous solutions of α-CD at room temperature, the solution became turbid and the complexes were formed as precipitates when the average molecular weight of PEG is higher than 300. β-CD, which is larger than α-CD, did not form complexes with PEG.

Table 2. Complex foramtion between CDs and polymers

	Polymer	MW	Yield(%)		
			α-CD	β-CD	γ-CD
PVA	-(CH₂CH)- OH	22,000	0	0	0
PAAm	-(CH₂CH)- CONH₂	10,000	0	0	0
PEG	-(CH₂CH₂O)-	1,000	92	0	trace
PPG	CH₃ -(CH₂CHO)-	1,000	0	96	80
PMVE	-(CH₂CH)- OCH₃	20,000	0	0	80

Instead, β-CD formed complexes with poly(propylene glycol) of molecular weight higher than 400 to give crystalline compounds in high yields although α-CD did not form complexes with PPG of any molecular weight [4]. We also found that γ-CD formed complexes with poly(methyl vinyl ether)(PMVE), although α- and β-CD did not form complexes with PMVE [5].

In the course of the experiments of the preparation of the complexes of α-CD with PEG, we found that the rates of the complex formation depend on the molecular weight of PEG. PEG of molecular weight 200 did not form complexes. PEG of MW 600 and 1,000 formed complexes instantancously. The rates of complex formation decreased with an increase in the molecular weight of PEG when the molecular weights are higher than 1,000.

The complexes were isolated by filtration or centrifugation, washed with water, and dried. **Fig. 2.** shows the yields of the complexes of α-CD with PEG as a function of the molecular weight of PEG. α-CD did not form complexes with the low molecular weight analogs, ethylene glycol [6], diethylene glycol, and triethylene glycol. α-CD formed complexes with PEG of molecular weight higher than 300. The yields increased with the increase in the molecular weight. The complexes were obtained almost quantitatively with PEG of molecular weight over 1,000. β-CD did not form complexes with PEG of any molecular weight.

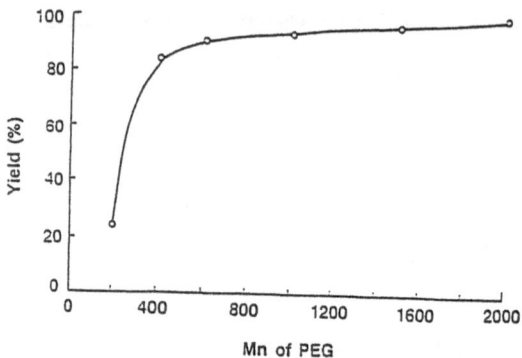

Fig. 2. Yields of complexes of α-CD with PEG as a function of the MW of PEG

Complex formation between α-CD and PEG with various end groups was tested. PEGs with small end groups, such as methyl, dimethyl, and amino groups form complexes. PEG carrying bulky substituents such as 3,5-dinitrobenzoyl group and 2,4-dinitrophenyl group at both ends of the PEG, which do not pass through the α-CD cavity, did not give any complexes with α-CD.

Fig. 3 shows a proposed structure of the complex of poly(ethylene glycol) with α-CD. The inclusion complex formation of PEG in α-CD channel is entropically unfavorable. However, formation of the complexes is thought to be promoted by hydrogen bond formation between cyclodextrins.

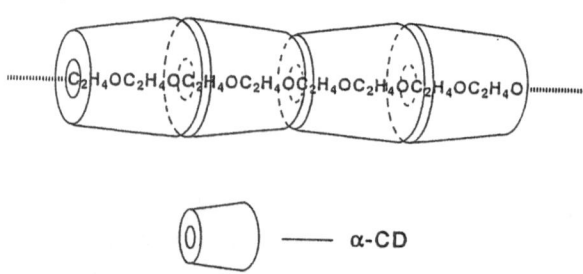

Fig. 3. Proposed structure of the complex of PEG with α-CD

However, β-CD formed complexes with PPG of molecular weight higher than 400, although α-CD did not form complexes with PPG of any molecular weight. The complex was obtained almost quantitatively with PPG of molecular weight 1,000 and the yields decrease with the increase in the molecular weight of PPG. γ-CD also forms complexes with PPG in high yields even when the molecular weight of PPG is low (400-725).

Stoichiometries of the Complexes

The stoichiometries of the complexes were estimated by the continuous variation method and ^{1}H NMR spectra of the complexes. The stoichiometries were found to be 2:1 for α-CD-PEG and β-CD-PPG complexes . γ-CD-PMVE complex shows 3:1(three monomer units and one CD). The length of two ethylene glycol units(or two propylene glycol units) or three methyl vinyl units corresponds to the depth of the CD cavity (6.7 Å).

Properties of the Complexes

The complexes of α-CD with PEG of low molecular weight (-1,000) are soluble in a large amount of water. The complexes of PEG of high molecular weight can be dissolved in water by heating. The addition of an excess amount of benzoic acid, which is a low molecular weight guest, to the suspension of the complex resulted in solubilization of the complex when the molecular weight of PEG was low (-1,000). The formation of the complex is reversible.

Binding Modes of the Complexes

The X-ray powder patterns of the complex of α-CD with PEG show that the complexes are crystalline and the patterns are different from those of the complexes with small molecules, which have a cage structure, and similar to those of the complexes of valeric acid or octanol, which have been reported to have a channel structure. These results indicate that the complexes of α-CD and PEG are isomorphous with those of channel type structure.

Molecular model studies show that PEG chains are able to penetrate α-CD cavities, while the PPG chain cannot pass through the α-CD cavity due to the hindrance of the methyl group on the main chain. These views are in accordance with our results that α-CD formed complexes with PEG but not with PPG. β-CD did not form complexes with PEG. β-CD cavity is too large to fit PEG. Instead β-CD forms complexes with poly(propylene glycol) as we reported previously in communication. Model studies further indicate that the single cavity (depth 6.7 Å) accommodates two ethylene glycol (or propylene glycol) units. 6.6 Å)

POLYROTAXANES

The importance of non-covalent interactions in biological systems motivates much of the current interest in supramolecular assemblies. A classic example of a supermolecule is provided by the rotaxanes, in which a molecular rotar is threaded by a

74

linear axle. Previous examples have included cyclic crown ethers threaded by polymers [8], paraquat-hydroquinone complexes [9] and cyclodextrin complexes [10]. we have succeeded in preparation of compounds in which many cyclodextrins are threaded on a single PEG chain and are trapped by capping the chain with bulky end groups. We call this supramolecular assembly a molecular necklace. A route of the preparation of the polyrotaxane is shown in **Scheme 1**.

Scheme 1.

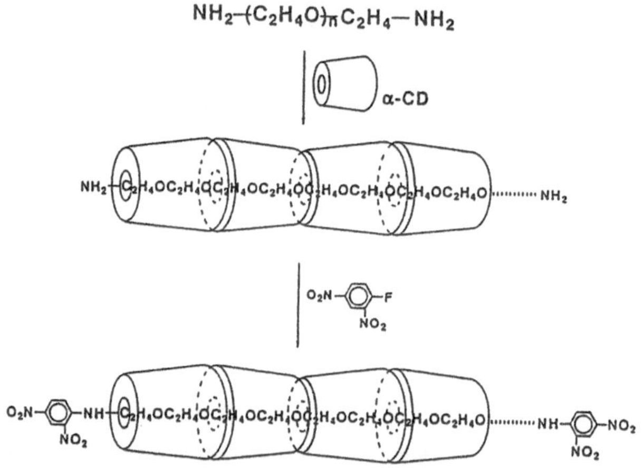

$$NH_2\text{-}(C_2H_4O)_n C_2H_4\text{-}NH_2$$

α-CD

$$NH_2\text{-}(C_2H_4OC_2H_4OC_2H_4OC_2H_4OC_2H_4OC_2H_4OC_2H_4OC_2H_4O \cdots NH_2$$

$$O_2N\text{-}\bigcirc\text{-}F$$
$$NO_2$$

$$O_2N\text{-}\bigcirc\text{-}NH\text{-}(C_2H_4OC_2H_4OC_2H_4OC_2H_4OC_2H_4OC_2H_4OC_2H_4OC_2H_4O \cdots NH\text{-}\bigcirc\text{-}NO_2$$
$$NO_2 \qquad\qquad NO_2$$

We obtained inclusion complexes of α-CD with PEG bisamine(PEG-BA) (MW=3,350) by adding an aqueous solution of PEG-BA to a saturated aqueous solution of α-CD at room temperature, using a method similar to that used to prepare complexes of α-CD and PEG [3]. The resulting complex was collected and dried, We added to the complex an excess of 2,4-dinitrofluorobenzene, which is bulky enough to prevent dethreading, together with dimethylformamide, and stirred the mixture at room temperature overnight. The reaction mixture was poured into excess amount of ether. The precipitate was washed with ether to remove unreacted 2,4-dinitrofluorobenzene and with dimethylformamide to remove free α-CD, PEG-BA, and dinitrophenyl derivatives. The residue was dissolved in dimethylsufoxide(DMSO) and washed with water to remove unreacted α-CD, PEG-BA and water-soluble dinitrophenyl derivatives. The product was collected, washed with ether, and dried(yield 60 %). Finally, the product was purified by column chromatography on Sephadex G-50 using DMSO as solvent. The product. was found to be pure and contained no free α-CD, PEG-BA or dinitrophenyl derivatives.

The product is insoluble in water and dimethylformamide although each component, α-CD, PEG-BA, bis(2,4-dinitrophenyl)-PEG, and even α-CD-PEG-BA complexes are soluble in water. The product is, however, soluble in DMSO and 0.1 N NaOH. The hydroxyl groups of α-CD (pKa=12) may be ionized at this pH, so that α-CD becomes soluble in an aqueous medium. Neutralizing this solution by adding 0.1 M HCl instantly produced a precipitate. This reaction is reversible. The structure of the product was confirmed by elemental analyis, infrared and ultraviolet spectroscopy, X-ray diffraction, and proton and ^{13}C nuclear magnetic resonance(NMR). Proton and ^{13}C NMR spectra of the product are consistent with a compound that contains both components α-CD and PEG-DNB, although some broadening was observed. Gel chromatography on Sephadex G-50(exclusion limit 30,000) using DMSO as eluent shows a single peak close to the void volume. A chromatogram of a mixture of product, α-CD and PEG-DNB shows three peaks. The first peak, which could be detected by both ultraviolet spectroscopy and optical rotation, was identified as the product. The second, which could be detected only by ultraviolet, was identified as PEG-DNB. The third, which could be detected only by optical rotation, was identified as α-CD. The average molecular weight was determined by ^1H NMR to be 23,000, which indicates that 20 α-CD units are captured in a polymer chain, and by ultraviolet spectroscopy to be 26,400, indicating 23 α-CD units in a single molecule. We consider these to be the same within experimental error. X-ray powder diffraction shows that the complexes are crystalline, and the patterns are similar to those of complexes of α-CD with valeric acid or octanol, which differ from those of α-CD complexes with small molecules, such as propionic acid or propanol. These results indicate that the product is isomorphous with those of channel-type structure.

At present we have not determined how adjacent cyclodextrins are oriented. But the fact that the product is insoluble in water but soluble in 0.1 N NaOH shows that strong hydrogen bonds exist between α-CDs, suggesting that they alternate as shown in **Scheme 1.**

MOLECULAR TUBE: SYNTHESIS OF A TUBULAR POLYMER FROM THREADED CYCLODEXTRINS

Recently, much attention has been focused on designing and fabricating nano-scale structures. Carbon nano-tubes(4-30 nm), for example, has been constructed by arc-discharge evaporation method by similar method to that of fullaren synthesis. In order to design and construct smaller tubes, sub-nano tubes, it is more convenient to construct them chemically. One of the most promising method for the construction of sub-nano-tubes chemically is to pile up ring molecules. Among the ring molecules, cyclodextrins are the most suitable candidates for this purpose, because they have cavities with diameters of 0.45 nm for α-CD, 0.7 nm for β-CD, and 0.85 nm for γ-CD, respectively. Moreover, the cavities of CDs have some depth(0.7 nm) and CDs have many hydroxyl groups on the both sides of the rings, which might enable CDs to link together to form a channel.

We have succeeded in preparing a sub-nano tubes by crosslinking neighboring CDs of polyrotaxane and removing large substituents at both ends and the polymer chain as shown in **Scheme2**. We call these sub-nano tubes a "molecular tube" [11].

Scheme 2.

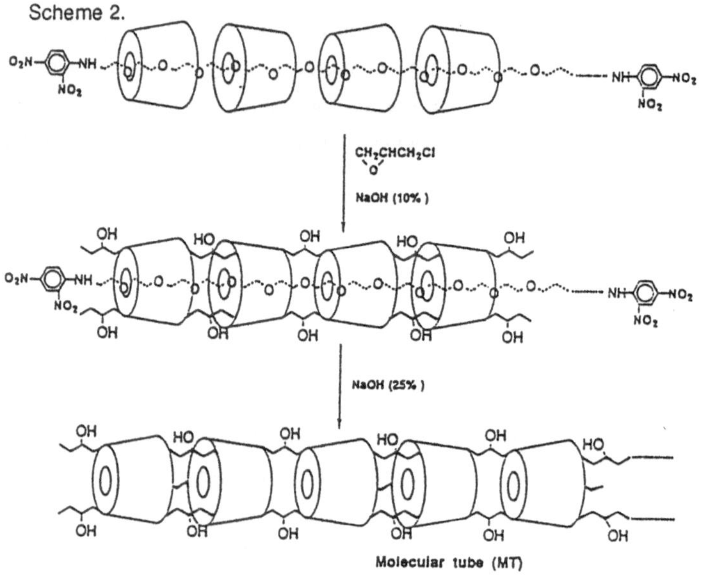

Molecular tube (MT)

Polyrotaxanes were prepared as described previously. In this case poly(ethylene glycol) of molecular weight 1,450 was used, because α-CD forms complexes with PEG of molecular weight 600-2000 most efficiently. The complexes are nearly stoichiometric (two ethylene glycol units: a single CD), that is, α-CDs are close packed from end to end of the polymer chain. The polyrotaxane(22.5 mg) was dissolved in 10% NaOH solution and epichlorohydrin(3.84 mmol) was added to the solution. The solution was stirred at room temperature for 36 hours. The reaction mixture was neutralized by the addition of HCl. The yellow solid was precipitated from ethanol. In order to remove stoppers at both ends, the product was treated with strong base (25% NaOH) at 45 °C for 24 hours. The reaction mixture was cooled and then neutralized with HCl. **Fig. 4** shows the elution diagram of the reaction mixture of polyrotaxane with 25 % NaOH (a), that of the crosslinked polyrotaxane (b), and that of the reaction mixture of the crosslinked polyrotaxane with 25 % NaOH (c). The crosslinked product was eluted at the void volume(b), the same as the polyrotaxane. After the polyrotaxane was treated with 25 % NaOH and the solution was neutralized before crosslinking, two peaks were observed: one could be detected only by UV (360 nm), which is assigned to dinitrophenyl group(DNP), and the other, which could be detected only by optical rotation, is identified as the product, the molecular tube. The second peak, that was detected only by UV(360 nm), is assigned as a dinitrophenyl group. The molecular weight of the molecular tube has been estimated to be about 18,000 by GPC, which is consistent the fact that the molecular weight of the molecular tube prepared from PEG of molecular weight 1,450 is about 17,000. Therefore, intra-chain crosslinking took place predominantly. The yield of the final product is 92 %.

The product was soluble in water, DMF, and dimethylsulfoxide (DMSO), although polyrotaxanes are insoluble in water and DMF and soluble in DMSO. The product was characterized by ^{1}H NMR, ^{13}C NMR, IR and UV spectra and GPC.

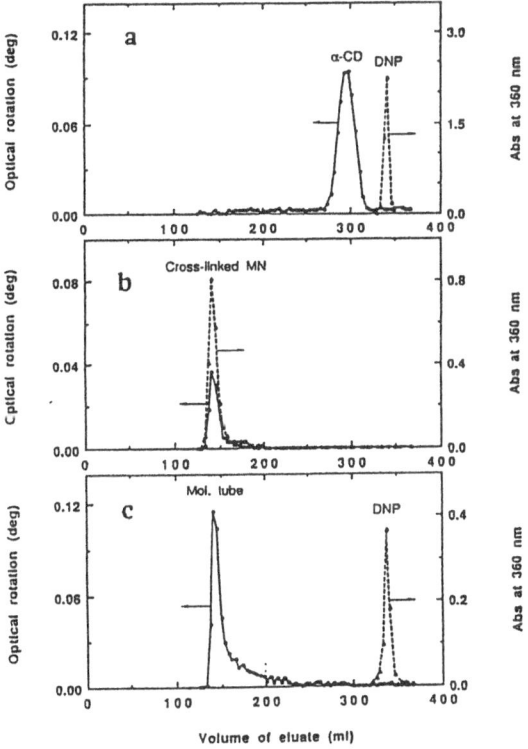

Fig. 4. Elution diagrams of the reaction mixture of polyrotaxane with 25% NaOH (a), that of the crosslinked polyrotaxane (b), and that of the reaction mixture of the crosslinked polyrotaxane with 25% NaOH (c). A Sephadex G-25 column (2.2 x 93 cm) with water eluate was used.

References:

1 Harada A, Li J, Kamachi M (1993) *Proc. Japn Acad.*, **69**: Ser. B, 39

2 Szejtli J (1982) Cyclodextrins and Their Inclusion Complexes, Akademiai Kiado, Budapest

3 Harada A, Kamachi M (1990) *Macromolecules* **21**:2821

4 Harada A, Kamachi M (1990) *J. Chem.Soc., Chem. Commun.*, **1990**:1322

5 Harada A, Li J, Kamachi M (1993) *Chem. Lett.*, **1993**:237

6 Harada A, S. Takahashi (1984) *Chem. Lett.*, **1984**:2089

7 Gibson HW (1991) *Makromolc. Chem., Makromolec. Sympo.***42/43**:395

8 Ashton PR (1991) *Angew. Chem., Int. Ed. Engl.* **30**:1042

9 Isnin R, Kaifer AE (1991) *J. Am. Chem. Soc.*, **113**:8188

10 Harada A, Li J, Kamachi M (1992) *Nature* **356**:325

11 Harada A, Li J, Kamachi M (1993) *Nature* in press

Case II Diffusion in Polystyrene Fiber: *In-situ* Measurement of a Moving Sharp Front

Mitsuhiro FUKUDA

Textile Materials Science Laboratory, Hyogo University of
Teacher Education, Yashiro-cho, Hyogo 673–14, Japan

Abstract

In-situ measurement of a moving sharp front formed in a polystyrene fiber- organic vapor system was presented by using optical microscopy and light scattering technique. The diffusion of acetone vapor with the relative vapor pressure of 0.92 into the physically aged polystyrene fiber of 25 μm diameter was measured at 25°C. The system showed following characteristics; (1) an induction time, (2) formation of a sharp front after the induction time, (3) nearly constant rate of the sharp advancing front, (4) step-like distribution of the penetrant concentration at the sharp front, (5) uniform state of swelling behind the sharp front. These findings strongly suggested Case II diffusion. The effect of the physical aging on the polymer glass to get front formation was also discussed based on Thomas and Windle model.

Introduction

Diffusion of low molecular weight substance in a polymer is of great interest to the food and electronic packing industry. Experimental technique for measuring diffusion in polymers requires good spatial resolution, methods such as Rutherford backscattering sepectrometry, forced Rayleigh scattering, laser interferometry and NMR imaging have recently become available for this purpose [1–4]. In this paper, we present a simple method for accomplishing this objective. To follow the real-time solvent diffusion in polymer, we have chosen a glassy amorphous fiber of small diameter in this study. To gain a complete picture of the diffusion process, one wishes to follow as a function of time, both the fiber diameter and the radial concentration gradient. As we will show, the diffraction of a laser-light beam from the fiber offers a simple and sensitive probe for following these variables. In addition, the analysis of the diffraction pattern is straightforward because the scattering from a cylinder with a radial refractive index gradient has been solved.

Case II diffusion is observed in the diffusion of organic solvent in a glassy polymer recognized by Alfrey et al.[5] and has the following characteristic aspects; (1) A sharp advancing front can be observed, separating a glassy region where the solvent concentration is negligible from a swollen rubbery shell with a high solvent content. (2) The rubbery shell region has a homogeneous distribution of solvent. (3) The sharp front moves throughout the polymer linearly with time. Therefore the sorption curve increased linearly in the case of the film sample.

The objective of the present study is first to confirm that three characteristics as men-

A. Teramoto, M. Kobayashi, T. Norisuje (Eds.)
Ordering in Macromolecular Systems
© Springer-Verlag Berlin Heidelberg 1994

tioned above can be applied to the glassy polymer fiber, and second to consider the conditions to observe the front formation.

Materials and Methods

Polystyrene (PS) fiber with 25 μm diameter was supplied by Toyobo Co. Inc., Japan as multifilament yarn. As measured by gel permeation chromatography, the weight–average molecular weight and polydispersity of the fiber are 178,000 and 1.5, respectively. Using optical microscopy with a Berek compensator, the birefringence of the fiber was found to be negligible ($\Delta n < 0.0005$). The PS fiber which was physically aged for 18 hr at 60 °C was used throughout the study unless otherwise mentioned. To get reproducible results physical aging was significant. The cross section of the fiber thus conditioned was found to be fully circular.

The sample chamber was constructed in order to measure the fiber monofilament under the given temperature and given vapor pressure. The block diagram of the apparatus is shown in Figure 1. The PS monofilament of 30 mm long was mounted on a sample chamber without any tension. After the fiber monofilament was vacuum dried for 24 hr, the acetone vapor at given vapor pressure at 25°C was induced into the sample chamber. The change of the shape in the monofilament exposed to acetone vapor was observed by optical microscopy.

Sorption rates were measured gravimetrically using a helical quartz spring at 25°C. The details of the apparatus are referred to elsewhere [6].

Time–resolved light scattering apparatus was used to measure both the diameter and the concentration distribution of the organic penetrant in the monofilament. The sample chamber used in optical microscopy was also applied to the light scattering experiments. The incident

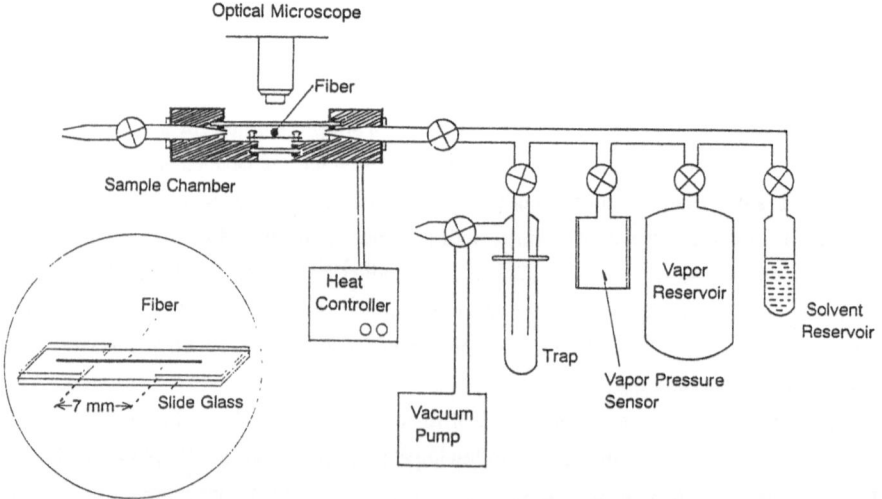

Figure.1 Experimental arrangement to keep a fiber at given organic vapor pressure, and method of mounting a fiber (in the circle).

radiation is provided by a 15 mW He–Ne laser with a wavelength of 632.8 nm oriented with its plane of vibration along the fiber axis. The intensity of the diffracted radiation is measured as a function of the scattering angle from $\theta=0°$ to $\theta=35°$ by using a dynamic image analyzer (DIANA, HASC Co.Ltd., Kyoto).

Results

Optical Microscopy

Figure 2 shows the microphotograph of PS monofilament after exposed in acetone vapor at 25°C. The relative vapor pressure, p/p_s, was kept at 0.92, here p_s means the saturated vapor pressure. Note that we could not observe a sharp front at p/p_s lower than 0.8. The moving sharp front can be clearly found after 220 sec and it advanced to the center of the fiber (Figures 2–(b) – 2–(d)). The interface at the front seemed to be somewhat rugged presumably due to the small crack or craze generated in the glassy media in response to the swelling stress by solvent penetration. When the sharp front was just disappeared at 350 sec as shown in Figure 2–(e), the fiber become completely rubbery state and began to shrink. The shrinkage seemed to be caused by a kind of relaxation process in the rubbery state.

It is important to note that in the as–spun fiber without heat treatment the sharp front was not appeared even at $p/p_s=0.92$.

Sorption Curve

Figure 3 shows the sorption rate, when the volume fraction of acetone, $\phi_{acetone}(t)$, is plotted against $\sqrt{t}$ (minutes) at 25°C. The relative vapor pressure p/p_s is kept at 0.92. The sorption curve seems to be complicated shape and didn't obey Fickian diffusion. If the velocity of the advancing sharp front observed in optical microscopy is assumed to be constant, then the sorption curve for the cylindrical material, $\phi_{acetone}(t)$ should be proportional to $\sqrt{t}$ not to t. The straight line drawn in the time range from 200 sec to 350 sec showed almost linear.

The fiber remarkably shrank at the last stage of the sorption and the amount of the vapor sorbed in the fiber rapidly decreases at that point to reach equilibrium. The arrow indicates the time when the fiber start to shrink. It is noted that the point indicated by the arrow corresponded to the time when the sharp front just disappeared in optical microscopy.

In the initial stage of the diffusion the sorption curve showed non–Fickian type and the sharp front also didn't appear. In Case II diffusion, the induction time is frequently observed due to insufficient time or space in the fiber to develop the step concentration profiles depending on the sample thickness or diameter.

Light Scattering Study

The results obtained from optical microscopy and sorption curve that (1) the sharp advancing front was formed, (2) the front velocity is almost constant and (3) existence of the induction time, strongly suggested Case II diffusion for the present system. If we assume the glassy state ahead of the front and rubbery state behind it in the PS fiber, light scattering theory from a infinitely long coaxial cylinder with dual components can be applied.

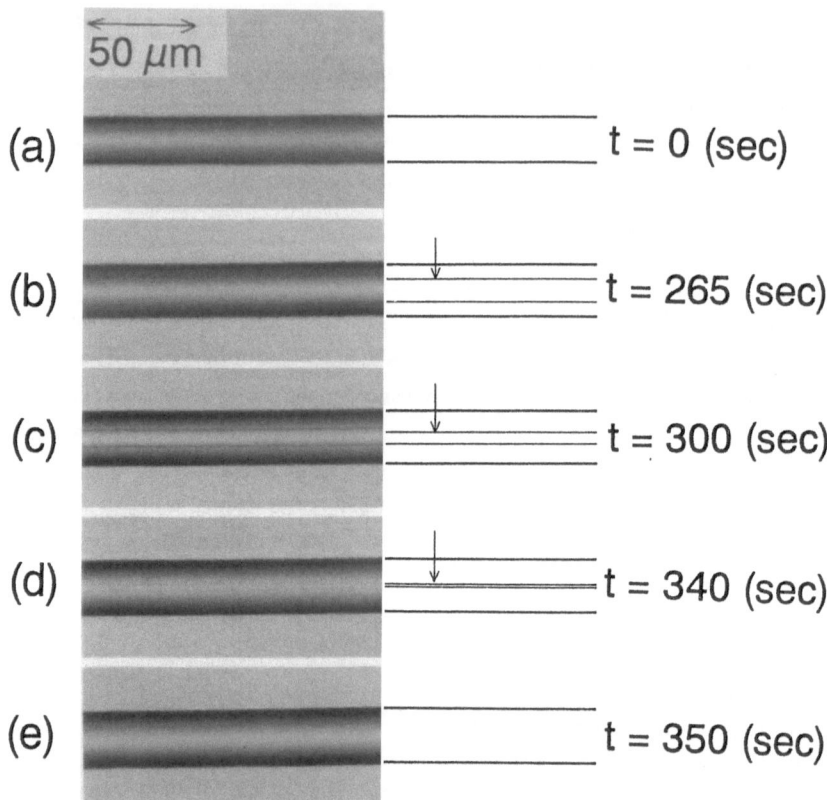

Figure.2 Optical microphotographs showing the advancing sharp front in PS fiber with time after exposed in acetone vapor of $p/p_s = 0.92$, at 25°C.

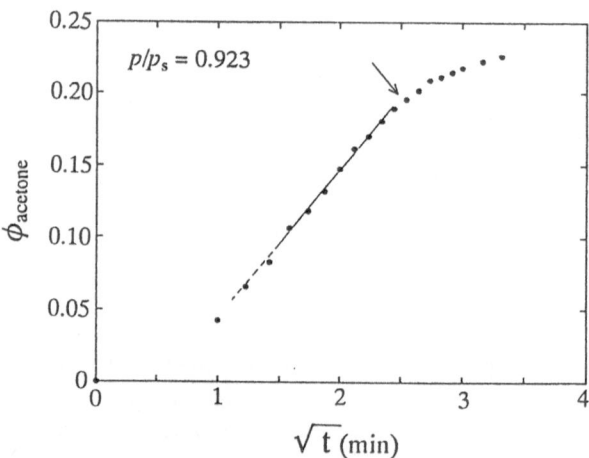

Figure.3 Sorption of acetone vapor in PS fiber at 25°C. The relative vapor pressure, p/p_s, is kept at 0.923. The arrow indicates the time when the fiber just began to shrink.

We first evaluated the refractive index of the rubbery swollen PS fiber (m_2) by assuming the linear dependence on the volume fraction of the penetrant as follows:

$$m_2 = \phi_{acetone} n_{acetone} + (1 - \phi_{acetone}) m_1 \tag{1}$$

where $\phi_{acetone}$ is the equilibrium volume fraction of acetone sorbed in the PS fiber. The refractive index of the virgin fiber (m_1) was 1.589 and the values of $n_{acetone}$ and $\phi_{acetone}$ are 1.357 and 0.202, respectively. Given these values for m_1, $n_{acetone}$, and $\phi_{acetone}$, the refractive index of the rubbery swollen fiber (m_2) was calculated as 1.542.

The scattered intensity distribution from the infinite length of a coaxial cylinder was calculated from a model refractive index profile consisting of a glassy core and a rubbery shell as shown in Figure 4. For an incident ray with its electric–field vector parallel to the fiber axis, the angular distribution of scattering intensity, $I(\theta)$, relative to the incident intensity, I_0, is given by [7];

$$I(\theta)/I_0 = (\lambda /\pi^2 r) \ \left| \ b_0 + 2 \sum_{n=1}^{\infty} b_n cos(n\theta) \ \right|^2 \tag{2}$$

where r is the fiber–to–detector distance, θ is the angle between the scattered and incident beams, and λ is the wavelength of the incident beam.

For a coaxial cylinder of infinite length, b_n in Eq.(2) is given by [7];

$$b_n = \frac{\begin{vmatrix} J_n(m_3\alpha_2) & H_n(m_2\alpha_2) & J_n(m_2\alpha_2) & 0 \\ m_3 J_n'(m_3\alpha_2) & m_2 H_n'(m_2\alpha_2) & m_2 J_n'(m_2\alpha_2) & 0 \\ 0 & H_n(m_2\alpha_1) & J_n(m_2\alpha_1) & J_n(m_1\alpha_1) \\ 0 & m_2 H_n'(m_2\alpha_1) & m_2 J_n'(m_2\alpha_1) & m_1 J_n'(m_1\alpha_1) \end{vmatrix}}{\begin{vmatrix} H_n(m_3\alpha_2) & H_n(m_2\alpha_2) & J_n(m_2\alpha_2) & 0 \\ m_3 H_n'(m_3\alpha_2) & m_2 H_n'(m_2\alpha_2) & m_2 J_n'(m_2\alpha_2) & 0 \\ 0 & H_n(m_2\alpha_1) & J_n(m_2\alpha_1) & J_n(m_1\alpha_1) \\ 0 & m_2 H_n'(m_2\alpha_1) & m_2 J_n'(m_2\alpha_1) & m_1 J_n'(m_1\alpha_1) \end{vmatrix}} \tag{3}$$

where m_1 and m_2 are the refractive indices of the core and the shell, respectively, and m_3 is the refractive index of the surrounding medium. J_n is the Bessel function of order n, and H_n is the Hankel function of the second kind. J_n' and H_n' are the derivatives of J_n and H_n, respectively. In our experiments, the fiber is surrounded by vapor and therefore m_3 is ≈ 1. R_1 and R_2 denote the radii of the inside and outside cylinders. The parameters α_1 and α_2 are $2\pi R_1/\lambda$ and $2\pi R_2/\lambda$, respectively.

The change of the scattered intensity distribution from the PS fiber after exposed to ace–

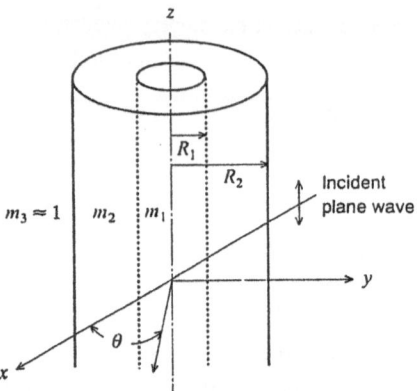

Figure.4 Schematic drawing of a coaxial fiber with dual components, showing coordinates.

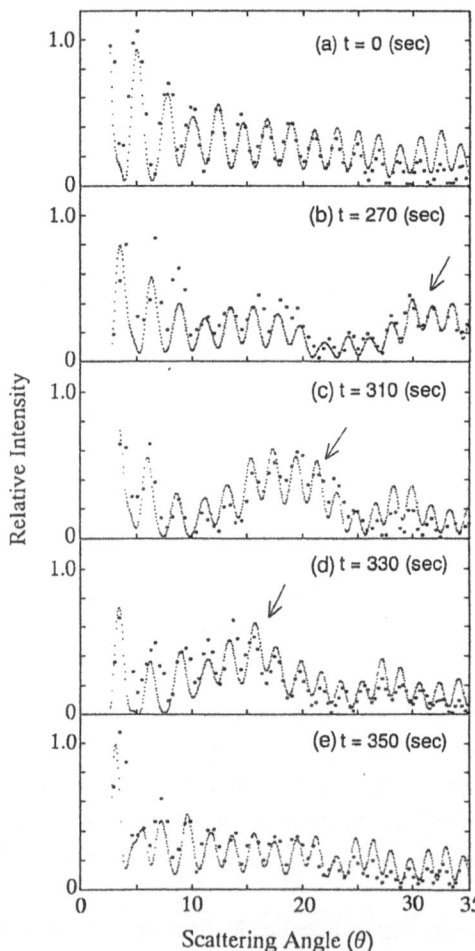

Figure.5 Changes of the scattering intensity distribution (■) with exposure time. The dotted curves (······) indicate the theoretical calculation according to Eq.(2) which fit to the observed curves. Arrows indicate the broad fringes which are characteristic of a coaxial fiber.

tone vapor is shown in Figure 5. For analyzing the intensity curve from Figures 5–(b) to 5–(d), we must determine four parameters, R_1, R_2, m_1 and m_2. In the determination of these four parameters from the experimental intensity distribution, another analysis from the ray tracing methods as described below are also applied. The difference of the optical pass length between the diffracted ray passing only through the shell and that through both the shell and core make the make the broad fringes as reported by Watkins [9]. The fringes are characteristic of the coaxial fiber which remarkably appears in the lower scattering angle region as the core diameter decreases. Although the detailed methods is not shown here [10], we can determine the four parameters by combining the rigorous theory in Eq.(2) and ray tracing methods. The calculated results showed fitness to the observed ones as shown in Figure 5.

In Figure 6 the front position from the surface of the swollen fiber $(R_2 - R_1)$ with time is plotted. The result obtained from the light scattering method showed in good agreement with that from optical microscopy. The front position proceeded almost linearly after detecting the front.

Discussion

Thomas and Windle model based on the osmotic pressure theory must be the most successful one to explain several essential aspects of Case II diffusion [11]. The model was constructed for the diffusion processes occurring at the front in terms of the viscous response of

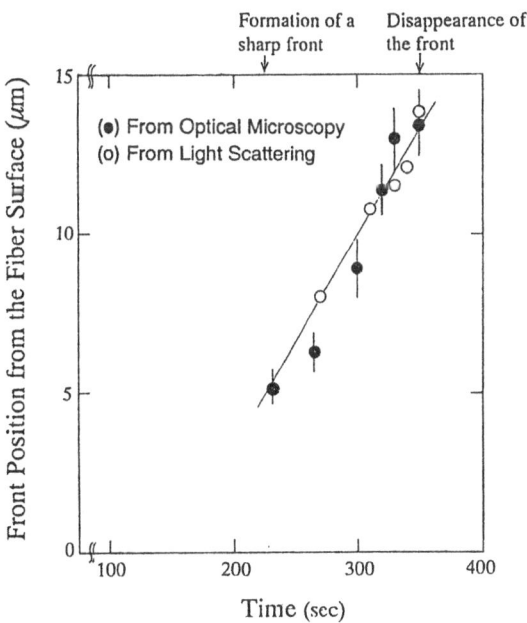

Figure.6 The relation between the front position from the fiber surface and time for PS– acetone system, at $p/p_s = 0.92$ and at 25°C.

swelling element of glass to the osmotic pressure of the penetrant. Two basic parameters, diffusivity, D, and the flow viscosity of the glassy polymer, η_0, both greatly depend on the concentration of the penetrant, are significant to characterize Case II diffusion. In the very thin element of the glassy polymer, the rate of the normalized volume fraction of the solvent, $\overline{\phi}$, ($\overline{\phi}$ is the ratio of volume fraction of penetrant, ϕ, to the equilibrium volume fraction, ϕ_e) is proportional both to the osmotic pressure, P, and to viscous flow rate, $1/\eta$, and expressed as;

$$\frac{\partial \phi}{\partial t} = \frac{P}{\eta} = -\frac{kT}{V_1 \eta_0} \exp(M\overline{\phi}) \ln(\frac{\overline{\phi}}{a}) \tag{4}$$

where k is Boltzmanns constant, T is temperature, V_1 is the partial molar volume of the penetrant, η_0 is the viscosity of the unswollen polymer, a is penetrant activity and M is a constant to be determined. The Fick's first law is the relation of the flux, J, and diffusion coefficients, D, which dependent on the volume fraction of the penetrant is given by

$$J = -D(\phi) \frac{\phi}{a} \frac{da}{dr} \tag{5}$$

Numerical solution by combining Eqs.(4) and (5) give the distribution of volume fraction, penetrant activity and pressure in the polymer. Thomas and Windle showed that various types of the diffusion can be obtained by varying D (from 0.5×10^{-14} to 2×10^{-14} (m^2/s)), $1/\eta_0$ (from 2×10^{-15} to 1×10^{-14} (m^2/MNs)) and M (from 5 to 20) for PMMA–methanol system [11].

We showed that the moving sharp front was observed in the physically aged PS fiber but was not in the as–spun fiber. The relation between the volume of the polymer and temperature is often represented as in Figure 7. When the glassy polymer is prepared by quenching the polymer melt, the polymer in glass is initially nonequilibrium state and contains much excess volume. Heat treatment below glass transition temperature undergo structural recovery and accelerate the reduction of the excess volume to reach quasi–equilibrium state. In this process the packing of the polymer chain become denser, which bring about the increase in viscosity of the polymer glass. As the viscosity of the wholly swollen polymer is independent on that in the initial glassy state with or without any heat–treatment, the parameter M in Eq.(4) may also be decreased by physical aging. From the calculation based on Thomas and Windle theory, both smaller viscosity and smaller M value in polymer glass inhibit the formation of the sharp front. This is because typical aspect of Case II diffusion didn't observed in the as–spun PS fiber.

The diffusion of organic vapor in a glassy fiber is schematically represented for both physically aged and as–spun fiber in Figure 8. Supposed volume fraction of penetrant and swelling pressure distribution are also revealed. Hui et al. developed Thomas and Windle theory and applied to the experimental results obtained for the diffusion of iodohexane in a PS film of 4 μm by using Rutherford–backscattering technique [12]. The PS film they used was physically aged for 24 hr at 50 °C. Although they did not refer to the interfacial thickness at the sharp front as indicated by l in Figure 8, the thickness lies in the range from 100 nm to 300 nm. Further study

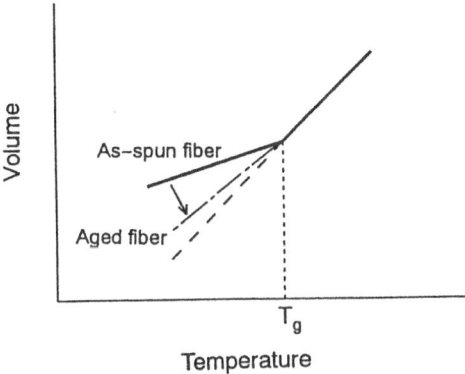

Figure 7. Effect of physical aging on the volume of the glassy polymer.

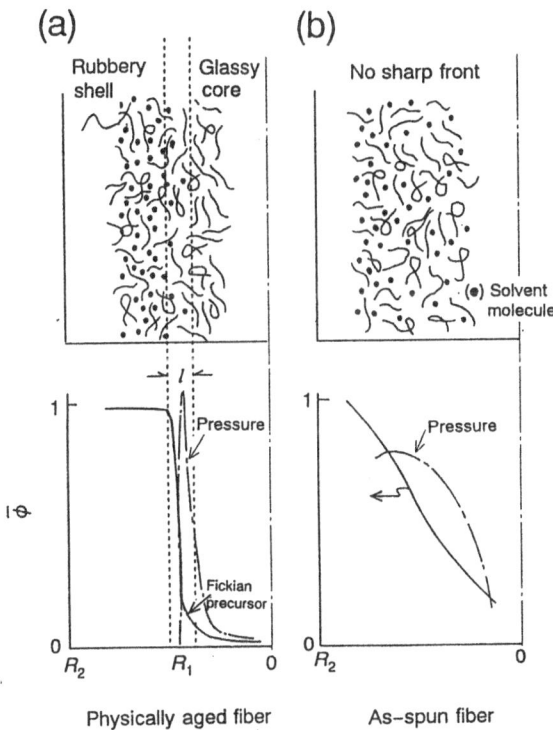

Figure 8. Schematic representation showing the penetrant distribution in (a) physically aged fiber at near the sharp front and in (b) as-spun fiber. Corresponding normalized volume fraction and the swelling pressure are also indicated.

on the interfacial thickness based on the molecular dynamics will be needed.

The common features of the Case II diffusion includes the induction time and the exist-ence of Fickian precursor just ahead of the front. The coaxial cylinder model with dual compo-nents used in this study showed good fitness to the observed intensity distribution. Taking the presence of Fickian precursor into account, the assumption that the refractive index ahead of the front is just the same as the glassy PS is maybe incorrect. Nevertheless, as the determination of the small refractive index distribution in the fiber is impossible only from the scattered intensity curve, the assumption that the step-like distribution of the refractive index may be accepted in the present stage.

According to Thomas and Windle model, the swelling pressure of radial direction showed maximum just ahead of the sharp front. Alfrey et al.[5] showed that the tensile stress in the glassy core increased as the sharp front advanced for the cylindrical PS sample, while compres-sive stress worked on the rubbery shell. When the sharp front (or core) has just disappeared, the tensile stress and radial swelling stress become suddenly down to zero in the whole region of the fiber. This is the reason the PS fiber shrank just after the sharp front has disappeared.

Acknowledgment

The author gratefully acknowledge Professor Richard S.Stein, University of Massachu-setts for his encouragement and suggestion to accomplish this work. We also wish to thank Dr.Kazuyuki Yabuki, Toyobo Co. Inc. for preparing polystyrene fiber.

References

1. Lasky,R.C., Kramer,E.J. and Hui,C.-Y., *Polymer* **29**, 673 (1988)
2. Kim,H., Waldow,D.A., Han,C.C., Qui,T.-C. and Yamamoto,M., *Polymer Commun.* **32**, 108 (1991)
3. Saenger,K.L. and Tong,H.M., *Polym.Eng.Sci.* **31**, 432 (1991)
4. Weisenberger,L.A., and Koenigh,J.L. *Macromolecules*, **23** 2445 (1990)
5. Alfrey,T., Gurnee,E.F. and Lloyd,W.G. *J.Polym.Sci.*(C) **12**, 249 (1966)
6. Fukuda,M. and Kawai,H., *Text.Res.J.* **63**, 185 (1993)
7. Van de Hulst,H.C., "Light Scattering by Small Particles" Chap.15, John Wiley & Sons, Inc., New York. (1957)
8. Kerker,M. and Matijevic,E., *J.Opt.Soc.Am.*, **51**, 506 (1961)
9. Watkins,L.S., *J.Opt.Soc.Amer.*, **64**, 767 (1974)
10. Fukuda,M in preparation
11. Thomas,N.L. and Windle,A.H. *Polymer* **23**, 529 (1982)
12. Hui,C.-Y., Wu,K.-C., Lasky, R.C. and Kramer,E.J., *J.Appl.Phys*, **61**, 5129, 5137 (1987)

"Liquid Crystallization" Mechanism of Liquid Crystalline Polymers

M.Hikosaka, K.Mabuchi, K.Yonetake, T.Masuko
G.Ungar* and V.Percec**

Faculty of Engineering, Yamagata University, Yonezawa, 992 Japan
*The University of Sheffield, Sheffield, UK ** Case Western Reserve University, Ohio, USA

Abstract   Morphology of the liquid crystals of thermotropic main chain liquid crystalline polymers isothermaly formed at atmospheiric pressure from the melt and their formation mechanism was studied. Materials used were a copolyether based on 1,2-Bis (4-hydroxy phenyl) ethane (BPE-8/12) and poly [bis (trifluoro ethoxy) phosphazene], (PBFP). Morphology of liquid crystals observed by polarizing microscopy and that of the quenched liquid crystals into crystalline phase oberved by transmission electron microscopy was very similar to that of extended chain crystals (ECCs) of crystalline polymers, such as polyethylene (PE). Therefore it is concluded that extended chain liquid crystals (ECLCs) are formed from the melt, which suggests that lamellar thikening growth is fast within the liquid crystalline phase. The lateral growth rate (V) of an ECLC showed the same degree of supercooling (ΔT) dependence as has been observed on crystalline polymers, that is $V \propto \exp(-\frac{C}{\Delta T})$ where C is a constant. This indicates that the lateral growth of the liquid crystals is a nucleation controlled process, as has been proposed by Papkov et al. Combination of these two results leads to a conclusion that the liquid cyrstals are formed by the coupling of the thickening growth and the nucleation controlled lateral growth, which we proposed to term **"liquid crystallization"** mechanism.

INTRODUCTION

One of authors (MH) and Keller et al found a new mode of polymer crystal growth, namely "lamellar thickening growth" of an isolated extended chain single crystal (ECSC) (Fig.1) on several polymers, such as polyethylene (PE)[1], poly (vinylidene fluoride /trifluoroethylene) [P(VDF/TrFE)][2], poly (chlorotrifluoro ethylene)   [PCTFE)][3] and poly(trans-1.4butadiene) [PTBD][4]. Our work is an extention of well known Wunderlich et al's well known work on exteded chain crystals (ECCs) of PE.[5] MH presented a "sliding diffusion theory", which showed that the thickening growth rate is fast when sliding diffusion within crystals is easy, while it is slow when sliding diffusion is difficult (Fig.2).[6] In the theory, we showed that polymer crystals are formed by coupling of two different grwoth mechanism, one is long known lateral grwoth and the other is the lamellar thickening growth.[6] This scheme successfully explains Bassett's well known finding on PE that "crystallization from the melt into the hexagonal phase gives ECCs and crystallization from the melt into the orhtorhombic phase gives folded chain crystals (FCCs),"[7]

A. Teramoto, M. Kobayashi, T. Norisuje (Eds.)
Ordering in Macromolecular Systems
© Springer-Verlag Berlin Heidelberg 1994

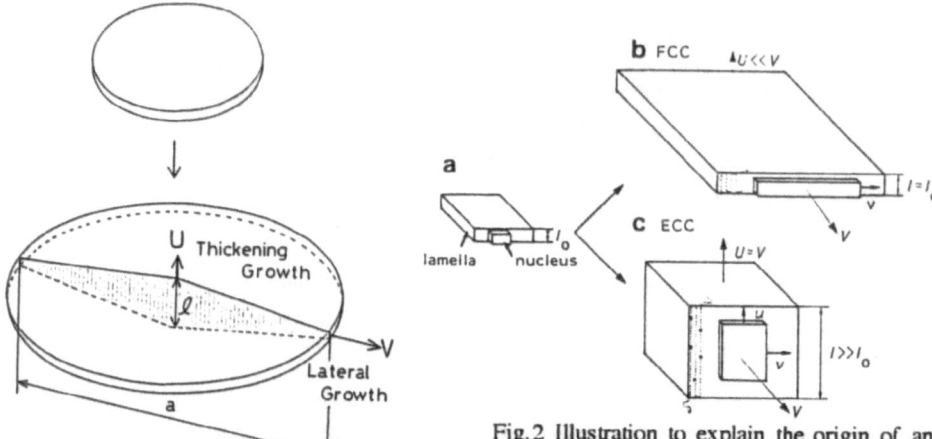

Fig. 1 Schematic perspective of an extended chain single crystal (ECSC) of PE.

Fig. 2 Illustration to explain the origin of an ECC and a FCC. a) the primary folded chain nucleus grows into b) a FCC or c) a ECC when sliding diffusion is diffucult or easy.[6]

since the sliding diffusion is easy in the mobile hexagonal phase, whereas it is difficult in the orthorhombic phase. The important role of the high chain mobility which is the same as the high sliding diffusion in "condis crystalline state" was first discussed by Wunderlich et al.[8]

The sliding motion must be much easier in the liquid crystalline phase than in any crystalline phase. (Here we will limit this work to thermotropic main chain liquid crystalline polymers.) Therefore it can be expected that the thickening growth is much faster in the liquid crystalline phase, which will result in formation of extebded chain liquid crystals (ECLCs). The ECLCs will be able to change into ECCs when the sample will be cooled below †he liquid crystal to crystal transition temperature (T_{LCC}). The first purpose of this contribuiton is to show an evidence that **ECLCs and ECCs are formed via "liquid crystallization" and liquid crystal-crystal phase transition** in the case of some liquid crystalline polymers, which has been already shown in our preliminary report.[9] Papkov, Godvsky et al found that liquid crystals of poly(diethyl siloxane) [PDES] show very similar optical morphology to ECCs of crystalline polymers.[10] They estimated by optical microscopy the lamellar thickness of the liquid crystals to be about $2 \mu m$ thick. Although it is difficult to have a final conclusion without eivdences by transmission electron microscopy, this strongly suggested that the liquid crystals may be ECLCs.

If we could have ECLCs, the interesting problem is what is the formation mechanism of the ECLCs. One possible answer has been already suggested by Papkov, Godvsky et al on PDES.[10] They showed an interesting result that the lateral growth rate (V) of the liquid crystals obeys the following experimental formula, $V \propto \exp(-C/\Delta T)$, where C is a constant. This indicates that the lateral growth rate is mainly controlled by nuclearion process of the secondary nucleus. The second purpose of this paper is to show that **the nucleation controlled** (lateral growth) **process is one of the universal mechanism in liquid crystal formation process.** Here it is to be noted that the nucleation process is only related to the lateral growth.

From combining the above results, the new liquid crystal formation process coupled with the thickening growth and nucleation controlled lateral growth will be presented, which **we would like to propose to term "lliquid crystallization"** mechanis. The universality of the coupling the thickening growth and nucleation controlled lateral growth in both liquid crystal formation and crystallization processes will be discussed. This is interesting not only for understanding the nature of liquid crystallization and crystalization mechanisms but also for finding a new method to cotrol the strucutre and morphology of polymer materials.

EXPERIMENTALS

Materials used in this study were a copolyether based on 1,2-Bis (4-hydroxy phenyl) ethane (BPE-8/12) (Fig.3a)[12] and poly [bis (trifluoro ethoxy) phos- phazene], (PBFP), $Mw=15.4*10^4$, $Mn=8.6*10^4$, $Mw/Mn=1.8$ (Fig.3b). Both polymers are thermotropic main chain liquid crystalline polymers. All the expeiment was carried out at atmospheric pressure. Sample was put in a hand-made hot stage and the "liquid crystallization" processes from the melt were directly observed by polarized optical microscopy (Olympus POM). The liquid crystallization behavior was recorded by camera (Olympus PM-10ADS) and ultra high sensitive video camera system (Flovel, HCC-1950). Range of the degree of supercooling (ΔT) where this experiment was carried out is illustrated in the DSC traces (Fig.4 a and b), where ΔT is defined by $\Delta T = T_m^0 - T_c$ where T_m^0 and T_c are equilibrium melting temperature and crystallization temperature. T_m^0 was estimated by Hoffman-Weeks plot.[13] Thin thermocouple (0.2mm in diameter) was directly inserted into the sample in order to measure the acurate sample temperatrure. The lateral growth rate (V) was directly obtained by observing the change of the lateral size (a) of an isolated liquid crystal with time (t) by applying the definition, $V = \frac{1}{2} \frac{da}{dt}$. At some stage of "liquid crystallization" where isolated liquid crystals are formed, the sample was quenched into ice water and then replica of fracture surface was prepared in order to observe the morphology by transmission electron microscopy (TEM, JEM100CX). These procedure are the same as has been carried out in the study of crystallization.[1]

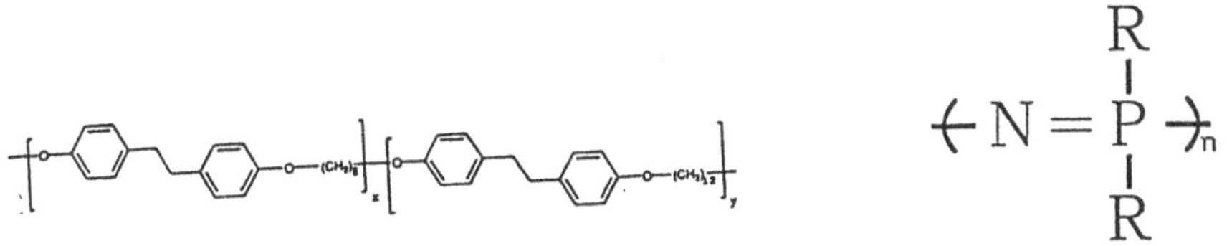

$$\begin{array}{c} R \\ | \\ \left(N = P \right)_n \\ | \\ R \end{array}$$

$$R : OCH_2CF_3$$

Fig.3 Structural formulae of molecules. (a) BPE -8/12 and (b) PBFP (a)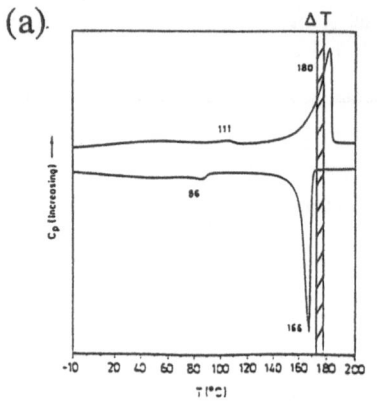

(b)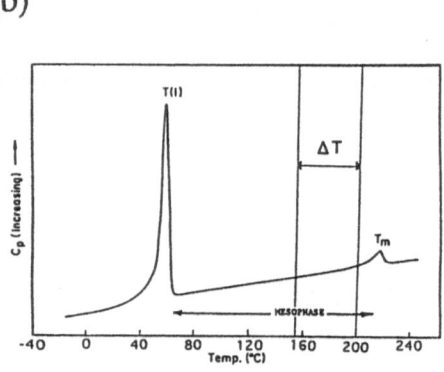

Fig.4 DSC traces of (a) BPE-8/12[12] and (b) PBFP. Range of degree of supercooling is illustrated.

RESULTS

1) Extended Chain Liquid Crystals (ECLCs) and ECCs

Fig.5a and Fig.6a represent polarizing optical micrographs of isolated liquid crystals of BPE and PBFP resectively. They showed lens like or cigar like shape which was very similar to ECSCs of PE, P(VDF/TrFE), PCTFE and PTBD.[1-4] The shape is also similar to that observed by Papkov et al on PDES.[10] This suggests that they are "single liquid crystals". There was no change in optical morphology of isolated liquid crystals of BPE and PBFP between before and after the liquid crystal to crystal phase transition with quenching, from which it was considered that we could guess the

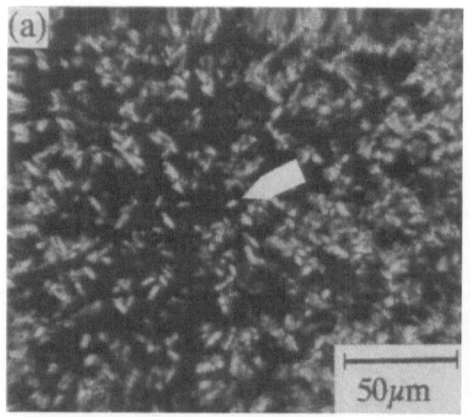

Fig.5 Morphology of a liquid crystal of BPE-8/12 isothermaly grown at $\Delta T= 4K$. (a) polarized optical micrograph, the arrow shows a lens like shape and (b) transmission electron micrograph, showing thick stacked extended chain crystals (ECCs).

morphology of the single liquid crystals much more in detail by using transmission electron microscopy (TEM).

Transmission electron micrograph of the quenched samples of BPE (Fig.5b) showed thick stacked lamellae 0.2-0.4μm thick with striation, which is the same as the morphology of stacked ECCs of the above polymers.[1-4] Therefore they must be stacked ECCs. In the case of PBFP, thick isolated lamella with striaton was seen, 1.2μm thick (Fig.6b), which is similar to the morphology of an isolated **extended chain single crystals (ECSCs)** of the above polymers.[1-4] Therefore it must be an isolated ECSCs of PBFP. Under the same condition, another huge cigar-like liquid crytals were seen by optical microscopy (Fig.7a). The electron micrograph of them showed regularly stacked thick lamellae with striation, 0.4μm thick, (Fig.7b). This indicates that the huge liquid crystals must be stacked ECLCs, formed by overgrowth, perhaps due to screw dislocation.

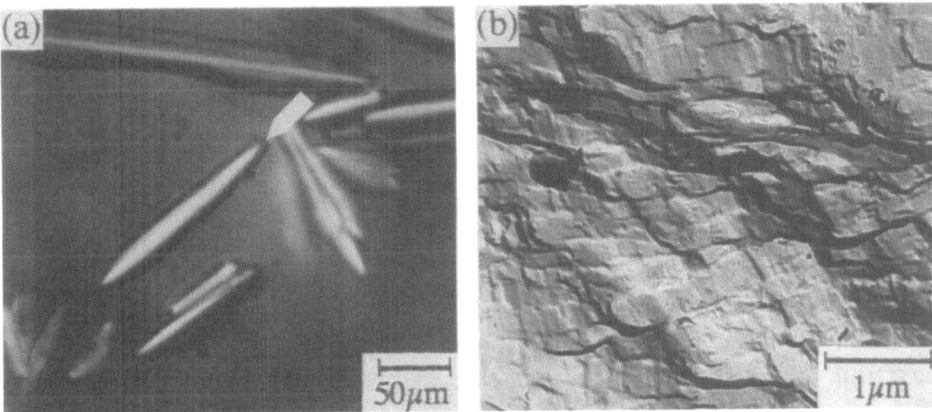

Fig.7 Morphology of a liquid crystal of PBFP isothermaly grown at ΔT= 16K. (a) polarized optical micrograph, the arrow shows a huge cigar like liquid crystals and (b) transmission electron micrograph, showing thick stacked ECCs.

Fig.6 Morphology of a liquid crystal of PBFP isothermaly grown at ΔT= 16K. (a) polarized optical micrograph, the arrow shows a disk shape and (b) transmission electron micrograph, showing an isolated extended chain single crystal (ECSCs).

The size and shape of isolated ECSCs or ECCs observed by TEM well corresponded to liquid crystals directly observed by polarizing optical microscopy. Therefore we may conclude that the ECSCs and/or ECCs of BPE and PBFP were ogininaly obtained from the extended chain single liquid crystals (ECSLCs) and/or the extended chain liquid crystals (ECLCs) through the liquid crystal to crystal phase transition. In another word, **the first conclusion is that ECSLCs (or ECLCs) of BPE and PBFP are formed directly from the isotropic (i.e., melt) phase and transform into ECSCs or (ECCs) by quenching** into a temperature below the $T_{I\text{-}C\text{-}C}$. Here the single liquid crystal is a new concept to show close resemblance to the single crystals. So far as the authors know, the formation of an ECSLC (or ECLCs) from the melt is found at the first time. This finding is ΔT interesting in understanding the mechanism of liquid crystal formation.

2) Nucleation Controlled Lateral Growth

Length of growing isolated ECSLCs (a) of BPE and PBFP increased linearly with time (t) (Fig.8a and b). This indicates that the lateral growth of the ECSLCs is a steady process, which is the same as has been shown on ECSCs of PE, P(VDF/TrFE), PCTFE and PTBD.[1-4] From the gradient of a-t lines, the lateral growth rate (V) was easily obtained. Logarithmic V vs. inverse of degree of supercooling (ΔT^{-1}) gave a straight line (Fig.9a and b), hence the following experimental formula was obtained.

$$V \propto \exp(-\frac{C}{\Delta T}) \qquad (1)$$

where C is a constant. This is again the same as has been shown on ECSCs of PE, P(VDF/TrFE) and PCTFE.[1-3] We have alredy shown that the lateral grwoth rate of these crystalline polymers is mainly controlled by nuclearion process of the two dimensional nucleus.[1-3] Therefore **the second conclusion is that the lateral growth of exteΔTnded chain single liquid crystal (ECSLCs) of BPE and PBFP is mainly controlled by the nuclearion process of the**

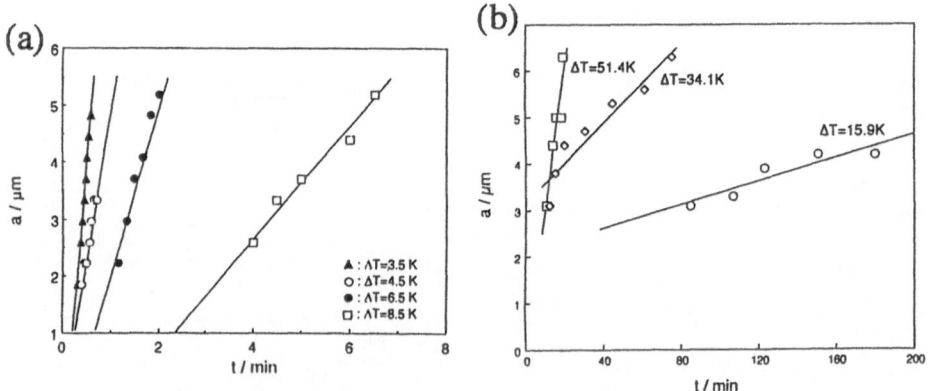

Fig.8 Linear increase of lateral size (a) of an ECLC with time (t) observed by optical microscopy. (a) BPE-8/12 and (b) PBFP.

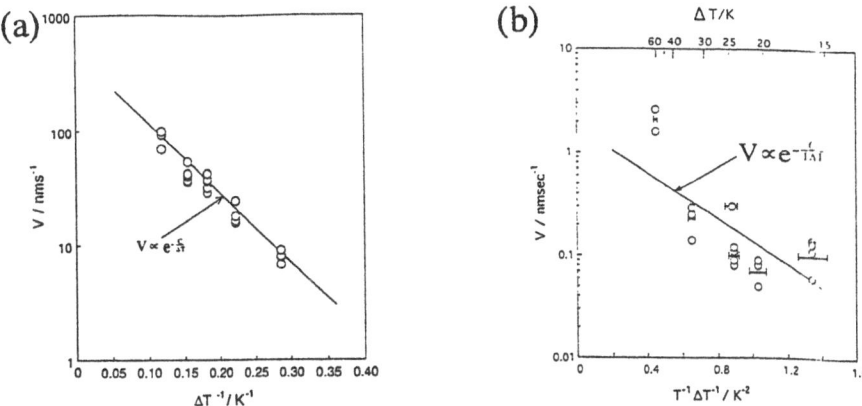

Fig.9 logV vs inverse of degree of supercooling ΔT^{-1} (a) BPE-8/12 and (b) PBFP.

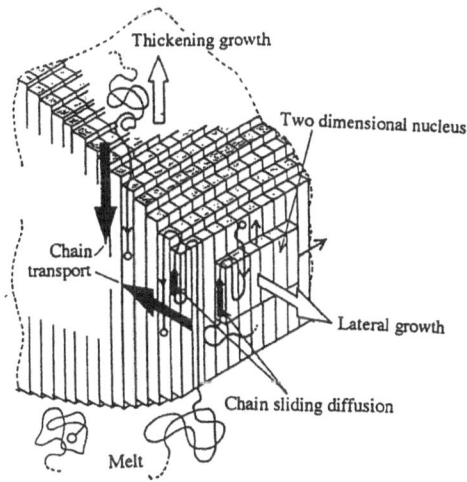

Fig. 10 Schematic illustration of the two dimentional nucleus on the flat side surface of an ECLC.

two dimensional nucleus. The two dimensional nucleus on the side surface of liquid crystals is schematically illustrated in Fig. 10. This is similar to Papkov et al's conclusion obtained on the liquid crystals of PDES.[10] The difference between us and Papkov et al is small, that is, the former is obtanined on ECSLCs and the latter on some single liquid crystals which morphology is not sure as to whether chain is extended or not within the liquid crystal. Thus the nucleation controlled lateral growth of single liquid crystals is probably one of the universal mechanism.

3) Proposal of a New "Liquid Crystallization" Mechanism

Based on the obtained new results, we will propose a new liquid crystal formation mechanism, that is, a single liquid crystal will be formed by coupling of thickening growth and nucleation controlled lateral growth, which we would like to term **"liquid crystallization"** mechanism. **We may predict that the single liquid crystal will grow into an extended chain singl liquid crystal (ECSLC)** due to high chain sliding diffusion which results in fast thickening growth. This predicition is an extended one presented on crystalline polymers.[6]

DISCUSSIONS

Is it reasonable to consider the two dimensional nucleus on the side surface of liquid crystals? is a probable question to our proposal. The side surface of a single liquid crystal must be definite, flat and smooth interface if the temperature is below so called thermal roughening temperature, because all chains are at least parallel within the liquid crystals, whether it is in the smectic or nematic phase, wheras random within the melt. In this case it is general in normal nucleation theory to consider the two dimensional nucleus on the flat and smooth surface, which is an answer to the above question.

The important role of chain sliding diffusion (or chain mobility) in formation of so called meso-phase and/or condis crystal or, more generally, any crystals is one of the important topics in polymer science. It has been widely discussed by Wunderlich[8] and recently restudied by Keller, MH, Rastogi et al by connecting with the important role of stability of phases[14,15] There are many other interesting experimental results which have close relation to this study. One example is Mitchell et al's work on some random copolyesters,[16] where thick lamellae are obtained by annealing. Another one is shown on PDES by Moler et al. They showed that ECCs are formed when it is annealed in β_2 crystalline phase which was shown to be chain mobile phase,[11] which is essentially the same as has been seen on extended chain crystals of PE. Further interst question is if PDES or similar siloxane polymers would show the same "liquid crystallization" mechanism and form ECLCs or not.

CONCLUSIONS

1) Formation of extended chain single liquid crystals (ECSLCs) or extended chain liquid crystals (ECLCs) of BPE and PBFP from the melt is sugested at the first time. This is suggested from the fact that extended chain single crystals (ECSCs) or extended chain crystals (ECCs) are confirmed on the quenched liquid crystals which is isothermally formed.

2) Degree of supercooling (ΔT) dependence of the lateral growth rate (V) of an isolated ECSLC , $V \propto \exp(-\dfrac{C}{\Delta T})$ where C is a constant, showed that the lateral growth of the liquid crystal is a nucleation controlled process.

3) We proposed a new **"liquid crystallization"** mechanism where a single liquid crystal will be formed by coupling of thickening growth and nucleation controlled lateral growth. **We predict that the single liquid crystal will grow into an "extended chain singl liquid crystal " (ECSLC)** due to high chain sliding diffusion which results in fast thickening growth. This predicition is an extended one presented on crystalline polymers in the sliding diffusion theory.[6]

REFERENCES

1) M.Hikosaka, S.Rastogi, A,Keller and H.Kawabata, J.Macromol.Sci.,Phys., 1992,B31,87,

2) M.Hikosaka, K.Sakurai, H.Ohigashi and T.Koizumi, Jpn.J.Appl.Phys. 1993,32, 2029 and ibid, 1993.32.2780

3) M.Hikosaka, et al, Polym.Prepr.Jpn,1989,30.307

4) S.Rastogi, M.Hikosaka, A.Keller and G.Ungar Progr.Coll.Polym.Sci.,1992. 87.42

5) B.Wunderlich and L.Mellilo, Macromol.Chem., 1968,118.250

6) M.Hikoska, Polymer. 1987.28. 1257 and ibid, 1990,31,458

7) D.C.Bassett, S.Block and G.J.Piermarini, J.A.P., 1974,45,4146

8) B.Wunderlich and T.Arakawa, J.Polym.Sci.,1964,A2,3697

9) T.Itaya et al, Polym.Prepr.Jpn., 1992

10) V,S,Papkov, V,S,Svistunov, Yu.K.Godovsky and A.A.Zhdanov,J.,Polym.Scil., 1987,25, 1859

11) G.Kögler, K.Loufkis and M.Möller, Makromol.Chem. Macromol.Symp., 1990, 34,171

12) G.Ungar, J.L.Feijo, V.Percec and R.Yord, Macromolecules, 1991, 24, 953

13) J.D.Hoffman and J.J.Weeks, J.Res.NBS, 1962.66A.13

14) A.Keller, M.Hikosaka, S.Rastogi, A.Toda,P.J.Barham and G.Goldbeck-Wood, submitted to J.Mat.Sci.,

15) S.Rastogi, M.Hikosaka, H.Kawabata and A.Keller, Makromol. Chem., Macromol. Symp., 1991.48/49.103

16) R.H.Dutton, D.C.Bassett and G.R.Mitcell, A Conf. to mark the retirementof A.Keller FRS. (Bristol, 1991)

Thermotropic Liquid Crystals in Polypeptides

Junji Watanabe

Department of Polymer Chemistry, Tokyo Institute of Technology,
Ookayama, Meguro-ku, Tokyo 152, Japan

Several types of the liquid crystal (LC) polymers have been synthesized and the mesomorphic properties and structures have been studied. Among these LC polymers, the polypeptide is of special interest since it has comprised a peculiar class of hard-rod polymers by pretending the stable α-helical conformation. By this reason, the mesomorphic behavior and structure have been extensively studied in the lyotropic solutions.

The further interest in the LC of the polypeptides has been accelerated by the recent discovery of the thermotropic polypeptides, which can be attained by the simple chemical modification [1-14]. Its first observation has been made in poly(γ-methyl D-glutamate-co-γ-hexyl D-glutamates) [2]. In this copolymer system, the thermotropic LC nature can be induced for copolymers with the intermediate copolymer contents of 30 to 70%. The similar phenomena can be observed in the other copolymer systems [1,5,6], leading to the conclusion that an appreciable difference in length of side chains of the two comonomers, that is a difference of five or more methylene units, is necessary for the induction of the thermotropic nature. The thermotropic LC behaviors are also observed in the homopolymer system. The typical example is poly(γ-alkyl L-glutamates) [3,4], in which the

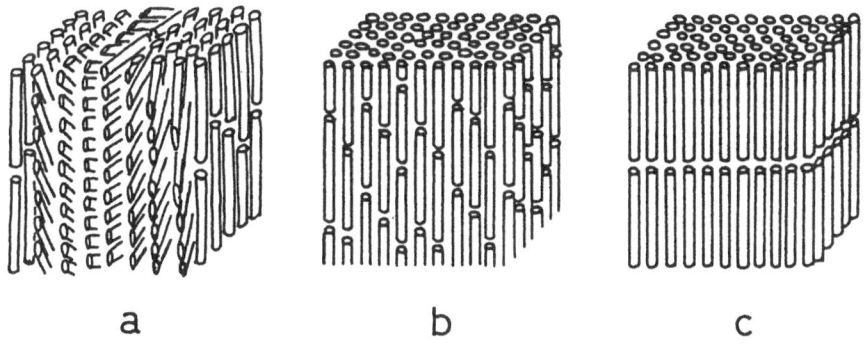

Figure 1. Illustration of (a) cholesteric, (b) columnar and (c) smectic A liquid crystals.

A. Teramoto, M. Kobayashi, T. Norisuje (Eds.)
Ordering in Macromolecular Systems
© Springer-Verlag Berlin Heidelberg 1994

thermotropic nature arises if the n-alkyl groups longer than decyl group are attached as side chains.

In both systems, the long flexible side chains are responsible for the induction of thermotropic LC such that they may play the role of solvent in the lyotropic LC [3,5,7,8]. These novel LC polymers having a mesogenic α-helical rod surrounded by flexible side chains, should be differentiated from the two familiar types of thermotropic LC polymers, so-called the side-chain and main-chain LC polymers, and can be classified into a third type, the rigid-rod main-chain LC polymer [5].

In this paper, I will review our experimental data so far collected in these polypeptides and present their thermotropic transition behaviors and mesophase structures.

1. Properties and Structure of Cholesteric Mesophase

Because of the chirality in main-chain α-helix, the typical liquid crystal observed is a cholesteric (see Figure 1a). Most representative materials forming the cholesteric liquid crystal are poly(γ-benzyl L-glutamate-co-γ-dodecyl L-glutamates) (BD-N: N, % dodecyl content) (refer to Figure 2) [5]. In these materials, the transformation of crystal to liquid crystal appears as a first-order transition on DSC thermogram. The markedly small transition entropies of 0.1 cal/mol K are estimated. Such small values may be reasonable since no conformational change of the main chain is included on this transition. In the X-ray method, one can observe that the crystalline pattern is replaced by a pattern displaying only diffuse hallos [6]. Simultaneously, the mechanical method exhibits an abrupt drop of storage modulus. Both show the liquid-like nature of the material.

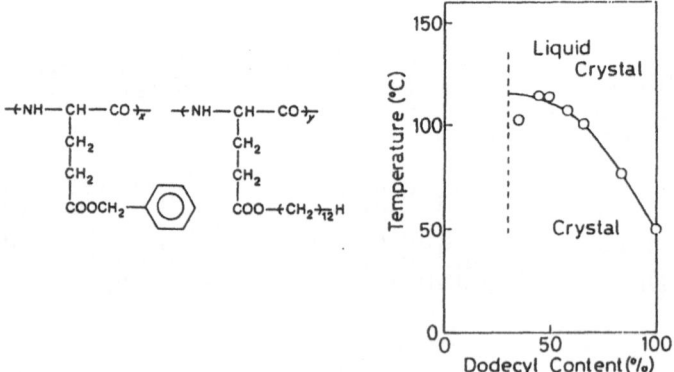

Figure 2. Variation of transition temperatures with the dodecyl content in BD copolymer system.

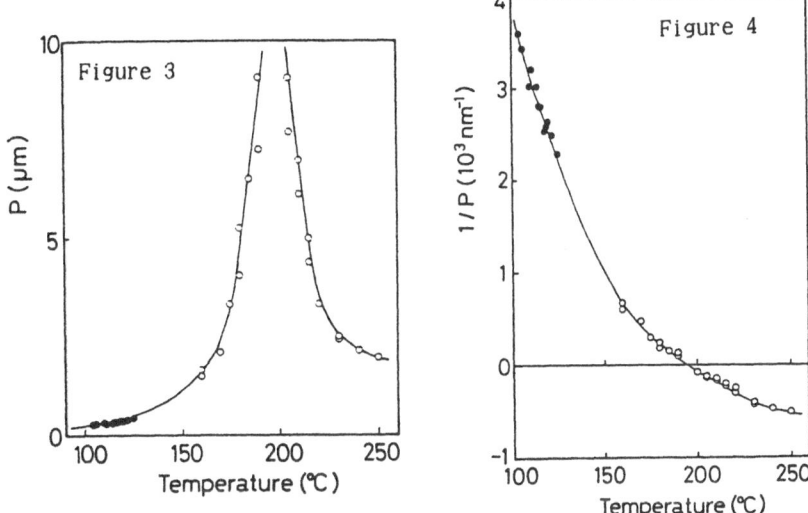

Figure 3. Variation of pitches with temperature for the cholesteric phases of BD-58. The closed and open circles incicate the data obtained by CD and microscopic observations, respectively.

Figure 4. Variation of the reciprocal pitches with temperature.

The transition temperatures thus determined are plotted against the dodecyl content in Figure 2. The thermotropic cholesteric LC phases are found to be induced for copolymers with the dodecyl contents more than 30%. The transition temperature is around 115°C for copolymers with dodecyl contents of 30 to 60 %, gradually decreases with increasing dodecyl content and finally reach 50°C for poly(γ-dodecyl L-glutamate). No isotropic liquid appears in the temperature range below the decomposition temperature of around 250°C and hence, the cholesteric phase is stably formed in the wide temperature range above the transitions.

One of the significant features in this cholesteric system is the strong temperature dependence of cholesteric pitches [9,11]. It should be noted here that hereafter any mesophases have been annealed to form the equilibrium phase structure. The annealing period is altered depending on the molecular weight and mesophase temperature, but generally several hours are required because of the high viscosity. Figure 3 shows the typical temperature dependence of cholesteric pitch which has been observed in BD-58 [9]. For this material, the cholesteric phase appears having the small pitches around 300 nm at the initial temperatures just after the transition. Such small pitches can be determined by the circular dichroism (CD) spectra due to the selective reflection of circularly polarized light (see later Figure 5). These are given as closed circles in Figure 3. One can find here that in the temperature range from 100 to 130°C, the pitches are so small to be detected by CD spectra, and increase with increasing temperature. This trend can be also recognized from the visual observation of reflection colors changing from blue to red on increasing temperature. With a further increase of temperature, the pitches become larger so that cholesteric helical structure can be detected as the fine striation lines under an optical microscope. By measuring the spacing between lines, the pitches were determined in the temperature region of 150 to 250°C. These are also plotted as open circles against temperature in Figure 3. Interestingly, the pitch increases substantially with the increase of temperature from 150°C, diverges at 195°C, and then decreases as the temperature is raised further.

This reverse temperature dependence of cholesteric pitches can be attributed to the sense inversion of cholesteric helix from the right-handed to the left-handed helix, which has been confirmed by the sign inversion in the induced CD for the phenyl groups of side chains and also in the form optical rotation due to cholesteric helix [9]. By consideration of this sense inversion, the reciprocal pitches proportional to the twist angle, are plotted against temperature in Figure 4. Here, the positive values correspond to the right-handed cholesteric helix while the negative to the left-handed. One can find that on this plot overall data points fall on a smooth curve passing through zero at 195°C (T_N). At lower temperatures, the remarkable variation of twist angle is observed whereas at higher temperatures the curve indicates less variation and seems to aproach a constant value. This type of temperature (T) dependence of reciprocal pitches (1/P) has commonly been observed for the other copolymers and can be described by the following empirical equation [5,11],

$$1/P = (1/P_0)(T_N - T)/(T - T_C)$$

It is worth stating here that the empirical equation has a similar form as the theoretical one which is proposed by Kimura et al.[15]. Also under the approximation of T = T_N, it can be deduced to the empirical equation which has been elucidated in the lyotropic cholesteric systems [16].

The easy preparation of Grandjean cholesteric films is a second striking feature in this system [10]. If the cholesteric film is prepared with a thickness less than 100 μm, it shows the uniform domain texture in the microscopic observation. In the conoscopic observation, the dark cross typical of monoaxial crystal can be seen. The use of a quarter-wave plate further dictates the feature of optically negative crystal. These are to be

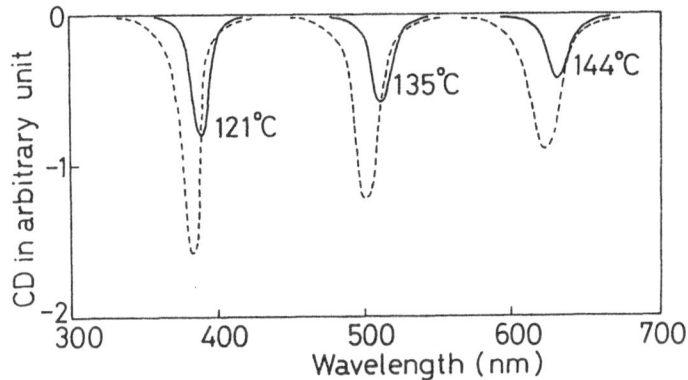

Figure 5. The CD spectra (solid curves) for cholesteric mesophases at different temperatures. The dashed curves show the CD spectra observed for the solid cholesterics which were prepared by quenching the mesophases.

expected for a Grandjean cholesteric in which the cholesteric helical axis lies uniformly in a direction perpendicular to the film surface.

When the mesophases are prepared to have the pitches comparable to the wavelength of the visible light, these exhibit the beautiful cholesteric colors as mentioned above. This color effect can be quantitatively analyzed from the CD spectra. The typical spectra are shown in Figure 5. Fairly sharp spectra are observed having a spectral width of 10 to 30 nm. These widths are only slightly larger than the values of 10 to 20 nm expected from the equation $\Delta\lambda = \Delta n\, P$, again leading to the uniform alignment of cholesterics [10].

The direct proof of the Grandjean structure is given by the transmittance electron microscopic (TEM) observation. Figure 6 indicates the TEM photograph observed for a thin microtomed film cut out perpendicularly to the film plane of solid cholesteric. Here, the solid cholesteric was prepared by quenching the mesophase into nitrogen. In this Figure, one can observe a set of black and white lines periodically spaced. The variation in transmissibility of electron beam from one pseudo-nematic layer to the other is attributable to an alternative appearance of black and white lines [10,17]. A black line means less transmissibility of an electron beam and its portion corresponds to the pseudo-nematic layers with their molecular directors almost parallel to the microtomed film plane while the white portion corresponds to the layers having the director nearly perpendicular to the film plane. It is obvious that the layers are continuously twisted up from one surface of film to another and there are no appreciable defects such as dislocations.

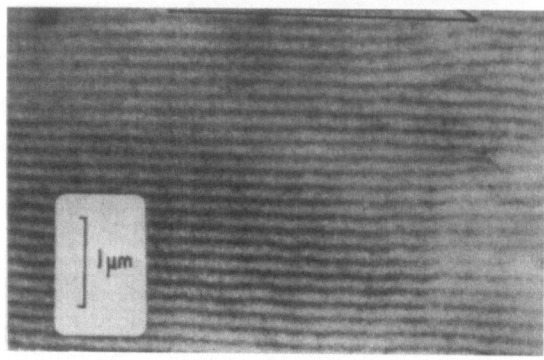

Figure 6. The TEM photograph of thin microtomed films cut perpendicularly to the film plane of solid cholesteric. The arrow indicates the direction of the plane of solid cholesteric film.

An easy preparation of the Grandjean texture in the present polypeptides is attributed to a preferential planar orientation of rigid-rod backbones with a large axial ratio. For rigid-rod mesophases, the director must be developed parallel to the surface, since condensing the ends of rods at the surface costs a lot of entropy. The high elastic constant for splay and bend deformations [18] may also promote the formation of monodomain structure. There is hence no serious condition for the preparation of the Grandjean texture in thermotropic polypeptides. Only some shear flow is needed to prepare a thin mesophase of less than 100 μm and also to promote a parallel alignment of hard-rod molecules to the surface.

In order to utilize the remarkable optical properties, the solid Grandjean films as prepared by quenching to a temperature below the crystal to mesophase transition are very useful. The dashed curves in Figure 5 show the CD spectra for such solid cholesterics and are interestingly compared with the solid curves observed in the mesophases. It can be found that the color effects are not seriously changed from those in the mesophases by solidification; there is no essential difference between spectra with respect to the maximum wavelength and can be seen only an increase of intensity. Thus, we can see the cholesteric colors in solid films similarly as in the mesophases. The retained colors are arbitrarily selected by changing the temperature of the prior existing mesophase and remain stable for a long time. The spectrum width is still so narrow as in the mesophase that nine or ten independent colors can be attained within a wavelength region of visible light. We can thus utilize these solid cholesteric films as optical filters, a performance of which is based on a selective reflection of circularly polarized light with desired wavelength.

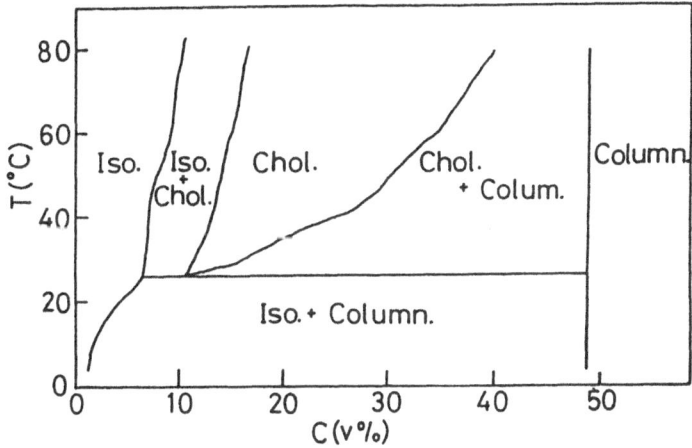

Figure 7. The phase diagram of poly(γ-phenylpropyl L-glutamate)—m-cresol system. Here, I, Chol and Colum mean the isotopic, cholesteric and columnar phases, respectively.

2. Properties and Structures of New Other Types of Mesophases

The cholesteric mesophase, thus, is typical of thermotropic mesophase in polypeptides and much less was known about other types of mesophases. More recently, however, the experimental data indicate that the rigid-rod polymers such as polypeptides may form liquid crystals which are neither nematic nor cholesteric. Livolant and Bouligand [19], and Lee and Meyer [20] have suggested through the microscopic textures that the poly(γ-benzyl L-glutamate) may form hexagonal columnar liquid crystals in its lyotropic solution. We have also observed the hexagonal columnar phase in poly(γ-phenylpropyl L-glutamate)—m-cresol system [21]. The phase diagram in this system as illustrated in Figure 7, indicates that the columnar phase appears in the polymer concentrations higher than 48 % after a narrow coexistence with the cholesteric phase. Strzelecka et al.[22], have reported that a smectic-like phase in addition to a cholesteric phase can be seen in aqueous solutions of DNA, and subsequently, Livolant et al.[23] have observed a columnar hexagonal phase in a similar DNA system. The smectic A phase has been also observed in the lyotropic solutions of tabacco mosais virus by Wen et al.[24]. On the theoretical side, Kimura et al.[25] and Frenkel [26] have predicted that in the hard-rod particles, smectic A and columnar liquid crystals may arise as the result of an excluded-volume effect of the hard rods. Thus, the search for new liquid crystals based on rigid-rod polymers and their identification are now proceeding in both experimental and theoretical aspects.

In the thermotropic polypeptides, we can also observe the several other mesophases, columnar and smectic phases (see Figures 1b and 1c) [12,13]. First, we will describe about the hexagonal columnar phase.

The columnar phase has been observed in poly(γ-octadecyl L-glutamate) [12]. For convenience, the polymer is abbreviated here as PG-18-200; 18 means the carbon number of alkyl side chain and next number is the degree of polymerization. In this polymer, the long octadecyl side chains are crystallizable at 45°C on cooling. On crystallization, the characteristic segregation structure of the main chain and side chain is formed such that the main-chain α-helices closely associate to form a layer and the side chains are placed in the space between the layers [3]. We have so far thought that in this system the cholesteric liquid crystal can be attained immediately after the melting of side-chain crystals as in a previous copolymer system; in fact, the sample has a fluidity and so has a liquid crystalline nature. However, the cholesteric phase can be detected only in the higher temperature range above 150°C, suggesting another thermotropic phase to exist in the lower temperature region from 60 to 150°C. The details of this lower temperature mesophase will be presented later. On further heating the cholesteric phase, the other phase transition takes place at around 200°C. This can be initially detected by the optical microscopic observation; the polygonal texture characteristic to the cholesteric is altered to the fine broken fan texture [12]. On this transition, the viscosity is substantially increased, and also the cholesteric helical structure simultaneously disappears. The latter can be unambiguously detected by the disappearance of the form optical rotation [12].

The phase transition can be also detected by X-ray diffraction method as in Figure 8. Curve a shows the X-ray diffraction profile from the cholesteric mesophase at 150°C. It includes two inner and outer reflections with the spacings of 20 Å and 4.7 Å. These are surely expected for a cholesteric mesophase; the inner reflection shows the disordered lateral packing of α-helices, and the outer reflection may result from the fifth layer line of α-helix and/or the alkyl side chain in melt. In contrast, the higher temperature mesophase exhibits a distinct X-ray profile (see curve b in Figure 8). It shows three sharp inner reflections although the outer broad reflection is invariably observed. The lattice spacings of the sharp reflections are in a ratio of 1, $1/\sqrt{3}$ and 1/2, indicating that α-helical rods are laterally packed into a regular two-dimensional hexagonal lattice. The denisity examination dictates that each α-helical molecule passes through each hexagonal unit cell. We thus conclude that the higher

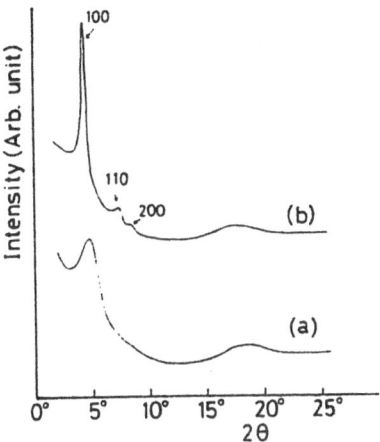

Figure 8. X-ray diffraction profiles observed for the mesophases of
PG-18-220: (a) cholesteric mesophase at 150°C and (b) columnar hexagonal
phase at 210°C.

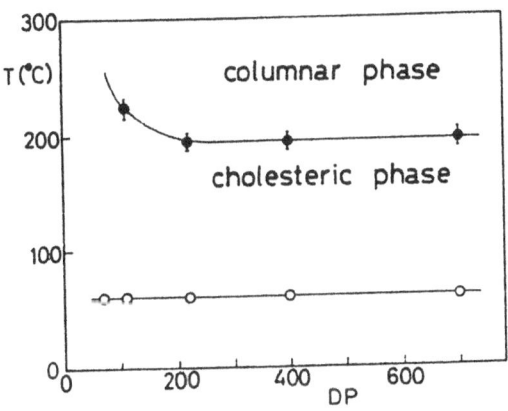

Figure 9. Variation of transition temperatures with the degree of
polymerization (DP) of PG-18. The open circles mean the melting temperature
of side-chain crystallites, and the closed circles represent the transition
temperatures from the cholesteric phase to the columnar hexagonal phase.

temperature mesophase is a hexagonal columnar phase as illustrated in
Figure 1b.

It is interesting to state here that the temperature of the cholesteric
to columnar phase transition depends on the degree of polymerization (DP).
Figure 9 shows the transition temperatures plotted against the DP. For the
materials with a DP more than 200, the transition temperature remains
constant at around 200°C, but it increases to 230°C for PG-18-110, and
finally it may move to a temperature beyond the decomposition temperature

for PG-18-70, the lowest molecular weight material employed here. It is thus obvious that the transition temperature decreases with an increase in the DP. In other words, the columnar hexagonal phase is more stable for the polymer with a higher DP. Here, we must emphasize the curiosity that the more ordered columnar phase appears in the higher temperature region than the less ordered cholesteric mesophase. This curiosity, however, cannot be explained now although it might give a clue for clarifying a mechanism of the transition.

We next refer to the smectic phases appearing in the lower temperature region than the cholesteric phase. For this study, we employed the PG-18-70 with the lowest molecular weigth and, hence, with the lowest viscosity of mesophases. As it is likely that the molecular weight and its distribution significantly affect the transition behavior and structure of smectic phases, the PG-18-70 was divided into four parts by the GPC fractionation [27]. Fractions of polymers were coded PG-18-70-I, II, III and IV in the order of increasing molecular weight. In Table I, the DP and the molecular weight distribution, Mw/Mn, of these samples are listed as measured by GPC on the standard of polystyrene. It is interesting to note that the markedly different phase behaviors were observed between the II and III samples. The I sample is similar to the II sample in the nature and the IV sample to the III sample. First, we shall refer to the II sample.

The PG-18-70-II polymer shows the unexpected phase behavior, which can be initially recognized from the microscopic observation. Figures 10a to 10c show the representative microscopic textures taken at 80°C, 120°C and 160°C. At initial temperature region of 60 to 80°C, the mesophase does not show the birefringence at all, as found in Figure 10a, and hence it is

Table I. Degree of polymerization and molecular weight distribution of the fractionated PG-18-70

Polymer	DP*	Mw/Mn
PG-18-70-I	30	1.59
PG-18-70-II	36	1.53
PG-18-70-III	45	1.45
PG-18-70-IV	60	1.60

* measured by GPC on the standard of polystyrene

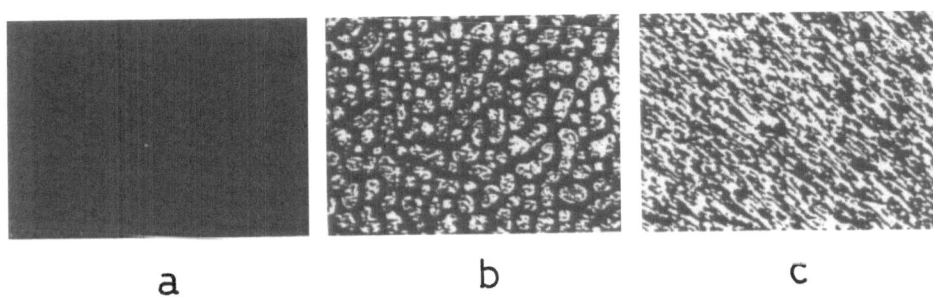

Figure 10. The optical microscopic textures of PG-18-70-II observed at (a) 80°C, (b) 120°C and (c) 160°C.

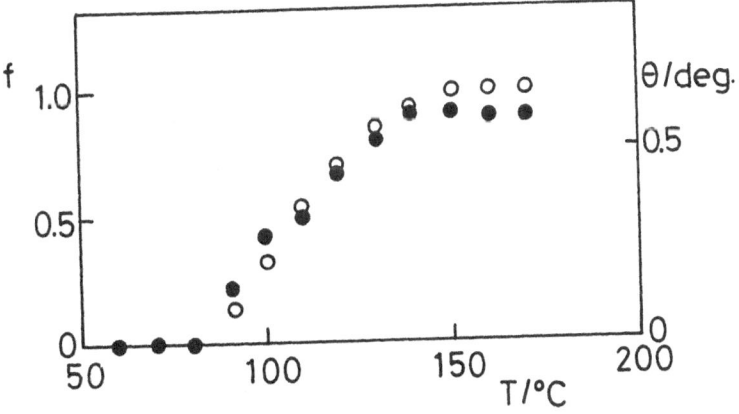

Figure 11. Temperature variation in the relative ratio of the anisotropic area to the isotropic area (f; open circles) as determined for the PG-18-70-II by microscopic observation and in the optical rotation, (θ; closed circles), as measured at 550 nm.

optically isotropic. On heating to the temperatures above 80°C, the birefringent cholesteric phase appears as spherulites in the dark domain. After the coexistence with the optical isotropic phase (Figure 10b), the anisotropic cholesteric phase spread over the whole domain at the temperatures higher than 150°C (Figure 10c). This phase behavior can be well recognized in Figure 11 where the relative ratios of anisotropic area to isotropic area are plotted as open circles against temperature.

This phase behavior can be also detected through the observation of the form optical rotation due to the cholesteric helix. In Figure 11, the optical rotations measured at 550 nm are also plotted as closed circles against temperature. The optical rotation cannot be observed at all for the isotropic phases at the temperatures below 80°C, appears gradually at 90°C and finally increases to a constant value in the temperature region above 150°C. The trend, thus, is obviously same as observed in the microscopic textures, leading to the conclusion that the uniform isotropic phase exists in the temperature region below 80°C and the uniform cholesteric phase appears in the temperature region above 150°C after the coexistence with the isotropic phase in the intermediate temperature region of 80 to 150°C.

The two phases can be also distinguished from each other through the crystallization behavior of side chains. In the DSC measurement, the isotropic phase exhibits a broad crystallization peak at 25°C on cooling while the cholesteric phase shows the sharp peak at 45°C. These results show that the crystallization of side chains occurs in a different manner between two phases and also that in the isotropic phase there is a poor organization of side chains into a crystalline form which may be caused by a lack of the uiaxial orientational order in a packing of α-helices.

With respect to the optically isotropic phase, we can consider two types of phases. One is a cubic isotropic mesophase and another is just an isotropic liquid. The cubic isotropic phases have been discovered in the certain lipids [28] and low molecular weight mesogenic materials [29]. They are called the cubic amphiphilic phase and the smectic D phase, respectively. Both have a net-work structure made up of jointed cylinders which are the rod-like micelles in the cubic amphiphilic phase or the rod-like columns in the smectic D phase. Both phases exhibit a striking resemblance in the X-ray patterns which contain several small-angle reflections and have been interpreted by the large cubic lattices. In the present material, the component molecule has a similar structural feature, a cylindrical shape surrounded by alkyl side chains, and hence can be expected to form the cubic phase. However, the observed X-ray pattern have offered no evidence for this, showing only a diffuse reflection with a spacing of 25 Å in a small angle region. On the other hand, the assignment of this isotopic phase to the isotropic liquid is hardly acceptable since it is very exceptional that the isotropic liquid is placed in the lower temperature region than the anisotropic cholesteric phase.

Although an ambiguity is thus remained to determine the phase structure, it is no doubt that the length of alkyl side chain is significantly correlated to the length of backbone for the formation of such a curious isotropic phase, since this phase can be observed only for the PG-18-70-I and PG-18-70-II with the lower DP. Possibly, the side chains in melt may attain the comfortable occupation in a space around α-helices by the cubic net-work association of α-helices rather than the parallel association in cholesterics. On this point, it is ineteresting to note that the segregation form with a cubic symmetry has been observed in the block copolymer system [30].

We next refer to the fractionated III sample, PG-18-70-III, which exhibits another interesting phase behavior. In contrast to the PG-18-70-II above mentioned, this material exhibits the anisotropic phases in the whole temperature region of 60 to 250°C. Also in this material, the cholesteric phases appear only in the temperature region higher than 170°C, leading to the existence of another different anisotropic phase in the lower temperature region. This phase can be differentiated from the cholesteric phase by the following characteristics.

1. It shows a microscopic texture with the focal conic domain.
2. It exhibits no form optical rotation.
3. It preserves the uniaxial orientation produced by shearing even after annealing for a long time.

These features can be recognized in Figures 12a and 12b; Figure 12a shows an aligned focal conic textures of the oriented sample and Figure 12b shows the X-ray diffraction pattern of the same sample, in which the equatorial reflection attributable to a lateral packing of α-helices can be observed with the strong intensity on the equator. Furthermore, it is interesting to note that in the oriented specimen the molecular axis lies perpendicular to the shear direction if the small shear deformation is applied. This orientation behavior is obviously opposite to that encountered in the cholesteric phase.

Overall features thus allow the unique assignment of the mesophase to the smectic A in which a smectic layer is constructed by a parallel side by

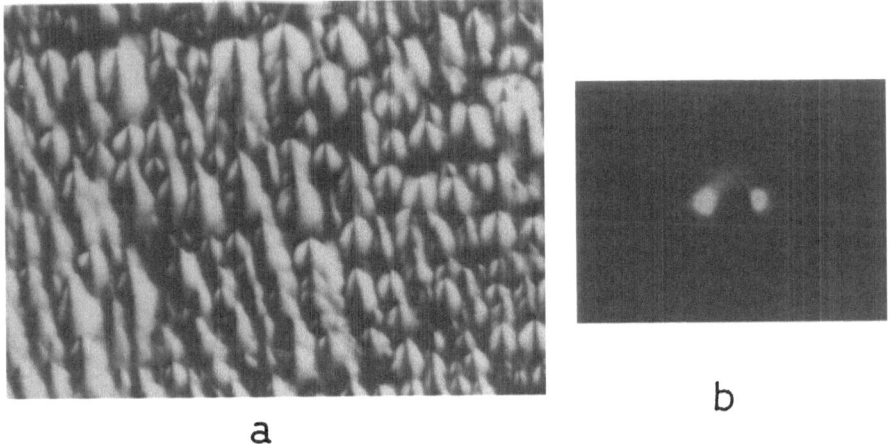

a

b

Figure 12. (a) Optical microscopic texture and (b) X-ray diffraction pattern observed for PG-18-70-III at 80°C. The sample was oriented by the shear applied in a horizontal direction.

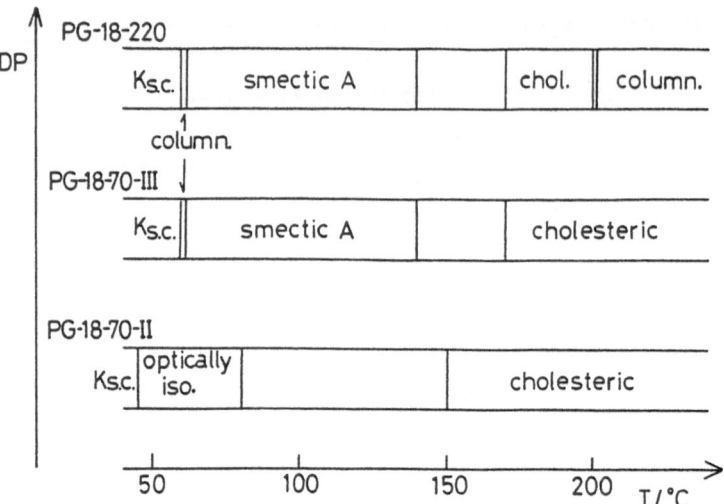

Figure 13. The schematic illustration exhibiting the phase behaviors of PG-18 polymers which appear different depending on the DP and temperature.

side packing of α-helical molecules as illustrated in Figure 1c, although the direct observation of the layered association cannot be made by the small-angle X-ray method. Possibly, this may be caused by the low contrast in electron densities with the layered fluctuation.

We have so far thought that the polypeptides form only a nematic or cholesteric phase, but the present study for the PG-18 polymers indicates the formation of several other liquid crystals. To conclude, the phase behavior is illustrated in Figure 13 as functions of the DP and temperature. If the DP is appreciably low around 40, optically isotropic phase possibly with the cubic symmetry appears in the lower temperature region than the cholesteric phase. With the increase of DP, the isotropic phase disappears and in place the smectic phase appears as the lower temperature mesophase. With further increase of DP, the columnar hexagonal phase appears in the higher temperature region than the cholesteric phase. Although we cannot give the space to present here, we have also observed the other type of columnar hexagonal phase in the narrow temperature span just below the smectic A phase [3]. In this columnar phase, the α-helices assume the two-strand coiled coil conformation and are packed into a two dimensional hexagonal lattice [4].

The present study is purely phenomenological and the identification of the mesophase structures is not yet satisfactory, but it is interesting that the phase behavior observed here, in one aspect, is similar to that expected theoretically through the excluded volume effect by Kimura et al.[25] and Frenkel [26]. The better understanding of the phase structure and the mechanism of phase transition may be done with the cooporation of theories.

References
1 Kasuya S, Sasaki S, Fukuda Y, Watanabe J, Uematsu I(1982) Polym Bulletin 7:241
2 Watanabe J, Fukuda Y, Gehani R, Uematsu I (1984) Macromolecules 17:1004
3 Watanabe J, Ono H, Abe A, Uematsu I (1985) Macromolecules 18:2141
4 Watanabe J, Ono H (1986) Macromolecules 19:1079
5 Watanabe J, Gotoh M, Nagase T (1987) Macromolecules 20:298
6 Watanabe J, Nagase T (1987) Polym J 19:781
7 Tsukahara M, Yamanobe T, Komoto T, Watanabe J, Ando I (1987) J Mol Cryst 159:345
8 Yamanobe T, Tsukahara M, Komoto T, Ando I, Watanabe J, Uematsu I (1988) Macromolecules 21:48
9 Watanabe J, Nagase T (1988) Macromolecules 21:171
10 Watanabe J, Nagase T, Itoh H, Ishii T, Satoh T (1988) Mol Cryst Liq Cryst 160:432
11 Watanabe J, Nagase T, Ichizuka T (1990) Polym J 22:1029
12 Watanabe J, Takashina Y (1991) Macromolecules 24:3423
13 Watanabe J, Takashina Y (1992) Polym J 24:709
14 Watanabe J, Tominaga T (1993) Macromolecules in press
15 Kimura H, Hoshino M, Nakano H (1979) J Phys 40:C3-174
16 Uematsu I, Uematsu Y (1984) Adv Polym Sci 59:37
17 Hara H, Satoh T, Toya S, Iida S, Orii S, Watanabe J (1988) Macromolecule 21:14
18 de Gennes P G (1982) In: Cifferri A, Meyer R D, Krigbaum W R (ed) Polymer Liquid Crystals. Academic Press, New York
19 Livolant F, Bouligand Y (1986) J Phys 47:1813
20 Lee S, Meyer R B (990) Liq Cryst 7:451
21 Watanabe J, Uemura N, Uematsu I to be published
22 Strzerecka T E, Davidson N W, Rill R L (1988) Nature 331:457
23 Livolant F, Levelut A M, Doucet J, Benoit J P (1989) Nature 339:724
24 Wen X, Meyer R B, Casper D L (1989) Phys Rev Lett 63:2760
25 Kimura H, Tsuchiya J (1990) Phys Soc Jpn 59:3563
26 Frenkel D (1989) Liq Cryst 5:929
27 Watanabe J, Takashina Y, Fukuda T, Miyamoto T to be published
28 Tardieu A, Luzzati V (1970) Biochi Biophys Acta 219:11
29 Tardieu A, Billard J (1976) J Phys 37:C3-79
30 Thomas E L, Alward D B, Henkee C S, Hoffman (1988) Nature 334:598

Study on the Orientational Order of the Main Chain Liquid Crystalline Polymers

Dong-Won Kim, Jung-Ki Park and Kwan-Soo Hong*

Department of Chemical Engineering, Korea Advanced Institute of Science and Technology, 373-1, Kusung-dong, Yusung-gu, Daejon, 305-701, Korea

*Physics Laboratory, Korea Basic Science Center, 966-5, Daechi-dong, Kangnam-ku, Seoul, 135-280, Korea

Abstract : The thermotropic liquid crystalline homo- or copolyesters based on 4-hydroxyacetophenone azine and diacyl chloride were synthesized. The liquid crystalline polymers(LCPs) were characterized by ^{1}H NMR, DSC, GPC, polarizing microscopy experiments. We analyzed the molecular orientational order in LCPs by changing the length of polymethylene spacers, or combining the two different aliphatic spacers in the magnetic field. An odd-even oscillation of orientational order parameters(S) measured from ^{1}H NMR spectra was observed in thermotropic homopolyesters, and the nematic order parameters associated with the mesogens or the spacers were found to be quite high, about 0.85~0.97 near solid-nematic transition temperature. The orientational parameters of copolyesters with the two polymethylene spacers in equal molar amounts decreased dramatically as compared with those of the corresponding homopolymers, and the copolymers connecting both even numbered spacers were shown to have a higher order than those connecting by even-odd or odd-odd number parity.

INTRODUCTION

Thermotropic liquid crystalline polymers(TLCPs) comprising rigid units connected by flexible spacer groups have been studied prominently in recent years[1-6]. For this class of thermotropic liquid crystalline polymers, it has been well known that the spacer groups play critical roles in determining the stability of mesophase and the ordering process in the liquid crystalline state. And the homologous series of LCPs usually also show odd-even effects in the isotropization tempertatue and molar entropic change[3,4,7]. These thermodynamic properties are closely related to the molecular order in TLCPs, and thus the knowledge of molecular parameters is of great importance for the understanding of their macroscopic properties. The ordering of anisotropic molecules can be characterized by the orientational order parameter, which is a very useful guide as to the quality of the orientation. For the TLCPs, it has been possible to measure the order parameter from ^{1}H NMR spectroscopy, because the pairs of proton will give rise to a doublet in the spectrum arising from intermolecular dipole interactions between the two protons[8-10].

In our study, we tried to demonstrate the nature of molecular order in nematic polyesters and its dependence on chemical structures of liquid crystalline polymers. The TLCP chosen for this study was main chain

A. Teramoto, M. Kobayashi, T. Norisuje (Eds.)
Ordering in Macromolecular Systems
© Springer-Verlag Berlin Heidelberg 1994

LCP(poly(oxy-α,α'-dimethylbezalazineoxyalkanoyl)), which was first investigated by Roviello and Sirigu[1]. This particular polymer was chosen since its transition temperature is reasonably low and its nematic range is broad. The order parameters of them were determined as a fuction of temperature, spacer parity, and in particular the influence of number of methylene units in aliphatic spacer was also investigated. From the systematic analysis, we could correlate NMR results to thermodynamic properties of homologous series of TLCPs.

EXPERIMENTAL

Monomer Synthesis

The 4,4'-Dihydroxy-α,α'-dimethylbenzalazine(DDBA) was synthesized according to the procedure of Blout et al.[11]. An ethanol solution of hydrazine monohydrate(0.10 mole) with 4-hydroxyacetophenone(0.22 mole) and a few drops of concentrated hydrochloric acid(0.5 mL) was refluxed under dry nitrogen for 8 hours. The DDBA separated out on cooling and was filtered off, and finally recrystallized from aqueous ethanol. The obtained DDBA was a yellow needle shape crystal with m.p = 226 ℃. Both the ^{1}H NMR spectrum and elemental analysis are in excellent agreement with the proposed structure, indicating that the compound is pure.
- ^{1}H NMR ;-CH$_3$(2.24 ppm), 1,4-phenylene(6.80, 7.74 ppm),-OH(9.61 ppm)
- Elemental Analysis ;
  C(71.62), H(6.01), O(11.93), N(10.44) : Calculated
  C(71.44), H(6.27), O(11.84), N(10.45) : Found

Polymer Synthesis

The thermotropic liquid crystalline homo- or copolyesters based on 4-hydroxyacetophenone azine and diacyl chloride(ClOC[-CH$_2$-]$_n$COCl, n=3~8, 10) were prepared via solution polycondensation at room temperature in glass reactor equipped with a nitrogen inlet, a reflux condenser, an additional funnel and a mechanical stirrer, which was a similar method to that used by Cimecioglu et al.[12]. The DDBA(50 mmole) was dissolved in 250 mL chloroform with a small amount of triethylamine(30 mL) and NMP (N-methyl-2-pyrrolidone)(5 mL). A solution of slight molar excess of diacyl chloride(52 mmole) in chloroform(60 mL) was added to reactor dropwise with a vigorous stirring. The reaction mixture was vigorously stirred for 1 hour and the product was separated by precipitation into ethanol which was filtered, and further washed with methanol, methanol/water and water, and finally dried at 80 ℃ in a vacuum for 36 hours. A light yellow powder was obtained in 80~ 90 % yield. The representative ^{1}H NMR and elemental analysis results of LCP10 are given below.
- ^{1}H NMR ; α-methylene(2.58 ppm), β-methylene(1.77 ppm),
              other methylene(1.30-1.43 ppm), methyl(2.3 ppm)
              para-substituted phenyl(7.06-7.15, 7.92-7.94 ppm)
-Elemental Analysis ;
  C(72.47), H(7.50), O(14.24), N(5.79) : Calculated

C(72.26), H(7.54), O(14.41), N(5.79) : Found

It is noted that the calculated element composition is based on the assumption that both end groups of the polymer chain are ethyl ester and DP is about 13.1 from ^{1}H NMR. For brevity, LCPs will be designated as LCPn in this paper, where n indicates the number of methylene units in the aliphatic spacer group.

Characterization

^{1}H NMR spectra were obtained in CDCl$_3$ solvent on a Bruker-AM-300 NMR spectrometer(300 MHz) with tetramethylsilane as an internal standard. The number average molecular weights of LCPs were also determined by end group analysis using ^{1}H NMR spectroscopy. The aliphatic chain ends were confirmed to be converted to the ethyl ester, and it was titrated by measuring the peak intensity of -OCH$_2$CH$_3$ at δ = 4.13 ppm. Elemental analyses were performed with W. C. Heraeus elemental analyzer. Textures of the materials were studied with a Leiz Ortholux polarizing microscope equipped with a Mettler hot stage. Differential scanning calorimetry(DSC) experiments were performed on a Du Pont 9900 instrument at a heating rate of 20 ℃/min and a cooling rate of 5 ℃/min. To assure that all samples had equivalent thermal histories, the results of the second heating and first cooling scans have been reported throughout. Gel permeation chromatography (GPC) was carried out using a Waters CV-150 instrument equipped with three μ-Styragel columns(10^3,10^4,10^5 Å) and the system was calibrated with polystyrene standards.

Measurement of Order Parameter

The NMR experiments were performed on Bruker-MSL 200(200 MHz) spectrometer. Since we were obliged to run our measurements below 180 ℃ (i.e, the upper limit of the NMR apparatus), the samples were heated to the isotropic state without the NMR magnet, and then slowly cooled by steps of 5 ℃ in the magnetic field. Due to the positive diamagnetic susceptibility of the rod-like liquid crystalline polymers, they aligned with the director parallel to the magnetic field axis. The order parameter(S) of the mesogenic group and the spacer can be derived from the NMR data in the same way as reference 9, and an assessment of the reproducibility of the S measurement was made by measuring a successive series of NMR spectra on one sample. It was found that the measured values of S were reproducible to within limits of ±0.015.

RESULTS AND DISCUSSION

The molecular weights measured by NMR and GPC are summarized in Table 1. As seen from the data in Table 1, the molecular weights of the polymers were relatively low, and the polydispersity indexes estimated by GPC ranged from 2.0 to 3.2. A little differences of number average molecular weights between GPC and NMR were probably due to the differences in the hydrodynamic volume of LCP and polystyrene.

Table 1. Molecular Weight and Molecular Weight Distribution

n	GPC[a]			NMR[b]	
	M_n	M_w	M_w/M_n	M_n	DP
3[c]	--	--	--	--	--
4[c]	--	--	--	--	--
5	5000	12000	2.4	5990	15.3
6	5300	17000	3.2	7530	18.5
7	5900	15900	2.7	6460	15.4
8	9100	25800	2.8	10340	23.8
10	8000	23100	2.9	6070	13.1
7/3[d]	4300	10100	2.4	5710	14.5
7/4	4700	11700	2.5	5570	13.9
7/5	5400	10800	2.0	5740	14.1
7/6	5100	12800	2.5	5650	13.7
7/8	5300	13400	2.5	4850	11.3
7/10	5600	14700	2.6	6970	15.8
3/10	5100	15100	3.0	5740	13.7
4/10	5800	13100	2.3	6200	14.7
5/10	5900	15000	2.5	6560	15.3
6/10	6100	16100	2.6	7820	18.0
8/10	6700	19000	2.8	7490	16.7

a Polystyrene standard, eluent : chlroroform
b Calculated on the assumption that both end groups of the polymer chains are ethyl ester
c These LCPs were not readily soluble for GPC and NMR analysis
d Copolymer containing 50 mol % of two aliphatic spacer groups

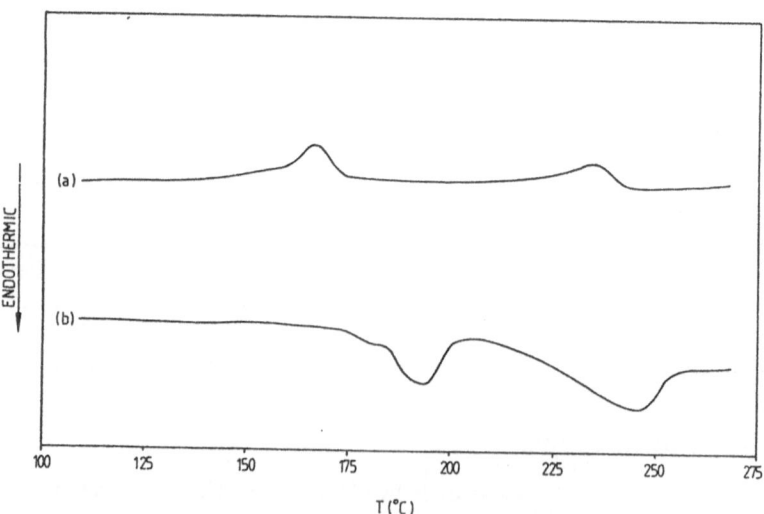

Fig. 1. DSC thermograms of LCP8
(a) On first cooling scan
(b) On Second heating scan

Figure 1 illustrates the typical DSC thermogram of LCP, and the phase transitions of all LCPs recorded by DSC upon heating and cooling are listed in Table 2. The isotropic transition of LCP4 was obscured by thermal decomposition and thus could not be determined by DSC. All the LCPs except for LCP4 showed the mesophase behavior with solid-nematic and nematic-isotropic transitions on heating and cooling. As is usual in other TLCPs, the supercoolings of T_{IN} and T_{NK} are observed. It is also found that copolymers have broader mesomorphic ranges($\triangle T$) than the homopolymers, and both the T_{KN} and T_{NI} are lower than for the homopolymers.

Table 2. DSC Results of TLCPs

n	Second Heating			First Cooling		
	T_{KN}(℃)	T_{NI}(℃)	$\triangle T$ $(T_{NI} - T_{KN})$	T_{KN}(℃)	T_{NI}(℃)	$\triangle S_{NI}$ (J/mru.K)
3	228	256	28	204	223	3.26
4[a]	231	--	--	--	--	--
5	193	225	32	145	205	4.69
6	202	267	65	163	258	6.37
7	175	219	44	147	213	6.23
8	192	244	52	167	237	6.98
10	193	218	25	172	212	8.31
7/3[b]	182	236	56	--	--	--
7/4[b]	143	250	107	--	--	--
7/5	157	234	77	101	205	2.79
7/6	167	246	79	126	216	4.38
7/8	156	214	58	126	198	4.93
7/10	139	217	78	97	192	6.60
3/10	191	230	39	107	212	4.26
4/10	140	232	92	111	222	5.00
5/10	149	215	66	107	201	4.63
6/10	155	251	96	138	230	5.84
8/10	145	228	83	108	212	7.71

a T_{NI} could not be observed by DSC
b The crystallization exotherm peaks were not appeared in the DSC thermogram, which may be due to a much less ordered structure of them

In Figure 2, the phase transition temperatures of homopolymers are plotted as a fuction of the number of the methylene units(n) in the aliphatic spacer. It is seen that they exhibit an odd-even effect with n. Thus T_{NI} of LCP with an even number of n is higher than those of an odd number of n. It has been widely accepted that the odd-even effects observed for TLCPs with flexible spacer group could be explained in terms of spatial orientation of the mesogens linked by the polymethylene spacers[4,5]. The mesogenic groups are approximately parallel to the chain axis in the polymer chains of even-numbered methylene units, while the mesogens linked by the odd-numbered methylene units are tilted from the same direcotor, leading to a less ordered state of the polymeric liquid crystals. From the Figure 2, $\triangle S_{NI}$ is

120

also found to be steadily increased as the length of the flexible spacer increased. This result suggests that the polymers with longer polymethylene spacer have higher degrees of order in their mesophases than those with shorter chain spacers.

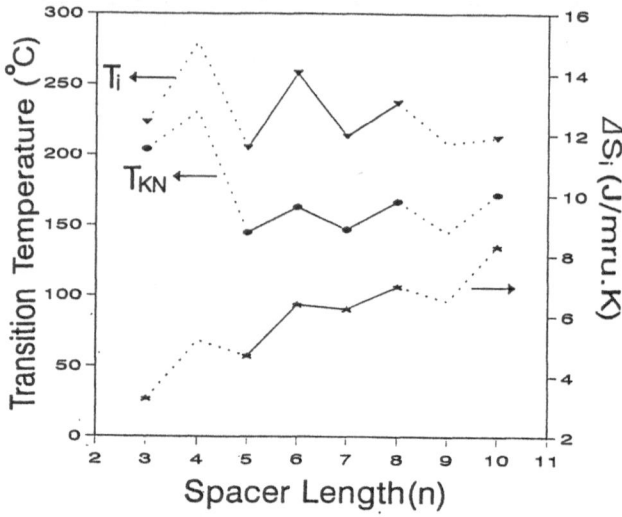

Fig. 2 Melting, isotropization temperatures and molar isotropization entropies as a fuction of spacer length for TLCPs

A Schlieren texture observed for LCP6 in Figure 3-(a) was indicative of the nematic texture, and we also observed similar textures for another LCPs. A more regular ordering of the microcrystals shown in Figure 3-(b) can be induced by shearing the liquid crystalline phase on cooling.

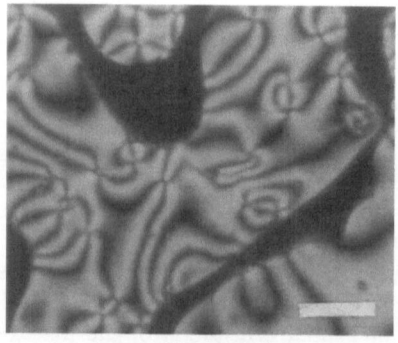 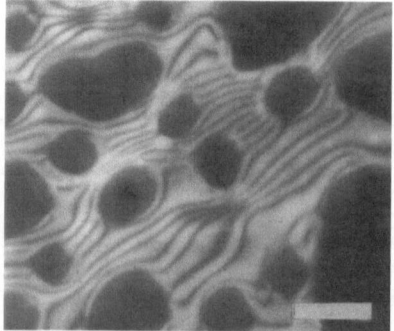

Fig. 3. Schlieren textures of LCP(n=6) seen between crossed polarizer on cooling scan(The bar in the lower right indicates 50 μm)
(a) At T=233 ℃
(b) At T=250 ℃(after shearing between glass plates)

In order to obtain the information on molecular orientation of the LCPs, NMR measurements were made on a series of polyesters in their nematic state except for LCP3 and LCP4. Sample alignment of LCP3 and LCP4 in spectrometer field could not be achieved due to the temperature limit of NMR apparatus. Representitative ^{1}H NMR spectra of LCP8 in the nematic range are shown in Figure 4.

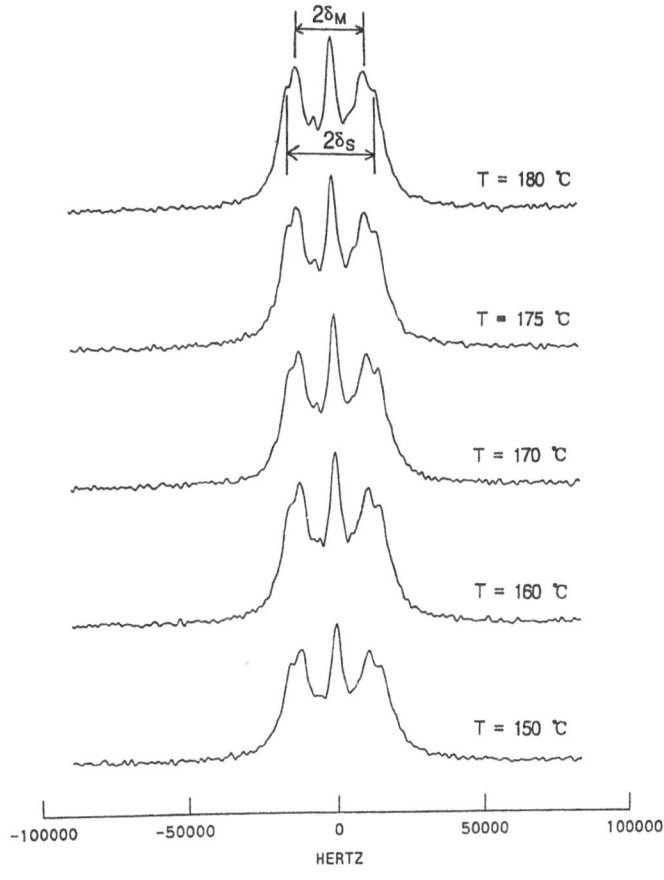

Fig. 4. Representative ^{1}H NMR spectra of LCP8 at various temperature ($2\delta_M$: splitting due to the dipolar interactions between mesogen (ortho phenylene) protons, $2\delta_S$: splitting due to the dipolar interactions between alkyl spacer protons)

The order parameter associated with the mesogenic moiety or spacer group was calculated from the dopolar splitting shown in NMR spectra. Martins et al. already proposed the models for the interpretation of NMR spectra of nematic phase, and their simulated and observed spectra were also shown to be split up into mesogenic moiety and spacer contributions in the thermotropic polyesters based on the 4,4'-dioxy-2,2'-dimethyl azoxybenzene moiety[9]. The order parameters of homopolymers obtained from NMR measurement are listed in Table 3.

Table 3. Temperature Dependence of S for the TLCPs

LCP T(℃)	LCP 5	LCP 6	LCP 7	LCP 8	LCP 10
180	0.773[a] (0.773)[b]	0.911 (0.911)	0.851 (0.851)	0.958 (0.951)	0.931 (0.927)
175	0.783 (0.781)	0.920 (0.918)	0.869 (0.872)	0.963 (0.962)	0.947 (0.948)
170	0.788 (0.784)	0.925 (0.922)	0.882 (0.889)	0.968 (0.970)	0.951 (0.956)
160	0.824 (0.822)	0.930 (0.932)	0.887 (0.900)	0.977 (0.974)	0.951 (0.965)
150	0.856 (0.854)	0.952 (0.936)	0.882 (0.907)	0.967 (0.984)	0.953 (0.972)

a S measured from the dipolar splitting of mesogen(ortho-phenylene) protons
b S measured from the dipolar splitting of spacer(methylene) protons

It is noted that S of spacer group is thought to be a degree of order of an isolated pair of CH_2 protons aligned, on the average, perpendicularly to the long axis of the mesogenic moiety. The ordering tendencies in mesogenic moiety and spacer group of these molecules were to be found the same : the flexible spacer parts aligned in the magnetic field with a degree of order comparable to that of the mesogenic moiety, which means that the flexible spacers are rather extended conformation. Such a result is in qualitative agreement with previous reports which has been already described for the other main chain LCPs[9,13]. It has also been proved possible to achieve a degree of preferred orientation upto S = 0.97. To our knowledge, this may be the highest value among the order parameters of the nematic polyesters ever reported. The decline in S with increasing temperature is thought to be a consequence of an approach to the nematic–isotropic transition temperature, which is commonly observed for other liquid crystals. It is also interesting to note that lengthening the flexible spacer causes an increase in order parameter. This result may suggests that the orientation under a magenetic field will be hard to achieve for low n, but it will be increasingly easy as n increase. LCP5, therefore, possessed a lowest value of S for homopolyesters studied in the whole temperature range considered. This result indicates that LCP5 has a much less ordered structure in the nematic state than the other homopolymers, which is consistent with the lowest value of thermodynamic entropy change of isotropization($\triangle S_{NI}$). That is to say, a small value of $\triangle S_{NI}$ means a lower molecular ordering in the nematic state, if we assume that LCPs have the same values of entropy in the isotropic state.

The effect of the length of aliphatic spacer on orientation at a given reduced temperature is better represented in Figure 5. As expected, the orders of mesogen and spacer are higher for n=even at a given reduced temperature. This result is due to the fact that the megogen axes fall alternately in and out of alignment as n changes from even to odd as explained above. The odd-even oscillation of orientational parameter observed in Figure 5 is consistent with the oscillation of $\triangle S_{NI}$ shown Figure 2.

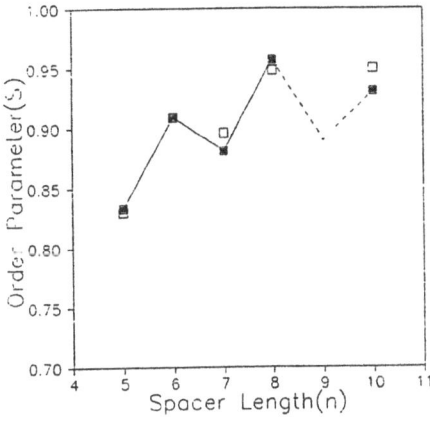

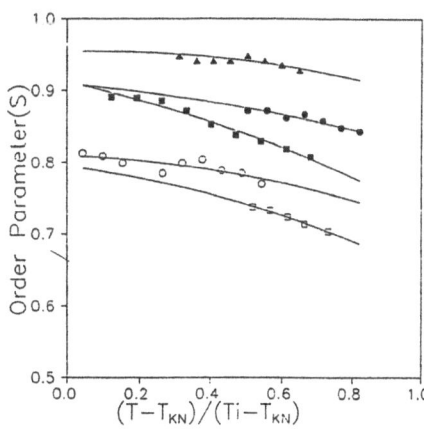

Fig. 5. Order parameter as a fuction of spacer length at $T^*_{red}=0.2$ (■:mesogen, □:spacer, $T^*_{red}=(T-T_{KN})/(T_i-T_{KN})$)

Fig. 6. Mesogenic order parameter versus arbitrary reduced temperature for copolyesters with two aliphatic spacers in equal molar amounts (□ : LCP(n=7,5), ○ : LCP(n=7,6), ■ : LCP(n=7,8), ● : LCP(n=7,10), ▲ : LCP(n=8,10))

The order parameters of copolyesters with two aliphatic spacers in equal molar amounts were shown in Figure 6 as a fuction of temperature. The copolymers containing equimolar amounts of flexible spacers of two different lengths were found to exhibit the lower orientational order as compared with those of the corresponding homopolymers, and the copolymers connecting both even numbered spacers were shown to have higher order parameters than those connecting by even-odd or odd-odd number parity. This effect arises from the conformational disorder introduced along the polymer chain in a statistically random manner. From the Figure 6, it is also found that the copolymers connected by the two longer aliphatic spacers have the higher values of S than those linked by the two shorter spacers as in case of homopolymers. Among the copolymers examined, LCP810, hence, possessed the highest value of S, indicating the best ordered structure of the copolymers studied in the nematic state. To conclude, the above NMR results of homo or copolyesters indicate that the length or type of flexible spacer plays an important role in controlling the orientation of the TLCPs.

CONCLUSION

The orientational ordering in the main chain TLCPs with mesogenic 4-hydroxyacetophenone azine moiety has been explored for the homopolyesters with different spacer length and the copolymers containing equimolar amounts of flexible spacers of two different lengths, and it could be rationalized on the basis of observed thermodynamic properties. The orientational order parameters of the homo-polyesters oscillated with number of methylene units at a given reduced temperature, and it also increased with increasing the length of aliphatic spacer. Comparable ordering of the mesogenic moiety and flexible spacer suggest that the aliphatic spacers should be fairly extended, and thus participate in the anisotropic intermolecular interaction in the nematic state. The copolyesters containing equimolar amounts of aliphatic groups of two different lengths had lower S as compared with the corresponding homopolymers, which was due to the destruction of molecular alignment by random insertion of flexible spacers of different lengths along the polymer chain. Our works propose that highly oriented materials such as high-strength fibers can be produced from thermotropic nematic polyesters studied, and this would also lead to better soluble and melt-processable materials without sacrifice of the directional properties associated with flexible aliphatic spacer.

REFERENCE

1. A.Roviello and A.Sirigu, *J.Polym.Sci.,Polym.Lett.Ed.*,13,455(1975)
2. S.Antoun,R.W.Lenz and J.I.Jin, *J.Polym.Sci.,Polym.Chem.Ed*,19,1901(1981)
3. A.Blumstein and O.Thomas, *Macromolecules*,15,1264(1982)
4. A.Abe, *Macromolecules*,17,2280(1984)
5. D.Y.Yoon and S.Bruckner, *Macromolecules*,18,651(1985)
6. P.Nieri, C.Ramireddy, C.N.Wu, P.Munk and R.W.Lenz, *Macromolecules*,25,1796 (1992)
7. A.Roviello and A.Sirigu, *Makromol. Chem.*,183,895(1982)
8. G.R.Luckhurst, *Mol.Cryst.Liq.Cryst.*,21,125(1973)
9. A.F.Martins, J.B.Ferreira, F.Volino, A.Blumstein and R.B.Blumstein, *Macro molecules*,16,279(1983)
10. R.B.Blumstein, E.M.Stickles, M.M.Gauthier and A.Blumstein, *Macromolecules*, 17,177(1984)
11. E.R.Blout, V.W.Eager and R.M.Gofstein, *J.Am.Chem.Soc*,68,1983(1946)
12. A.L.Cimecioglu,H.Fruitwala and R.W.Weiss, *Makromol.Chem.*,191,2329(1990)
13. A.Blumstein, S.Vilasagar, S.Donrathnam, S.B.Clough and R.B.Blumstein, *J. Polym.Sci.,Polym.Phys.Ed*,20,877(1982)

Statistical Theory of Liquid Crystalline Orderings in Hard Rod Fluids
- Nematic, Cholesteric, Smectic and Columnar Phases -

Hatsuo Kimura

Department of Applied Physics, Fuculty of Engineering,
Nagoya University, Nagoya 464-01, Japan

Abstract: A review is given of theoretical studies on systems of rod-like molecule with simple shape (cylinders, spherocylinders, square rods,...). The role of hard-core repulsion in liquid crystalline orderings are mainly discussed. It is shown that the excluded volume effect due to the hard-core repulsion stabilizes not only the orientational order but also positional ordering of the smectic A density wave in cylindrical molecules.

1. INTRODUCTION

Many types of orderings such as the nematic, cholesteric, smectic and columnar phases have been observed in several systems of rod like macromolecules as well as low molecular liquid crystals.
The nematic and cholesteric have long range orientational order but are positionally disorder as in the isotropic phase. The smectic and columnar phases have long range positional order.

In solutions of macromolecules, the transitions between these phases may be lyotropic, *i.e.* induced by changes of the concentration. While, in ordinary low molecular liquid crystals the phase transitions are thermotropic and brought about by temperature changes.

Physical origins of orientational order in the nematic phase are basically well known. The one is the hard-core repulsion between molecules. Onsager[1] showed, a half century ago, that the hard-core repulsion stabilizes the orientational order in systems of very long rod-like molecules. The isotropic to nematic phase transition observed in several solutions of macromolecules such as TMV, polypeptide, DNA, etc.. can be explained by this mechanism. On the other hand, Maier and Saupe[2] had shown that the anisotropic dispersion force can induce a thermotropic phase transition from the isotropic to nematic phase.

Though both theories treat the models in the simplest approximation, *i.e.* the 2nd virial approximation and a molecular field approximation, I believe these simple theories catch the physics of the orientational ordering correctly. Later, many sophisticated theories have been

A. Teramoto, M. Kobayashi, T. Norisuje (Eds.)
Ordering in Macromolecular Systems
© Springer-Verlag Berlin Heidelberg 1994

proposed on the same models, but they all support the essential conclusions of these theories.

In this paper, I will focus on two topics. Both of them are related to the hard-core repulsion. The first is the role of it in the ordinary thermotropic liquid crystals. Though in many theoretical analyses of low molecular liquid crystals the effect of hard-core repulsion is entirely neglected, we have shown up to now that many important physical properties of real thermotropic liquid crystals can be explained as a comprehensive result of the hard-core repulsion and attractive dispersion force. The second question is whether the positional orderings such as the smectic, columnar and crystalline phases can be stabilized by only the hard core repulsion or not.

2. ROLE OF HARD-CORE REPULSION IN ORIENTATIONAL ORDER

Let us discuss the first problem. The essence of the Onsager theory is *the excluded volume effect.*

One molecule excludes another in fluid state due to the hard-core repulsion. As the excluded volume is smallest for the parallel molecular orientations, we can easily see the packing entropy is larger in the parallel oriented nematic phase than in the orientationally disordered isotropic phase. On the other hand, the orientational entropy is of course larger in the isotropic phase. Therefore, if the density of molecules increases beyond certain critical value, the increase of packing entropy exceeds the decrease of the orientational entropy, and the system changes from the isotropic to nematic. This is the Onsager's mechanism.

In Fig.1 the results of several different theories on the Onsager's model are summarized. The systems of hard spherocylinder (cylinder of the length L-D and diameter D capped hemispheres of the same diameter on both ends) are studied. In the figure, the horizontal axis indicates the length to width ratio L/D of molecule, and the vertical axis denotes a quantity cL/D, where c is the density expressed by the volume fraction defined as ρv_m, ρ is the number density N/V and v_m the volume of a molecule.

The low density part of the figure shows the isotropic phase, and the upper part the nematic phase. Original Onsager theory predicted the system shows the 1st order isotropic to nematic phase transition, between the density c_i and c_n. At the density c_i the isotropic phase becomes unstable and the system goes to the nematic phase of higher density than c_n. Both c_i and c_n are proportional to the ratio D/L, so the values of c_iL/D and c_nL/D are constants respectively. The constants 3.3 and 4.2 are obtained by the 2nd virial approximation, so the results seem to be reliable only for very large values of L/D

Later, several theories have been proposed for shorter rods. All the theories conclude the 1st order lyotropic phase transition as the Onsager theory does, and give those values of c_iL/D, c_nL/D and the ratio c_i/c_n

which decrease continuously as the ratio L/D decreases. By a scaled particle theory, Cotter[3] calculated for rods of L/D $\leq$ 10. The method is considered to be valid for rather short rods. Kihara[4] calculated recently in the 3rd virial approximation which seems to be appropriate for moderately long rods: L/D $\geq$ 8. Results of a computer simulation by Frenkel, Mulder and McTague[5] for hard ellipsoid of revolution having the ratio L/D $\approx$ 3 are added in the figure.

According to these results, we can say the original Onsager theory correctly predicts the orientational phase transition in hard rod systems, and further this orienting mechanism is quite effective in ordinary low molecular liquid crystals which are composed of rather short rods in the range of the L/D = 3 - 5. For example, the scaled particle theory predicted the values c_i=0.51 and c_n=0.53 for L/D=3.

Therefore, a realistic theory of nematic should incorporate the hard-core repulsion as well as the attractive potential between molecules.

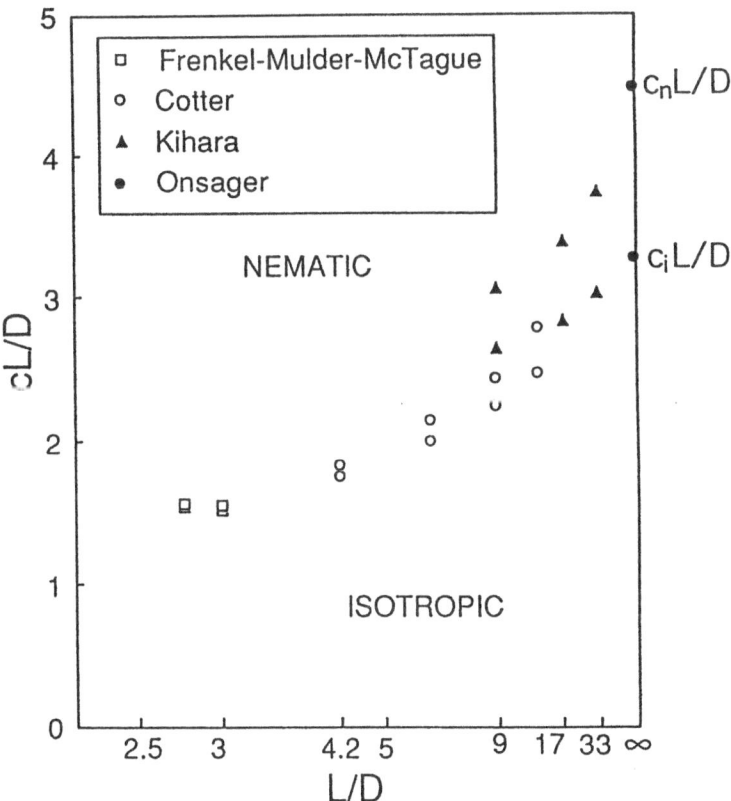

Fig.1: Nematic-isotropic transition densities in hard rods estimated by Onsager[1], Cotter[3], Kihara[4] and Frenkel, Mulder and McTague[5]

3.   HYBRID MODELS AND RESULTS

We have developed a simple theory of liquid crystals[6] adopting the following " *hybrid model* ".

The molecules are assumed to be simple cylinder or spherocylinder of the length L and diameter D.   The intermolecular potential ϕ_{ij} is assumed as the sum of the hard-core repulsion and the attractive dispersion force proposed by Maier and Saupe as

$$\phi_{ij} = \left\{ \begin{array}{ll} \infty & \text{if i and j intersect,} \\ \\ -C(r_{ij}) - A(r_{ij})\, P_2\!\left(\cos\theta_{ij}\right) & \text{otherwise.} \end{array} \right. \tag{1}$$

The distance and angle   between the molecules are denoted by r_{ij} and θ_{ij}, and $P_2(x)$ is the 2nd order Legendre polynomial.   We define the orientational order parameter S, which characterizes the nematic phase as

$$S = \left\langle \frac{1}{N}\sum_{i=1}^{N} P_2(\cos\theta_i) \right\rangle , \tag{2}$$

where θ_i is the angle between the long axis of the i-th molecule and the preferred direction of alignment, *i.e.* the director n. If we can calculate statistically mechanically the free energy of the system as a function of the order parameter S, we can predict equilibrium phases and phase transition among them.   Calculating the free energy of the present model using a 2nd virial theory, we can obtain the following results for the isotropic to nematic phase transition.

If the condition

$$\Gamma \equiv \frac{\widetilde{A}}{kT} + \gamma c > \Gamma_c \equiv 4.54 \tag{3}$$

is satisfied, the system goes to the nematic from isotropic phase through the 1st order phase transition, where

$$\widetilde{A} \equiv \rho \int A(r)\,dr \tag{4}$$

is mean value of the anisotropic part of the dispersion potential, γ expresses the effect of hard-core repulsion and is given by

$$\gamma \equiv \frac{5\pi DL^2}{16v_m} \tag{5}$$

This theory reduces approximately to Onsager's if we put $\widetilde{A}=0$, on the other hand putting c=0, we recover the popular Maier-Saupe theory of thermotropic liquid crystals. Therefore, this model is just the hybrid of the Onsager's hard-core model and the Maier-Saupe's attractive potential model. Using such model, we have studied various physical properties of real liquid crystals including systems of macromolecules. The model is very simple, but we have seen it works good and useful ; Frank elastic constants[7], surface tension[8], N-I interface[9], molecular orientations at hard wall[10] , etc...

4. TEMPERATURE DEPENDENT CHOLESTERIC PITCH

As is well known, chiral or optical active molecules form the cholesteric phase instead of the nematic phase. Polypeptide molecules with α- helical structure are the example. Such polypeptide molecules exhibit the cholesteric ordering in certain solvents or in the melt.

In the cholesteric phase, the molecular directions rotate spontaneously around the z-axis and form a helical arrangement. The pitch of the helix of this structure is known to depend strongly on temperatures and/or concentrations. To describe this phase ,we must somewhat generalize the model for nematic considering the chiral nature of the molecules. Though the definition of the order parameter S is the same as the nematic phase, now due to the helical arrangement, the director n depends on the position r as

$$\mathbf{n(r)} = (\cos qz, \sin qz, 0) \tag{6}$$

where the cholesteric pitch p is the inverse of the wave number q .

Due to the molecular chirality, the interaction potential contains extra asymmetric term, and the excluded volume has also an asymmetry. By considering these new factors, we can find such a temperature dependent expression for the cholesteric pitch[11]:

$$\frac{2\pi}{p} = \left(\frac{\widetilde{B}}{kT} - \kappa\right) / \left(\frac{\widetilde{A}}{kT} + \gamma c\right) \tag{7}$$

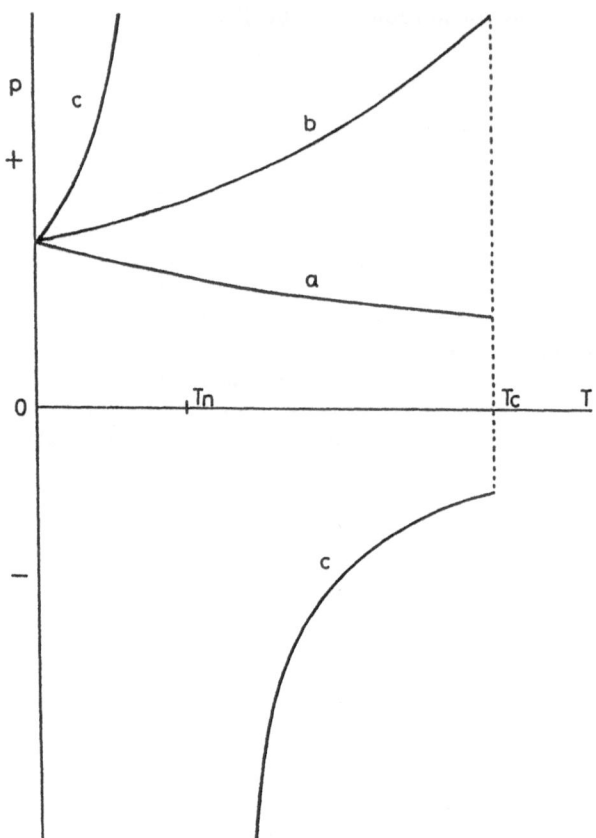

Fig.2: Temperature dependence of the cholesteric pitch p predicted by Eq.(7): (a) Decreasing, (b) increasing, and (c) the thermal conversion of the twisting.

Here, $\widetilde{B}$ expresses the strength of the chiral part of the dispersion force, κ the steric effect due to the coiled molecular shape. $\widetilde{A}$ and γ^c are similar quantities as Eqs.(4) and (5) for the nematic.

The temperature dependence of the pitch arizes from the combination of the effects of hard-core repulsion and attractive dispersion potential. Really, if we neglect either of the hard-core repulsion or the dispersion force, we must get a temperature independent constant cholesteric pitch. According to the values of $\widetilde{B}$ and κ, we can predict the possible three types of temperature dependence of p as shown in Fig. 2.

Though, all these types have been observed experimentally in real liquid crystals, the type (c) is the most remarkable. Such thermal conversion of the cholesteric helix has observed by Uematsu and Uematsu

[12] in various polypeptide solutions. In cases of L/D>>1, Eq.(7) can be simplified as

$$\frac{1}{p} = a\left(\frac{T_N}{T} - 1\right),$$

(8)

where a is a function of density c, and T_N is a " *nematic temperature* ". This simple theory could explain the experimental data very good[13].

More recent data by Watanabe and his coleagues[14] also show similar temperature dependence in polypeptide with dodecyl side chains. Though our formula (8) can be fitted near the temperatures T_N, the experimental temperature dependence is more strong and we cannot yet succeed to fit the overall dependence in observed entire temperature range.

5. CAN ROD-LIKE MOLECULES FORM SMECTIC PHASE WITHOUT ATTRACTIVE FORCES ?

To this problem, we have previously answered as "*yes* ". Our model was a system of N square rods of the length L and width D in a volume V. To simplify calculations, we assumed the molecules were allowed to orient only in three mutually perpendicular directions, x, y and z. The smectic A phase was assumed to be a square density wave along the z-axis as shown in Fig.3 (a), where P (z) is the density distribution function.

Now, however, we think this assumption of the square density wave was not good, because it is rather difficult structure to form in hard rod system. It was better to assume a sinusoidal density wave as Fig.3(b). Anyway, we explain the results of this model.

The amplitude τ of the wave is the order parameter of the smectic structure. The parameter τ and the wave length d are determined from the condition to minimize the free energy of the system. In the 2nd virial approximation, we found this system exhibits the nematic to smectic A phase transition when the mean density increases beyond a critical value[15]. The obtained phase diagram is shown in Fig.4(a). The lower critical line between the isotropic to nematic phases is the same as obtained by the Onsager theory. In more high density region, the nematic changes into the smectic A phase. The dashed part of the critical line indicates the 1st order phase transition, while the full line part shows the boundary of the 2nd order phase transition. The transition density for longer rods is 0.73 and nearly independent on the ratio L/D. The smectic period d is obtained to be 1.4L. After this calculation, we improved the approximation up to the 3rd order in the density (the 3rd virial theory), but assumed in this case the rods are all parallel to the z-axis, *i.e.* S = 1. By doing so, we found[16] a variety of phase sequence as shown in Fig.4(b).

The important results are (1) the transition density for the nematic to smectic A phase considerably decreases to 0.40 from the 2nd virial value

132

0.73. The period d is 1.36L in the 3rd virial theory. The optimum value of d seems to be not so sensitive to the degree of approximation.

(2) In the high density region, there appears more ordered phases: the columnar and the solid.   In the columnar phase, the rods are stacked to form liquid-like columns, and the columns constitute a two dimensional lattice in the xy-plane.   Though such structure has been observed in low molecular disc-like liquid crystals, it didn't be found yet in ordinary rod-like thermotropic   liquid crystals.

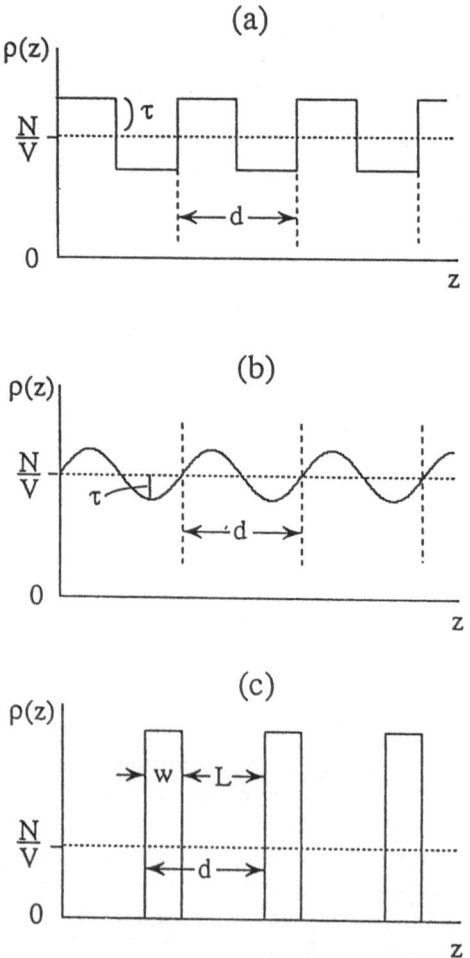

Fig.3: Models of the density function ρ (z) for the smectic A phase;
(a) Square density wave, (b) sinusoidal density wave, and (c) layered structure.

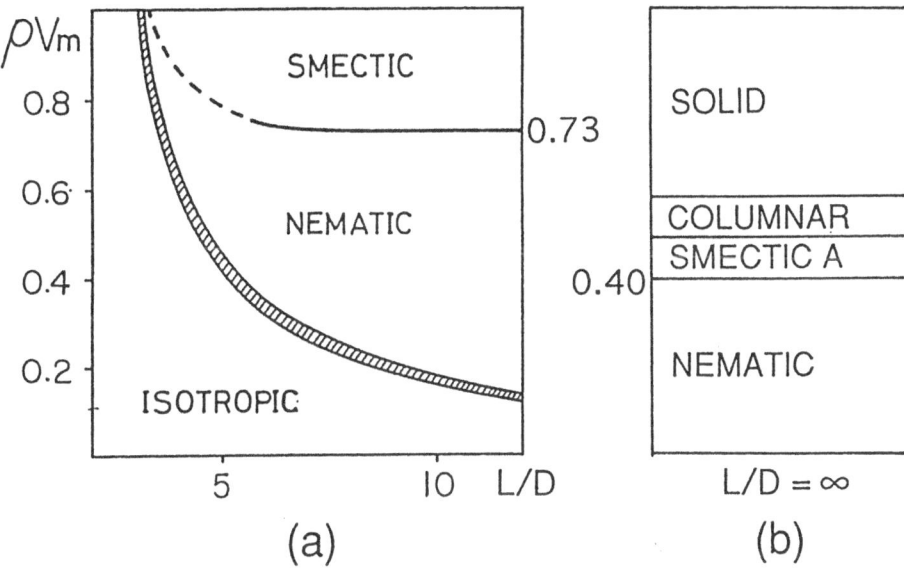

Fig.4: Phase diagram of square rods. (a) The 2nd virial theory; the square density wave is assumed for the smectic A phase. (b) The 3rd virial theory. Parallel alignment of molecules is assumed[15,16].

These results have attracted almost no attention at that time, perhaps because of the difficulty to see intuitively. To clear this point, let us consider a simple excercise. For the present parallel aligned hard rod system, we can express the free energy as

$$F \approx -NkT\log(V_a - \frac{1}{2}Nv_{ex}) \approx -NkT(\log V_a - \frac{N}{2V_a}v_{ex})$$ (9)

Here, V_a is an available volume for the molecular translation, $Nv_{ex}/2$ is the total excluded volume, v_{ex} denotes the excluded volume for a pair of molecules. Of course, this expression is correct only up to the 2nd order of molecular density. This free energy may be called as " *the packing entropy*" term.

Which has larger packing entropy the nematic or smectic A ? In the nematic phase, V_a is equal to the real volume V, and v_{ex} is given by $2\pi D^2 L$ for cylinders with the length L and diameter D. Now, let us imagine the smectic A phase is really a layered structure as shown in Fig.3(c), in which the centers of molecules can move only in narrow layers of width w separating with distance d = w + L to avoid collisions between molecules in different layers . Then, we have smaller values $V_a = Vw/d$ and $v_{ex} = \pi D^2 w$ than the nematic phase. Therefore, if the density c exceeds the value $\log(1+L/w)/(1-w/L)/2 \equiv c^*$, the packing entropy favours the layered structure than the nematic. However, as the quantity c^* has the minimum value 1.038 for $w/L=0.355$, the condition $c > c^*$ is physically impossible. Thus, the 2nd virial approximation tells us that the really layered square well type smectic structure cannot be stabilized by the hard-core repulsion. But we must investigate more carefully even in the 2nd virial approximation. Precisely , the free energy of the system is written in the 2nd virial approximation as

$$F/NkT = \text{const.} + \int \rho(r)\log\rho(r)dr - \frac{1}{2}\int \rho(r_1)\rho(r_2)b(r_1,r_2)dr_1dr_2$$

(10)

Here,

$$b(r_1,r_2) = -1 \quad \text{for molecules 1 and 2 intersect}$$
$$= 0 \quad \text{otherwise}$$

(11)

is the Mayer function for the hard-core repulsion. Therefore, we must find the distribution function $\rho(r)$ which minimizes the free energy. Assuming the square density wave as shown in Fig.3(a), we have found it becomes more stable than the nematic for $c > 0.73$. But we can solve exactly the equation $\delta F/\delta\rho = 0$ for the energy F given by Eq.(10) and find the true solution is the sinusoidal density wave Fig.3(b) and it is stabilized for $c > 0.57$ and $d = 1.40L$ [17].

Thus, the 2nd virial theory shows that the excluded volume effect stabilizes the sinusoidal density wave in hard rod system. This result seems to agree the notion that the smectic A phase in real liquid crystals is a density wave which was supported by x-ray experiments.

6. RECENT RESULTS ON POSITIONAL ORDERINGS

6.1. Computer Simulation

Recently, Stroobants, Lekkerkerker and Frenkel[18] made computer simulations for the system of 90 parallel hard spherocylinders. The pattern of the phase diagram obtained by them is almost identical with our previous results for the square rods. The system shows clearly the nematic to smectic A transition, and the critical density of the transition for the longest rods ($L/D = \infty$: this is equivalent to cylinders) was found as $c = 0.35$, and in more dense region the smectic A phase goes to the columnar phase. More recently, the simulation on 1080 parallel spherocylinders has been done by Veerman and Frenkel[19], and the phase diagram was somewhat revised (Fig.5). The essential difference is appearance of so-called AAA stacked crystalline phase on the high density side of the smectic A phase, where the columnar phase existed in the former diagram for 90 particles. On the other hand, the columnar phase entirely disappeared in the diagram for the 1080 particles. For cylinders, the nematic to smectic A phase transition occurs at the density c = 0.44, and the smectic A phase changes into the AAA crystal through the 1st order phase transition from c = 0.57 to 0.60.

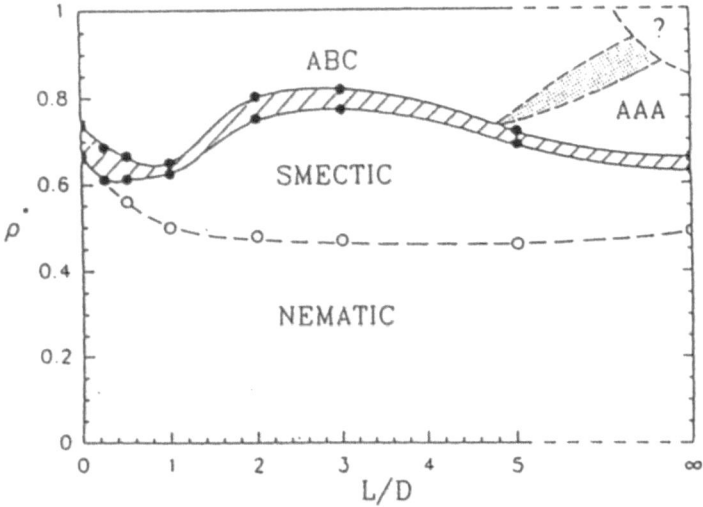

Fig.5: Phase diagram obtained by computer simulation on 1080 parallel hard spherocylinders[19].

6.2.   3rd Virial Theory

In order to discuss stabilities of these positional orderings, again we have made several calculations for systems of hard cylinders and square rods in the 3rd virial approximation. Up to the 3rd power of the density, we must add to the free enegy F/NkT of Eq.(10) the following term:

$$-\frac{1}{6}\int \rho(r_1)\rho(r_2)\rho(r_3)b(r_1,r_2)b(r_2,r_3)b(r_1,r_3)dr_1dr_2dr_3$$

(12)

6.2.1.   Cylinders[20,21,22]: By solving the condition to minimize the free energy, we found that in the system of hard parallel cylinders only the smectic A and the crystalline phases are thermodynamically stable. At the density c=0.36, the nematic to smectic A phase transition of the 2nd order occurs; then from 0.40 to 0.53 the smectic changes into a crystalline phase through the 1st order phase transition. As is shown in Fig.6, the calculated free energy of the columnar phase becomes lower than the nematic phase for c > 0.47, but where the stable states is either the smectic A or the crystal. So, the columnar phase is only a meta-stable state. These results agree quite well with those of the simulations on the recent 1080 particle system even numerically.

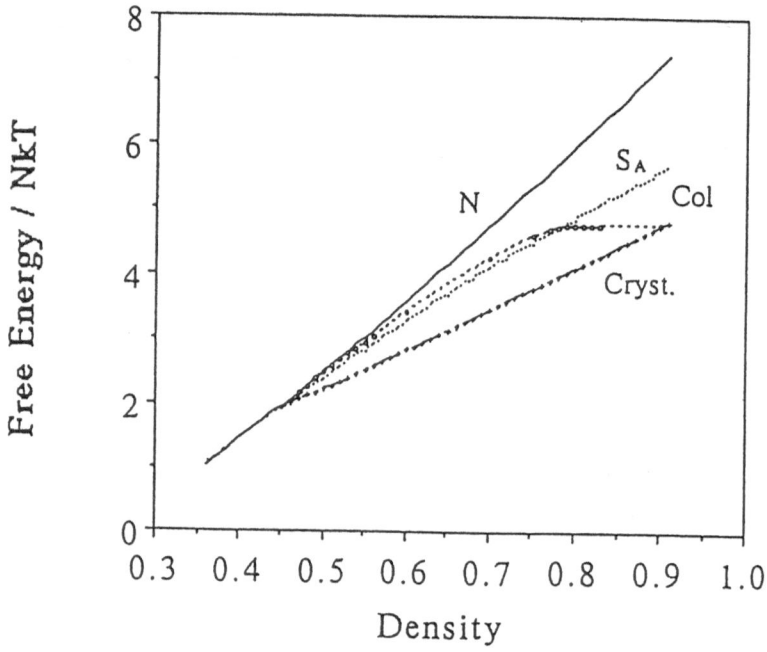

Fig.6: Free energies of parallel cylinders calculated by the 3rd virial approximation[22]. N: nematic, S_A: smectic A, Col: columnar, Cryst: crystal

6.2.2. Square rods[23]:   On the other hand, the calculation for the system of parallel square rods, shows that only the positional ordering which is stabilized by the excluded volume effect is the crystalline phase for c > 0.36, though the smectic A and columnar phases exist as meta-stable state in the same meaning for the columnar phase in cylinders. The behaviour of the squre rods differs from the case of cylinders. This indicates the free rotation around the long axis prevents real molecules from forming the crystalline ordering, so the molecules are forced to form a mesophase such as the smectic A.

7. CONCLUDING REMARKS

We have found that the positional ordering of the smectic A phase can be stabilized by the excluded volume effect in hard cylindrical rods. The similar conclusions have been obtained by others using different approximations[24,25]. The 2nd virial theory predicted that the smectic A phase is a sinusoidal density wave with period $d = 1.4L$ and the nematic to smectic A phase transition in parallel aligned molecules is the 2nd order. The columnar phase is predicted to be a meta-stable state. So, this phase may occur if the smectic A and crystalline phases are inhibited in the system for a certain reason. The stabilization of positional orderings due to the excluded volume effect depends delicately on the shape of molecules. In systems of perfectly oriented square rods, the smectic A phase cannot be stabilized and the system transits directly into the crystalline phase from the nematic phase.

The results are confirmed in the 3rd virial theory which agrees quite well with recent computer simulations.

Acknowledgements: The work reviewed in this article was done at Nagoya University where the author worked nearly 33 years. It has been the result of collaboration by Prof. H.Nakano and Dr. M.Hosino and young colleagues: M.Tsuchiya, A.Mori, T.Koda and H.Shindo. The author wish to express sincere thanks to them.

References :

1   Onsager L (1949) Ann NY Acad Sci 51:627
2   Maier W, Saupe A (1959) Z Naturforsch 14a:882
3   Cotter MA (1977) Phys Rev 66:1098
4   Kihara T (1990) Mol Cryst Liq Cryst 180B:329
5   Frenkel D, Mulder BM, McTague JP (1984) Mol Cryst Liq Cryst 104:1
6   Kimura H (1974) J Phys Soc Jpn 36:1280
7   Kimura H, Hosino M, Nakano H (1981) Mol Cryst Liq Cryst 74:55
8   Kimura H, Nakano H (1985) J Phys Soc Jpn 54:1730
9   Kimura H, Nakano H (1986) J Phys Soc Jpn 55:4186
10 Kimura H (1993) J Phys Soc Jpn 62:No 8
11 Kimura H, Hosino M, Nakano H (1979) J Phys Fr 40:C3-174
12 Uematsu I, Uematsu Y (1984) Adv in Polym Sci 59:37
13 Kimura H, Hosino M, Nakano H (1982) J Phys Soc Jpn 51:1584
14 Watanabe J, Nagase T, Ichizuka T (1990) Polym J 22:1029
15 Hosino M, Nakano H, Kimura H (1979) J Phys Soc Jpn 47:740
16 Hosino M, Nakano H, Kimura H (1982) J Phys Soc Jpn 51:741

138

17 Koda T, Kimura H, Doi M (1993) J Phys Soc J 62:170
18 Stroobants A, Lekkerkerker HNW, Frenkel D (1987) Phys Rev A36:2929
19 Veerman JAC, Frenkel D (1991) Phys Rev A43:4334
20 Kimura H, Tsuchiya M (1990) J Phys Soc Jpn 59:3563
21 Mori A, Kimura H (1991) J Phys Soc Jpn 60:2888
22 Mori A, Kimura H (1992) J Phys Soc Jpn 61:2703
23 Shindo H, Kimura H (1993) to be published in J Phys Soc Jpn
24 Somoza AM, Tarazona P (1988) Phys Rev Lett 61:2566
25 Holyst R, Poniewierski A (1989) Phys Rev A39:2742

Phase Diagram and Molecular Ordering of Dimer Liquid Crystals Dissolved in a Simple Nematic Solvent

- Identification of the Nematic Spacer Conformation

Akihiro Abe, Renato N. Shimizu, and Hidemine Furuya

Department of Polymer Chemistry
Tokyo Institute of Technology
2-12-1 Ookayama, Meguro-ku, Tokyo 152, Japan

ABSTRACT

The RIS analysis of the deuterium quadrupolar splitting data have been performed for ether-type dimer liquid crystals $NCPh_2O(CD_2)_nOPh_2CN$ (DLC) with n = 9 and 10 according to the simulation scheme previously established. The orientational as well as conformational characteristics of DLCs have been investigated under various conditions including binary mixtures with a low-molar-mass liquid crystal EtOPhCH=NPhCN. In the DLC systems studied, the intervening spacer $-O(CH_2)_nO-$ was found to take spatial arrangements characteristic of the nematic phase. The nematic conformation thus identified remains quite stable over a wide range of concentration and temperature.

INTRODUCTION

Polymer liquid crystals (PLC) comprising a rigid mesogenic core and a flexible spacer in the repeat unit often exhibit a distinct odd-even oscillation in various thermodynamic quantities at the nematic-isotropic (NI) transition temperature [1, 2]. In these systems, the orientational order parameter of the mesogenic core axis oscillates with the number of methylene units n, suggesting that the order-disorder transition of the mesogenic core is coupled with conformational changes of the intervening flexible spacers [1 - 3]. Dimer liquid crystals (DLC) comprising two mesogenic units jointed by a single spacer also exhibit similar odd-even characteristics [2, 4]. As easily shown by an inspection of a molecular model, the molecular axis does not generally coincide with the mesogenic core axis. In this respect, PLCs as well as DLCs are quite different from simple low-molar-mass (monomer) liquid crystals. The conformations in the nematic state have been elucidated for several examples of PLC and DLC from the relevant ^{2}H NMR data [2, 5 - 7]. Although the results reported from various laboratories seem to vary somewhat depending on the models adopted in the

A. Teramoto. M. Kobayashi, T. Norisuje (Eds.)
Ordering in Macromolecular Systems
© Springer-Verlag Berlin Heidelberg 1994

simulation of the spectra, all suggest that the flexible spacers prefer to take extended conformations in the nematic state.

In a series of papers [2], we have developed a rotational isomeric state (RIS) simulation scheme which enables us to estimate the conformational distribution in the nematic state by comparing the observed quadrupolar splittings with those calculated. Extensive studies were carried out for DLCs and PLCs such as

$$NC-\bigcirc-\bigcirc-O(CD_2)_nO-\bigcirc-\bigcirc-CN \qquad (DLC)$$

$$-[-\bigcirc-OC(O)-\bigcirc-O(CD_2)_nO-\bigcirc-C(O)O-\bigcirc-O(CH_2)_nO-]_x- \qquad (PLC)$$

An important consequence of the analysis was that the nematic conformation of the flexible segment was very similar between the dimer and polymer of a given n. The difference arises in the orientation of the molecular axis in the liquid crystalline domain. The order parameter of the molecular axis becomes higher in the polymeric systems. In more recent work [8], the results obtained from the RIS analysis of ^2H NMR data was successfully used to interpret the magnetic susceptibilities of the same liquid crystals determined by a SQUID magnetometer. These observations have led to a conclusion that the spacer is allowed to take various conformations in as much as they are compatible with the preferred orientation of the mesogenic cores. In this respect, the flexible spacer maintains liquidlike characteristics in the nematic state.

The molecular scheme described above requires an extension of the polymeric chain with a large persistence length. According to the recent results of the small-angle neutron scattering [9], the radius of gyration in the direction parallel to the nematic director is only several times larger than that in the perpendicular direction in the vicinity of the NI transition, suggesting that some folds or defects may occur quite frequently in the polymeric system. Such imperfections should be largely suppressed in DLCs. In this work, we have attempted to elucidate the orientational as well as conformational characteristics of DLCs under various conditions. Studies have been extended to include binary mixtures comprising a DLC and a low-molar-mass liquid crystal.

SAMPLES AND MEASUREMENTS

The preparation of α,ω-Bis[(4,4'-cyanobiphenyl)oxy]alkane (CBA-n) has been reported in our previous work [2]. The samples carrying perdeuterated spacers were obtained by starting from perdeuterated α,ω-dibromoalkanes,

Br(CD$_2$)$_n$Br with n = 9 and 10. Substitution of deuterons at the ortho-position of 4,4'-cyanobiphenyl group was easily accomplished by the standard method [4]:

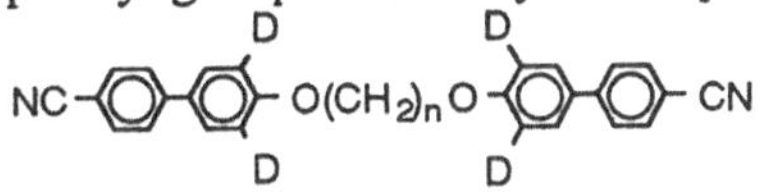

A partially deuterated 4'-Ethoxybenzilidene-4-cyanoaniline (EBCA) sample was prepared according to the following scheme [10, 11]:

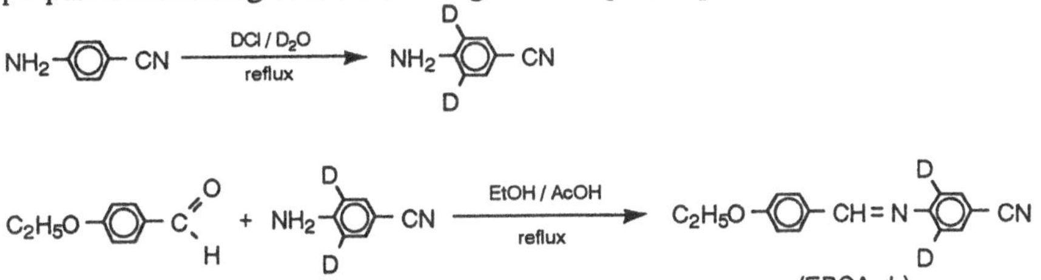

All samples exhibited enantiotropic nematic mesophase by the polarizing microscopic examination:

CBA-9	K	136.4 °C	N	174.9 °C	I
CBA-10	K	166.4 °C	N	186.9 °C	I
EBCA	K	105.9 °C	N	128.4 °C	I

The ^{2}H NMR spectra were recorded on a JEOL JNM-GSX-500 spectrometer operating at 76.65 MHz deuterium resonance frequency. Measurements were carried out under a complete decoupling and nonspinning mode. In the experiment, samples initially kept at a temperature above T_{NI} were cooled slowly to get into the nematic mesophase. The assignment of the individual splittings to the methylene groups of the spacer has been worked out previously by using partially deuterated DLC samples [2].

RESULTS

Shown in Figures 1 a and b are the phase diagrams for mixtures CBA-9/EBCA and CBA-10/EBCA, respectively. Differential scanning calorimetry (DSC) was used to determine the transition temperature and the associated enthalpy change. The ordinate indicates temperatures obtained on the cooling cycle. Both mixtures exhibit an appreciably wide range of nematic phase. For all binary mixtures, some broadening of the DSC peak was observed, indicating that a narrow biphasic equilibrium region exists around the NI transition temperature. The magnitude of the observed enthalpy change ΔH_{NI} was found to vary nearly linearly with the composition. The eutectic points appear at about 20 and 10 mol%, respectively, for CBA-9 and -10. In both diagrams, thermodynamic data associated with the phase boundary curve N/N+K were found to be reasonably compatible with the Schröder-van Laar expression [12]. The agreement has been slightly improved by adoption of the relation derived from the polymer solution theory. These results indicate that the CBA-n/EBCA solutions are practically ideal over the entire composition range.

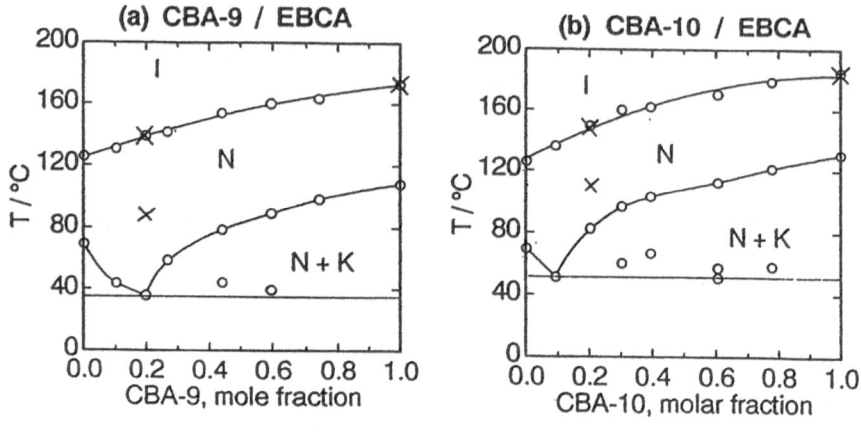

Figure 1.   Phase diagram of (a) CBA-9/EBCA and (b) CBA-10/EBCAsystems.   The symbol (x) indicates the points (concentration and temperature) at which RIS simulations of the spacer conformation were attempted (see Table 2).

Examples of the ^{2}H NMR spectra are shown in Figure 2 for mixtures containing 20 mol% of DLCs, where assignments of the observed quadrupolar Δv and dipolar splittings D_{HD} are also indicated.   The results of the ^{2}H NMR measurements carried out at various concentrations of DLCs are summarized in Figures 3 and 4, respectively, for CBA-9 and 10, where values of Δv and D_{HD} are plotted against the temperature ratio $T_r = T/T_{NI}$.   The corresponding variations for the nematic solvent EBCA are summarized in Figures 5 and 6.   Since signs of these couplings are not always definite in the present experiment, absolute values are listed in Figures 3 - 6.   In all measurements, the observed values of Δv and D_{HD} increase appreciably as temperature decreases.   The scattering of the dipolar coupling data tends to be pronounced at lower temperatures (cf. Fig. 3b and 4b). The analysis was therefore performed in the range above $T_r \cong 0.9$.

The observed values of Δv and D_{HD} plotted in Figures 3a-b and 4a-b are the quantities directly related to the order parameters of the mesogenic core axis.   It

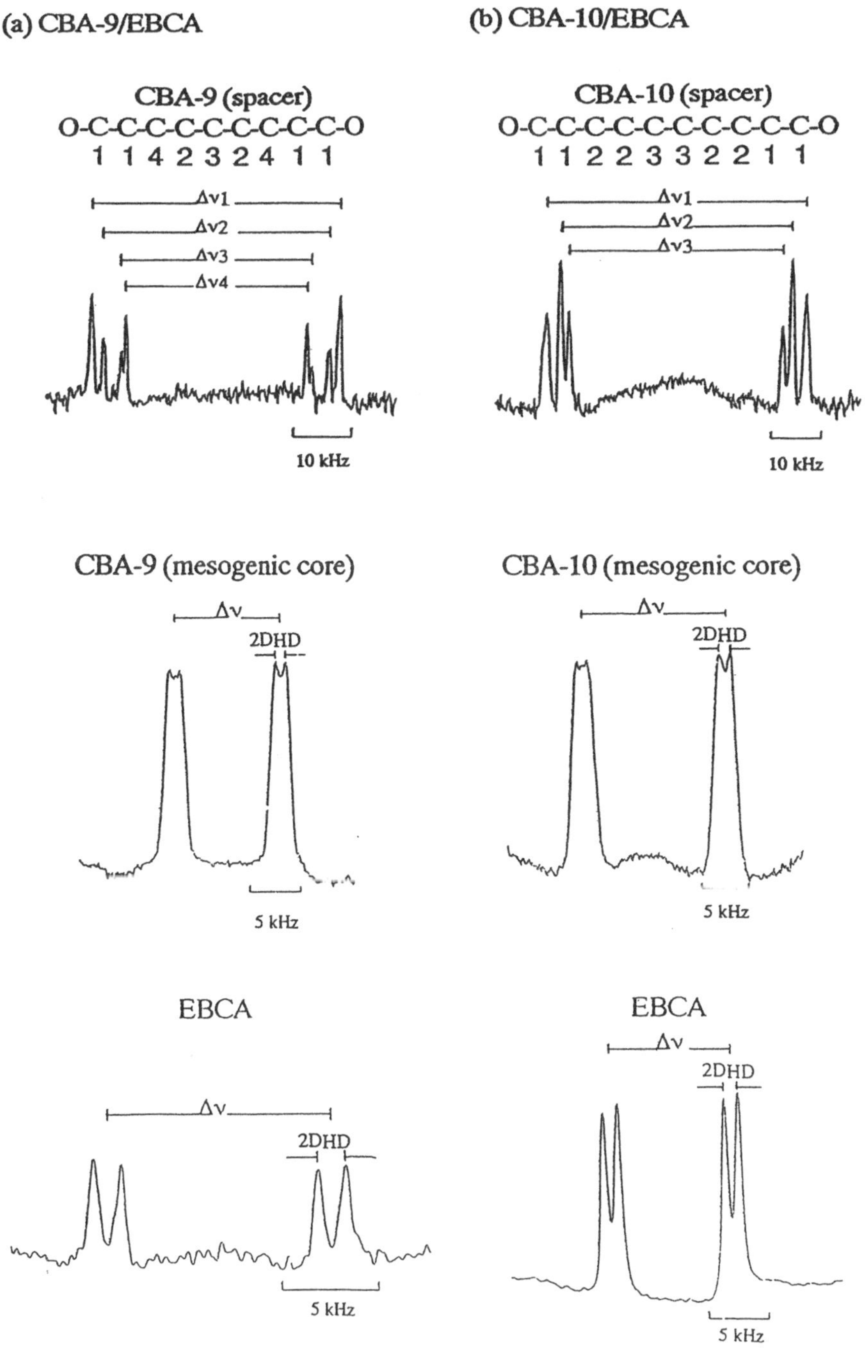

Figure 2.   Examples of ^{2}H NMR spectra observed for binary mixtures comprising 20 mol% of CBA-n: (a) CBA-9/EBCA at 130 °C and (b) CBA-10/EBCA at 130 °C.

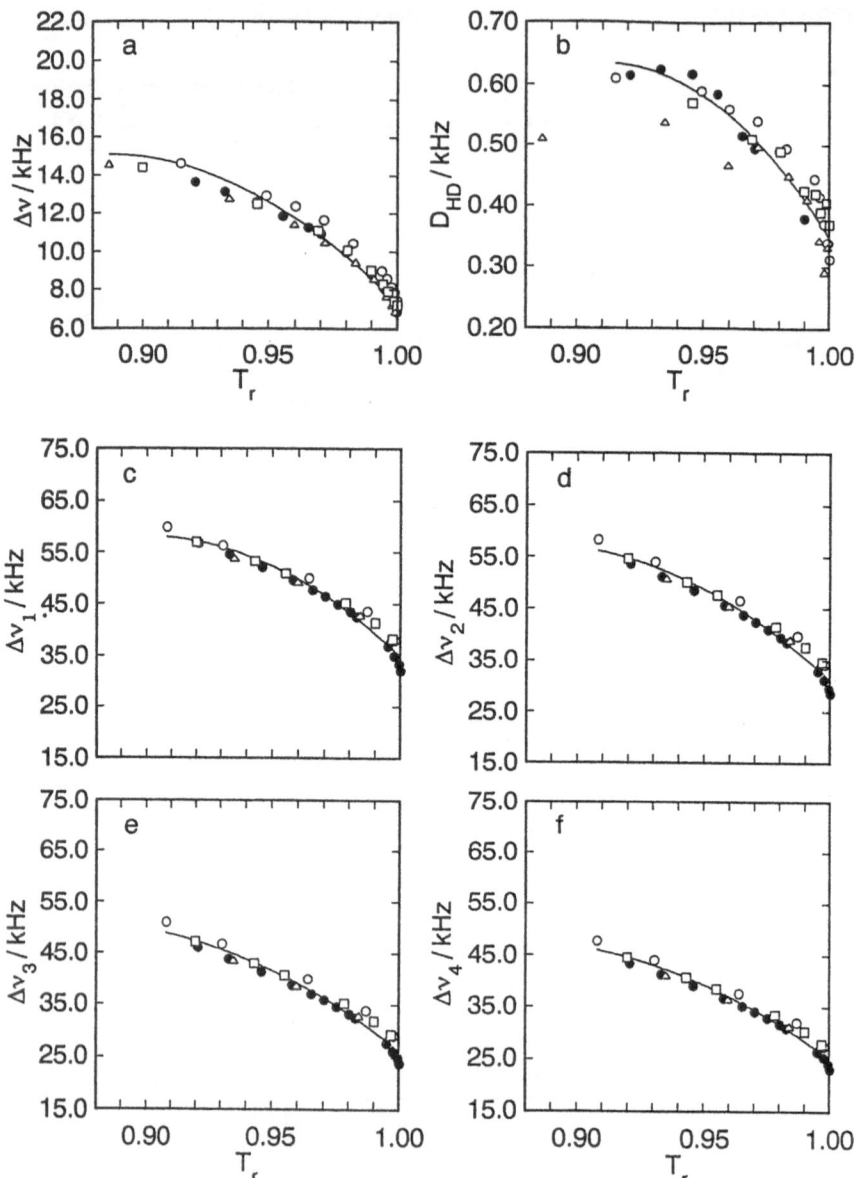

Figure 3.   Variation of $\Delta\nu$ and D_{HD} as a function of the reduced temperature T_r
= T/T_{NI}, observed for CBA-9 dissolved in EBCA: (a) $\Delta\nu$ and (b)
D_{HD} for the mesogenic core, and (c) - (f) for $\Delta\nu_i$ (i = 1-4) of the
spacer CD bonds.   Symbols:O 1.0, □ 0.6, △ 0.2, and ● 0.04,
expressed in CBA-9 mole fraction.

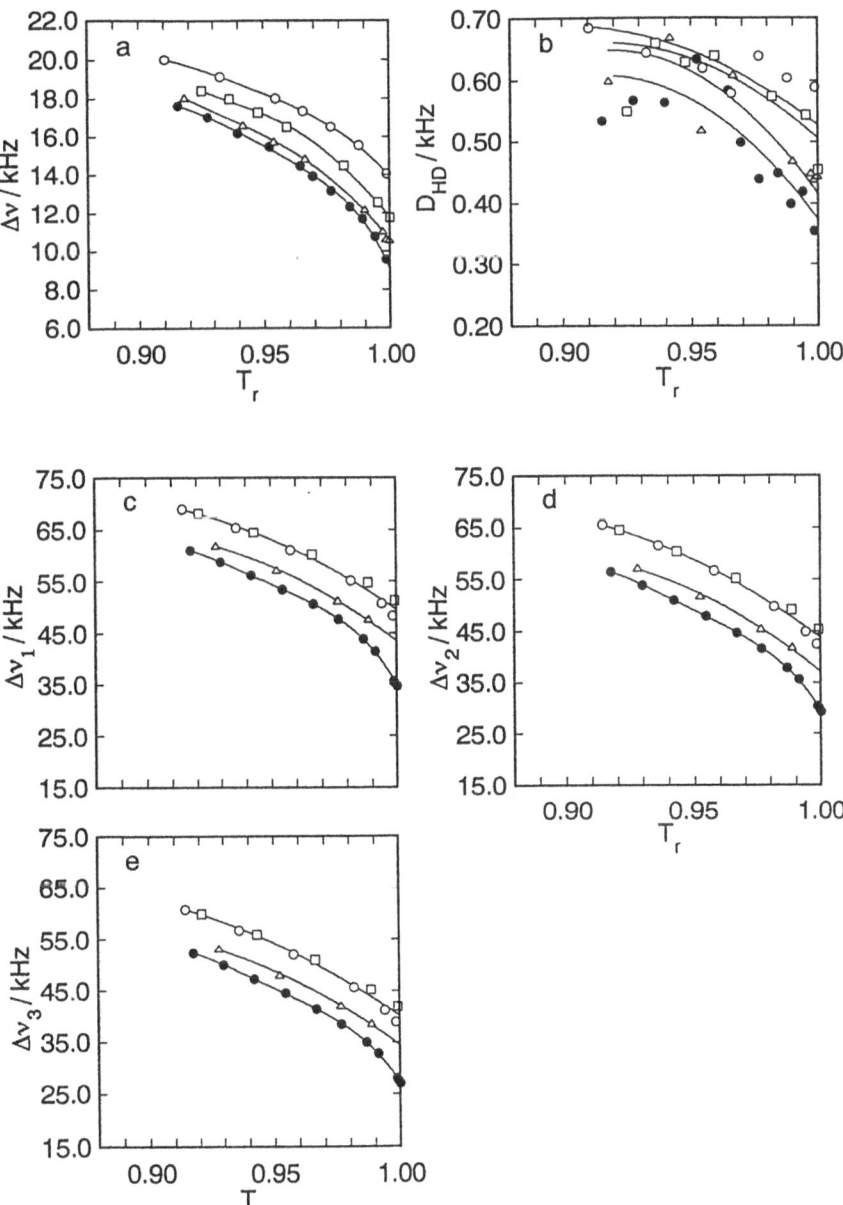

Figure 4. Variation of $\Delta\nu$ and D_{HD} as a function of the reduced temperature T_r = T/T_{NI}, observed for CBA-10 dissolved in EBCA: (a) $\Delta\nu$ and (b) D_{HD} for the mesogenic core, and (c) - (e) for $\Delta\nu_i$ (i = 1-3) of the spacer CD bonds. Symbols:O 1.0, □ 0.6, △ 0.2, and ● 0.04, expressed in CBA-10 mole fraction.

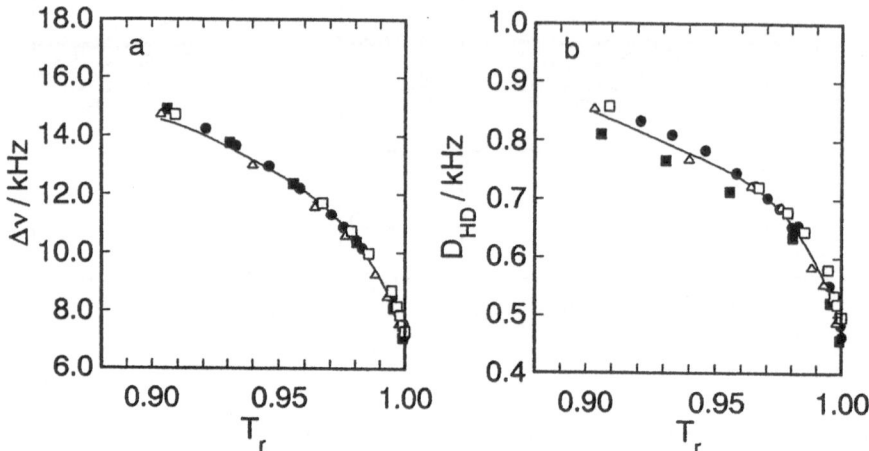

Figure 5.　Variation of $\Delta\nu$ and D_{HD} as a function of the reduced temperature T_r = T/T_{NI}, observed for the nematic solvent EBCA: (a) $\Delta\nu$ and (b) D_{HD}. Symbols: □ 0.6, △ 0.2, ● 0.04, and ■ 0, expressed in CBA-9 mole fraction.

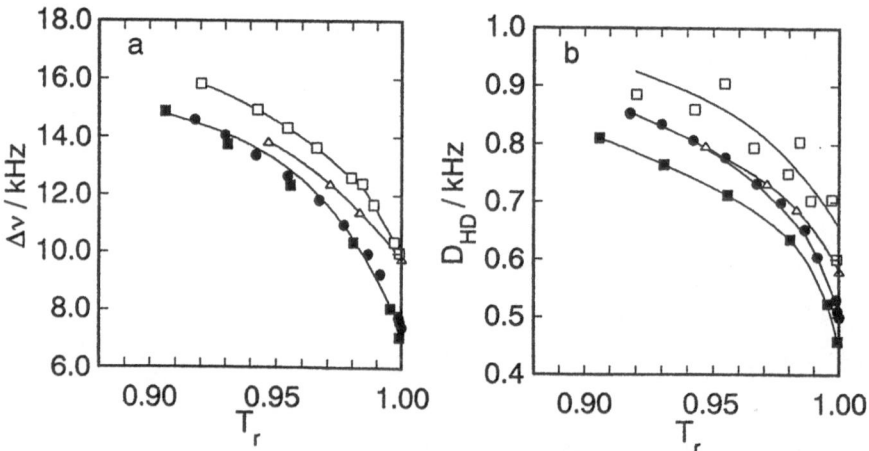

Figure 6.　Variation of $\Delta\nu$ and D_{HD} as a function of the reduced temperature T_r = T/T_{NI}, observed for the nematic solvent EBCA: (a) $\Delta\nu$ and (b) D_{HD}. Symbols: □ 0.6, △ 0.2, ● 0.04, and ■ 0, expressed in CBA-10 mole fraction.

has been reported that the order parameters determined for mixtures in which one component is a non-mesomorphic solute conform to a universal curve when plotted against T_r [13 - 15]. The results shown in Figures 3 and 5 suggest that a similar superposition principle holds for each component of the CBA-9/EBCA system over a wide range of concentration. The same argument applies to Figures 3c-f, where variations of Δv_i's of the spacer are indicated as a function of T_r. In contrast, the curves illustrated in Figures 4 and 6 for the CBA-10/EBCA mixture considerably deviate from the superposition rule.

CONFORMATIONAL ANALYSIS

The order parameters of the mesogenic core axis of the DLCs were first estimated from the observed D_{HD} and Δv according to the procedure previously described [16]. The geometrical parameters required for the calculation are listed in Table 1. Values of S_{zz}^R thus estimated vary approximately from 0.3 to 0.6 in both CBA-9 and -10 systems over the range $T_r = 1.0$ to 0.9. Contribution from the biaxiality term S_{xx}^R - S_{yy}^R remains small, being of the order of ca. 0.02 under all conditions examined.

We then attempted to analyze the conformation of the spacer on the basis of the deuterium quadrupolar splitting data collected for the individual CD bonds (cf. Fig. 3c-f and 4c-e). In the RIS simulation scheme previously established [2], the following expressions were adopted for the dipolar coupling (D_{HD}) due to the mesogenic core and the deuterium quadrupolar splittings (Δv_i) of the spacer:

$$D_{HD} = -\gamma_H \gamma_D h / \left(4\pi^2 r_{HD}^3\right) S_{zz} \left(3\langle\cos^2\psi\rangle - 1\right) / 2 \qquad (1)$$

$$\Delta v_i = (3/2)(e^2qQ/h) S_{zz} \left(3\langle\cos^2\phi_i\rangle - 1\right) / 2 \qquad (2)$$

where $\gamma_H = 2.6752 \times 10^8$ kg^{-1}sA, $\gamma_D = 4.1065 \times 10^7$ kg^{-1}sA, h = 6.6262×10^{-34} Js, $r_{HD} = 2.48$ Å and the quadrupolar coupling constant for the aliphatic CD bond $(e^2qQ/h) = 174$ kHz. Here, we assume that the molecular axis (z) lies in the direction parallel to the line connecting the centers of two neighboring mesogenic cores, and the molecules are approximately axially symmetric around the z-axis, and thus the orientation of these anisotropic molecules can be described by a single order parameter S_{zz}, the biaxiality of the system S_{xx} - S_{yy} being ignored for simplicity. In the equations above, ψ denotes the disorientation of the mesogenic core axis with respect to the molecular axis, and ϕ_i is the angle between the ith CD bond and the molecular axis. The bracket indicates statistical mechanical averages taken over all allowed conformations in the system.

According to Eq. 2, the ratio such as $\Delta v_i/\Delta v_1$ ($i > 1$) should solely depend on the spacer conformation, being free from the orientational order of the molecular axis S_{zz}. These ratios are plotted against T_r in Figures 7a and b, respectively, for CBA-9 and -10. In both diagrams, the ratios tend to increase somewhat as temperature decreases, indicating that the conformation of the spacer may be affected as well. The $\Delta v_i/\Delta v_1$ vs. T_r curves exhibit some small divergence (less than 4 % of $\Delta v_i/\Delta v_1$) as a function of concentration. As shown in Figures 8a and b, however, the superposition principle clearly holds for both CBA-n's when the ratio $\Delta v_i/\Delta v_1$ is plotted against Δv_1. These results immediately suggest that the conformational correlation along the spacer is governed by the same principle throughout the entire range of concentration and temperature studied.

In the present RIS simulation, the conformation map, on which spatial arrangements of the configurations allowed in the free state are identified by the inclination angles ψ_1 and ψ_2 of the terminal mesogenic core axes with respect to the molecular axis (cf. ref. 2), was used in the estimation of the nematic conformation. For a chosen ensemble of spatial configurations, the orientation of the constituent CD bonds was calculated in terms of $\cos^2\phi$. The average $\langle \cos^2\phi \rangle$ for given CD bonds may be estimated for a given conformer distribution. Statistical weight parameters defined for the individual skeletal bonds were so adjusted to minimize the difference between the calculated and observed results according to the relation given by Eq. 1 and 2. The best fit ensemble of configurations has been derived by repeating calculations iteratively. The results obtained around T_{NI} are as follows: $0 < \psi_1$, $\psi_2 < \psi_m$ with $\psi_m \cong 45°$ for CBA-9; $0 < \psi_1$, $\psi_2 < \psi_m$ with $\psi_m \cong 35°$ for CBA-10. Bond conformations of the spacer have been elucidated for several combinations of concentration and temperature chosen in consideration of the general feature of the phase diagram (Fig. 1a and b). The results of simulation are summarized in Table 2, where values of the trans fraction (f_t) are listed according to the bond order from the mesogenic terminal to the central bond of the spacer. The effect of concentration (c), expressed in terms of the mole fraction of CBA-n, can be examined by comparing the results for c = 1.0 and 0.2 at T_r = 1.0. The temperature was varied from T_r = 1.0 to 0.89 for CBA-9 and to 0.93 for CBA-10 by keeping the concentration at c = 0.2. The range of uncertainty was found to be alternatively enhanced along the chain (i.e. for bonds such as C_1-C_2, C_3-C_4, and C_5-C_6). As examination of a proper molecular model should reveal, the angular correlation between the terminal mesogenic groups are less sensitive to the rotation

Table 1. Structural Parameters Used in the RIS Analysis

bond	length, Å	bond angle	angle, deg	bond	rotational angle for gauche state, deg
O-C	1.40	∠CPhOC	120	CC-CC	±112.5
C-C	1.53	∠OCC	112	OC-CC	±117.0
		∠CCC	112	CPhO-CC	±100.0
		∠CCH	109		

Table 2. Bond Conformations of the Flexible Spacer Estimated from the Deuterium Quadrupolar Splitting Data, Sampling Points Being Indicated on the Phase Diagrams (Figures 1a and b)

(a) CBA-9/EBCA

T/°C (T_r)	c [a]	Bond Conformation, f_t				
		O-C_1	C_1-C_2	C_2-C_3	C_3-C_4	C_4-C_5
170.8 (1.0)	1.0	0.95±0.05	0.63±0.12	0.72±0.03	0.43±0.06	0.64±0.03
136.5 (1.0)	0.2	0.95±0.05	0.59±0.16	0.73±0.04	0.43±0.06	0.63±0.04
90.0 (0.89)	0.2	0.94±0.05	0.53±0.16	0.74±0.02	0.53±0.04	0.66±0.01

(b) CBA-10/EBCA

T/°C (T_r)	c [a]	Bond Conformation, f_t					
		O-C_1	C_1-C_2	C_2-C_3	C_3-C_4	C_4-C_5	C_5-C_6
184.2 (1.0)	1.0	0.97±0.03	0.56±0.18	0.90±0.03	0.44±0.09	0.84±0.04	0.56±0.22
139.7 (1.0)	0.2	0.95±0.05	0.50±0.22	0.88±0.03	0.46±0.14	0.81±0.06	0.45±0.24
110.0 (0.93)	0.2	0.95±0.05	0.36±0.13	0.92±0.04	0.40±0.09	0.85±0.06	0.36±0.12

[a] Expressed in CBA-n mole fraction

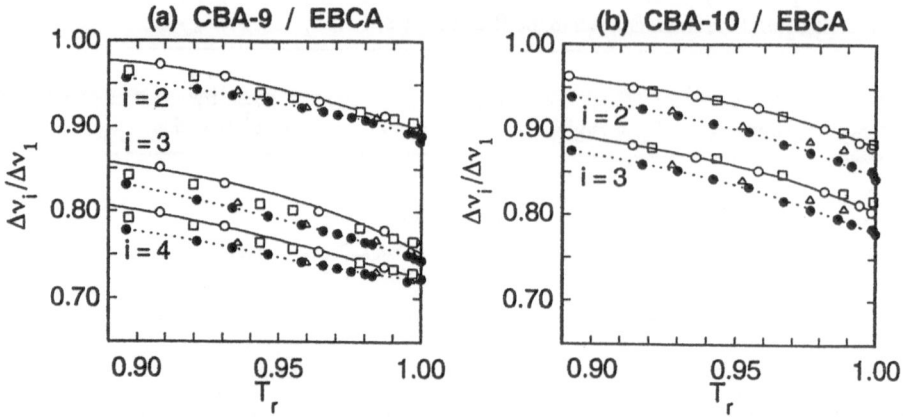

Figure 7. The quadrupolar splitting ratios $\Delta v_i/\Delta v_1$ observed for the spacer plotted against the reduced temperature $T_r = T/T_{NI}$: (a) CBA-9/EBCA and (b) CBA-10/EBCA. Symbols: —O— 1.0, □ 0.6, △ 0.2, and —●— 0.04, expressed in CBA-n mole fraction.

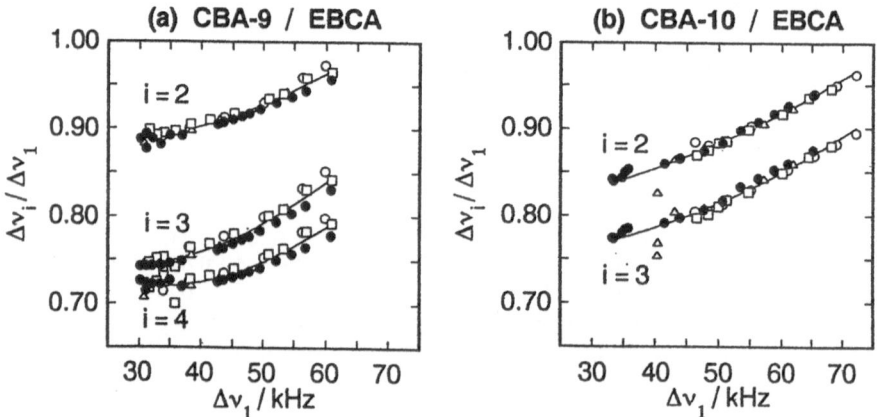

Figure 8. The $\Delta v_i/\Delta v_1$ vs. Δv_1 plot: (a) CBA-9/EBCA and (b) CBA-10/EBCA. Symbols: O 1.0, □ 0.6, △ 0.2, and ● 0.04, expressed in CBA-n mole fraction.

around these bonds.   Accordingly, the observed ^{2}H NMR profiles can be reproduced somewhat arbitrarily within the given range of bond conformation in the simulation.   Nevertheless, the odd-even alternation of the f_t value along the chain is marked for n = 10.   The alternation is slightly less distinct for n = 9.   As shown in Table 2, the bond conformations thus estimated are not much affected by the concentration and temperature as long as samples remain in the stable nematic phase.

DISCUSSION

The RIS analysis presented above simultaneously yields the order parameter S_{zz}, which characterizes the orientation of the long molecular axis of the DLC molecule. As indicated in Figures 9a and b, values of S_{zz} thus derived increase substantially as the reduced temperature T_r decreases. The range of uncertainty is indicated for the individual estimate by the error bar.   The effect of concentration is more marked for CBA-10/EBCA (Fig. 9b) than CBA-9/EBCA (Fig. 9a), being consistent with the observed trend in the D_{HD} vs. T_r plots shown in Figures 4a, b and 3a, b, respectively.   These results, in combination with the arguments presented in the preceding section, immediately suggest that the observed variation of Δv_i with temperature and concentration largely arises from the fluctuation of the molecular axis while the conformation of the spacer remains nearly unaltered. The experimental observations presented in this paper apparently support the previous view [1, 17] that a flexible spacer carrying mesogenic units on both terminals should take certain nematic arrangements, which are distinctly different in the conformational distribution encountered from that in the isotropic state.

Finally it may be interesting at this stage to examine the orientational correlation between the solute (DLC) and solvent (EBCA) molecules coexisting in the mixture.   The order parameter S_{zz}^M defined for the para axis of the EBCA molecule can be estimated easily from the ^{2}H NMR data given in Figures 5 and 6 according to the procedure described in the previous section. As in the case of DLCs, contribution from the biaxiality term was found to be very small at all concentration and temperature examined.   The orientational order parameters estimated independently for the individual components of binary mixtures are compared in Figures 10a and b.   In both CBA-9/EBCA and CBA-10/EBCA systems, correlations of the two order parameters derived under various conditions are reasonable.   These observations are compatible with the aforementioned statement on the molecular ordering of DLCs in the nematic phase.

The authors are grateful to the Asahi Glass Foundation for financial support of this work.

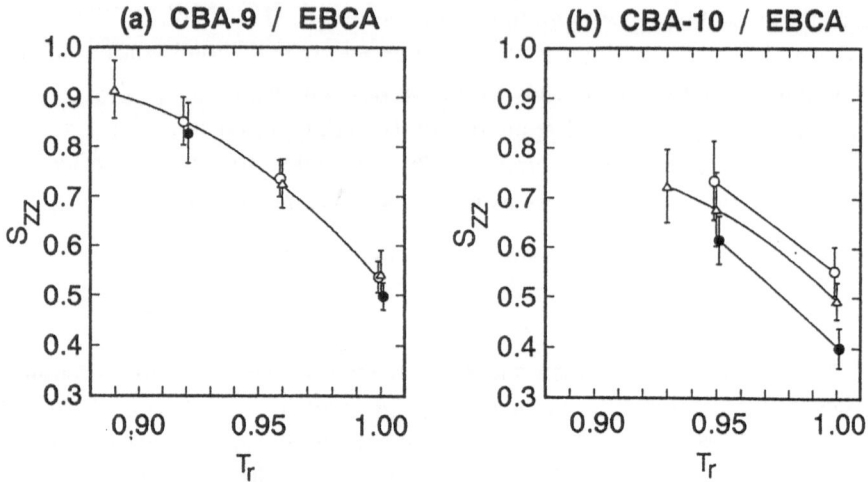

Figure 9.　Variation of the order parameter S_{ZZ} of the molecular axis for CBA-n as a function of the reduced temperature T_r: (a) CBA -9/EBCA and (b) CBA-10/EBCA.　Symbols: ○ 1.0, △ 0.2, and ● 0.04, expressed in CBA-n mole fraction.

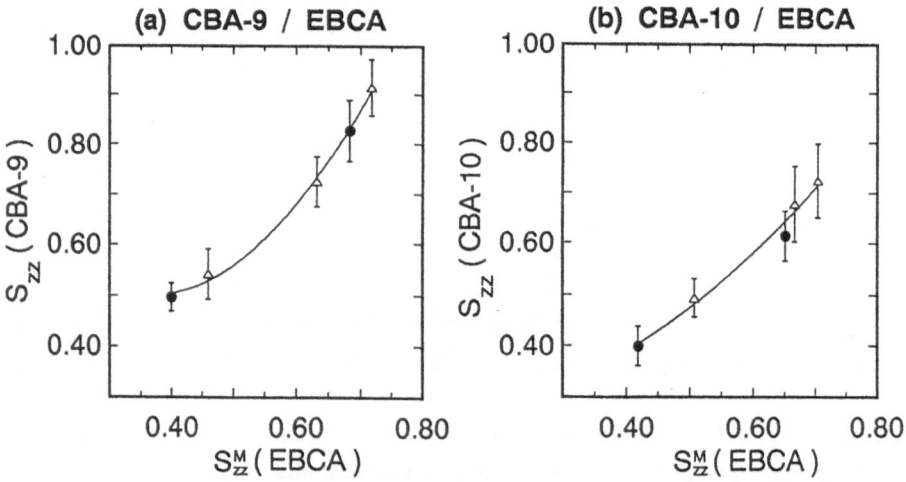

Figure 10.　The S_{ZZ} vs. S_{ZZ}^{M} plot, indicating the orientational correlation between solute (DLC) and solvent (EBCA) molecules: (a) CBA-9/EBCA and (b) CBA-10/EBCA.　Symbols: △ 0.2, and ● 0.04, expressed in CBA-n mole fraction.

REFERENCES

1   Abe A (1984) Macromolecules 17: 2280
2   Abe A, Furuya H (1989) Macromolecules 22: 2982   Furuya H, Abe A (1988)
    Polym Bull 20: 467   Abe A, Furuya H (1988) Polym Bull 19: 403   Abe A,
    Furuya H, Yoon DY (1988) Mol Cryst Liq Cryst 159: 151   Abe A, Furuya H
    (1988) Mol Cryst Liq Cryst (1988) 159: 99   Abe A, Furuya H (1986)
    Kobunshi Ronbunshu 5: 247   Furuya H, Asahi K, Abe A (1986) Polym J 18:
    779   Toriumi H, Furuya H, Abe A (1985) Polym J 17: 895
3   Blumstein A (ed) (1983) Polymer liquid crystals, Plenum, New York   Ciferri
    A (ed) (1991) Liquid crystallinity in Polymers -principles and fundamental
    properties, VCH, New York
4   Emsley JW, Luckhurst GR, Shilstone GN (1984) Mol Phys 53: 1023
5   Yoon DY, Bruckner S (1985) Macromolecules 18: 651   Yoon DY, Bruckner
    S, Volksen W, Scott JC, Griffin AC (1985) Faraday Discuss Chem Soc 79: 41
    Bruckner S, Scott JC, Yoon DY, Griffin AC (1985) Macromolecules 18: 2709
6   Samulski ET, Gauthier MM, Blumstein RB, Blumstein A (1984)
    Macromolecules 17: 479   Griffin AC, Samulski ET (1985) J Am Chem Soc
    107: 2975   Samulski ET (1985) Faraday Discuss Chem Soc 79: 7   Photinos
    DJ, Poon CD, Samulski ET, Toriumi H (1992) J Phys Chem 96: 8176
    Photinos DJ, Samulski ET, Toriumi H (1992) J Chem Soc Faraday Trans 88:
    1875
7   Emsley JW, Heaton NJ, Luckhurst GR, Shilstone GN (1988) Mol Phys 64: 377
    Cheung STW, Emsley JW (1993) Liq Cryst 13: 265
8   Furuya H, Dries T, Fuhrmann K, Abe A, Ballauff M, Fischer EW (1990)
    Macromolecules 23: 4122   Furuya H, Abe A, Fuhrmann K, Ballauff M,
    Fischer EW (1991) Macromolecules 24: 2999
9   D'Allest JF, Sixou P, Blumstein A, Blumstein R, Texeira J, Noirez L (1988)
    Mol Cryst Liq Cryst 155:581   D'Allest JF, Maïssa P, ten Bosch A, Sixou P,
    Blumstein A, Blumstein R, Teixeira J, Noirez L (1988) Phys Rev Lett 61: 2562
10 Zimmermann H (1989) Liq Cryst 4: 591
11 Pisipati VGKM, Rao NVS (1984) Z Naturforsch 39a: 696
12 Kelker H, Hatz R(eds) (1980) Hand book of liquid crystals, VCH, Weinheim
13 Chen DH, Luckhurst GR (1969) Trans Faraday Soc 65: 656
14 Kronberg B, Gilson DFG, Patterson D (1976) J Chem Soc Faraday Trans II
    72: 1673, 1686
15 Abe A, Iizumi E, Sasanuma Y (1993) Polym J, in press   Abe A, Iizumi E,
    Kimura N, in preparation
16 Kimura N, Abe A (1992) Makromol Chem, Theory Simul 1: 401
17 Abe A (1992) Makromol Chem, Makromol Symp 53: 13

Phase Equilibria in Liquid Crystalline Polymer Solutions: Theory and Experiment

Takahiro Sato and Akio Teramoto

Department of Macromolecular Science, Osaka University
Toyonaka, Osaka 560, Japan

Abstract: We compared various statistical thermodynamic theories with experimental results for isotropic–liquid crystal phase behaviors in many stiff-chain liquid-crystalline polymer solutions. Among the theories, the scaled particle theory for wormlike hard spherocylinders most successfully explained the experimental phase behaviors in neutral polymer systems, and the same theory considering the electrostatic interaction in a perturbative way was favorably compared with a phase diagram for a charged stiff polymer system. Good agreements between this theory and experiment were also obtained for the osmotic pressure and the orientational order parameter of poly(γ-benzyl L-glutamate) solutions.

1 INTRODUCTION

Liquid crystalline polymers can be roughly divided into four types according to their molecular architectures: (1) homopolymers with stiff backbones, (2) alternating copolymers with hard and flexible segments, (3) flexible polymers with mesogenic side groups, and (4) polymers with stiff backbones and long flexible side chains. Among them, the first type liquid crystalline polymers (stiff-chain polymers) are most well-characterized molecularly using dilute solution data on the basis of the wormlike chain model, and thermodynamic properties of their solutions have been extensively studied.

It is known that these stiff-chain polymers can form liquid crystalline (usually nematic or cholesteric) states in moderately concentrated solutions. They are distinct from flexible polymers in this property. Since most of stiff-chain polymers can dissolve only in good solvents, the liquid crystallinity of the solution must be induced by the hard-core repulsive interaction between polymers. Many statistical thermodynamic theories have been presented to explain the liquid crystallinity of stiff polymer solutions:

(1) Onsager [1] discussed the phase behavior of hard rod and charged rod solutions using the second virial approximation.

(2) Flory [2] proposed the lattice theory for hard rod solutions.

(3) Cotter [3,4] and others [5–7] applied the scaled particle theory to hard (sphero)cylinder systems in order to incorporate higher virial terms in the Onsager theory.

(4) Khokhlov and Semenov [8,9] extended the Onsager theory to semiflexible polymer solutions.

A. Teramoto, M. Kobayashi, T. Norisuje (Eds.)
Ordering in Macromolecular Systems

(5) Sato and Teramoto [10] incorporated higher virial terms into the Khokhlov–Semenov theory using the scaled particle approach and taking into account electrostatic interactions for semiflexible polyelectrolytes.

In this paper, we briefly summarize these theories and compared them with experimental phase equilibrium data of stiff-chain polymer solutions.

2 THEORIES FOR HARD ROD SOLUTIONS

2.1 Theories of Onsager, Flory, and Cotter

In order to calculate phase boundaries for a polymer solution, we need the excess free energy of the solution over that of the solvent. For hard rod solutions, the excess Helmholtz free energy F consists of the translational entropy term TS_{tr} and the orientational entropy term TS_{or}:

$$F = -TS_{or} - TS_{tr} \qquad (2.1)$$

For rods interacting each other only by the hard-core potential, no energy term appears in F. The orientational entropy S_{or} is represented by [1]

$$S_{or} = -nk_B \int f(\mathbf{a}) \ln[4\pi f(\mathbf{a})] d\mathbf{a} \qquad (2.2)$$

where n is the number of rods in the system, $f(\mathbf{a})$ the orientational distribution function of the rods, and $\mathbf{a}$ the unit vector parallel to the rod axis.

On the other hand, the formulation of the translational entropy S_{tr} is not an easy task. There are many statistical thermodynamic theories presented for this problem. Onsager [1] used the second virial approximation, and obtained

$$S_{tr} = -nk_B \left(\text{const.} + \ln c' + \tfrac{1}{2} c' \iint V_{ex}(\mathbf{a}, \mathbf{a}') f(\mathbf{a}) f(\mathbf{a}') d\mathbf{a} d\mathbf{a}' \right) \qquad (2.3)$$

where c' is the polymer number concentration and $V_{ex}(\mathbf{a},\mathbf{a}')$ is the mutual exclude volume between two rods oriented along $\mathbf{a}$ and $\mathbf{a}'$. If the rod has the length L much larger than the hard-core diameter d, $V_{ex}(\mathbf{a},\mathbf{a}')$ is given by

$$V_{ex}(\mathbf{a}, \mathbf{a}') \cong 2L^2 d |\sin \gamma| \qquad (2.4)$$

where $\gamma = \cos^{-1}(\mathbf{a} \cdot \mathbf{a}')$. Equations (2.3) and (2.4) indicate that the translational entropy is increased by the orientation of each rod to some common direction which reduces the angle γ and thus $V_{ex}(\mathbf{a},\mathbf{a}')$. However this orientation accompanies an orientational entropy loss. As a result, the relative stability of the isotropic and liquid crystal (nematic) states is determined by the competition between S_{or} and S_{tr}. The second virial term in Eq.(2.3) is negligible in F at very low concentration, but becomes significant with increasing polymer concentration. Thus the liquid crystal phase becomes stable at some high enough concentration. This is a qualitative explanation of the liquid crystallinity in the hard-rod system. Since the two competing terms are both entropic, the isotropic–liquid crystal phase equilibrium in hard rod solutions should be temperature independent. This makes a remarkable contrast with liquid-liquid phase equilibrium in flexible polymer–poor solvent systems.

Since Onsager neglected the third and higher virial terms in S_{tr}, the application of his theory to moderately concentrated solutions is questionable. Flory [2] proposed an alternate method to calculate S_{tr} using a lattice model, which is free from the second virial approximation. In this method, each rod is divided into X segments whose length is the same as the polymer diameter ($X = L/d$), and each segment of all rods is added to one lattice site without overlapping. At estimating the probability that a particular site where some segment should be added is vacant, Flory used a mean-field approximation, i.e., neglected the connectivity of the segments of the rods already existing in the lattice, just like Flory-Huggins theory [11].

Cotter [3,4] also considered the higher virial terms in S_{tr} for hard rod solutions using the scaled particle theory (SPT). The SPT introduces a hypothetical scaled particle (a hard spherocylinder), and calculates the following two quantities: the reversible work W of adding the scaled particle to a solution containing real solute particles at an arbitrary point under a constant (osmotic) pressure and the probability $P(a)$ of inserting the scaled particle oriented along the direction a at some arbitrary point of a solution without overlapping any other particles. The free energy F can be calculated from W or $P(a)$ for the scaled particle with the same size as the real particle.

The phase boundary number concentrations c_I' between the isotropic and biphasic regions and c_A' between the liquid crystal and biphasic regions can be calculated numerically from the phase coexistence equations. The three curves in Fig. 1 show the results for c_I' calculated from the three hard rod theories, where v is the molecular volume, and vc_I' represents the polymer volume fraction. All the three theories give vc_I' as a function only of the axial ratio X. The difference between vc_I' of the SPT and of the Onsager theory is observed only at considerably small X. (The same result is obtained for vc_A'.) This means that the higher virial terms neglected in the Onsager theory does not essentially affect the phase boundary concentration unless X is not too small. The Flory theory predicts appreciably higher c_I' than the SPT and the Onsager theories. This difference should be ascribed to the different models and approximations used in the theories.

2.2 Comparison with Experiment

The data points in Fig. 1 show the experimental phase boundary volume fractions vc_I' for several quasi-binary solutions consisting of one liquid crystalline (stiff) polymer sample and a good solvent; Table 1 lists the persistence length (the stiffness parameter) q and the diameter d for each polymer. The values of d (and the molecular volume v used in Fig. 1) were estimated from the partial specific volume of each polymer.

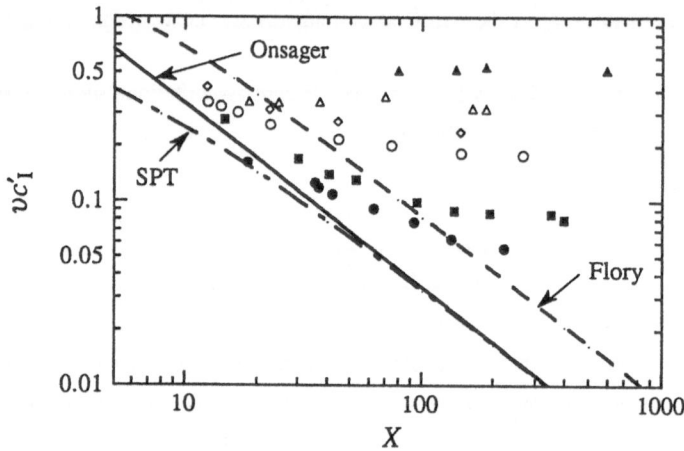

Fig. 1. Comparison between the hard rod theories and experimental results: (●) schizophyllan–water [12,13]; (■) PBLG–DMF [14-19]; (○)PHIC–toluene [20]; (◇) PHIC–DCM [20]; (△) HPC–water [21]; (▲) APC–DBP [22].

Table 1. Values of q and d of Liquid Crystalline Polymers

polymer	solvent	q/nm	d/nm
schizophyllan	water	200 [23]	1.68
poly(γ-benzyl L-glutamate) (PBLG)	dimethylformamide (DMF)	150 [24][a]	1.56
poly(n-hexyl isocyanate) (PHIC)	toluene (25°C)	37 [25]	1.24
	dichloromethane (DCM)	21 [25]	1.25
(hydroxypropyl)cellulose (HPC)	water	6.5 [21]	1.13
(acetoxypropyl)cellulose (APC)	dibutylphthalate (DBP)	5.9 [22]	1.19

[a] Analyzed by T. Itou.

None of the three theories can successfully predict the experimental results. Although these theories predict that $\upsilon c_I'$ is a unique function of X, the data points for the polymers with different stiffness do not follow any universal curve, but shift toward higher concentration with decreasing chain stiffness. This result indicates an important role of the chain flexibility in the isotropic–liquid crystal phase equilibrium in semiflexible polymer solutions, which was not paid due attention until recently. We consider this flexibility effect in the next section.

3 CHAIN FLEXIBILITY EFFECT

3.1 Khokhlov–Semenov Theory [8,9]

For semiflexible polymer solutions, the orientational state of the system must be specified by the orientational distribution function $f(a;l)$ of the unit vector a tangent to the polymer chain axis at the contour point l $(0 \leq l \leq L)$. In a nematic state, the tangent vector at each contour point should more or less align along some common direction. This means that the semiflexible polymer must take a stretched conformation in the nematic phase. Therefore the formation of the nematic phase is accompanied by a conformational entropy loss, which does not appear in the nematic phase formation in straight rod solutions.

Khokhlov and Semenov formulated this conformational entropy loss, applying the method of Lifshitz [26,27] to the wormlike chain model. Assuming $f(a;l)$ averaged along the chain contour to be represented by the Onsager trial function [1], they obtained the conformational entropy loss S_{conf} in the nematic phase as

$$S_{conf}/nk_B = \begin{cases} \ln \alpha - 1 + N(\alpha - 1)/3 & (N \ll 1) \\ N(\alpha - 1)/2 + \ln(\alpha/4) & (N \gg 1) \end{cases} \tag{3.1}$$

where α is the degree of orientation parameter in the Onsager trial function, and N the number of the Kuhn statistical segments $(\equiv L/2q)$; the expression of S_{conf} at intermediate N was not given. It is noted that S_{conf} includes the effects of both stretching and orientation.

The free energy of the wormlike chain solution can be written as

$$F = -TS_{conf} - TS_{tr} \tag{3.2}$$

If the persistence length q of the wormlike chain is much larger than the hard-core diameter d, S_{tr} of the chain can be approximated by that of the rod with the same contour length. Khokhlov and Semenov used Eq.(2.3) with Eq.(2.4) for S_{tr} (the second virial approximation), and calculated the phase boundary concentrations for semiflexible polymer solutions. Their results can be expressed in the following interpolation formulas

$$\frac{v c_I'}{\tilde{d}} = \frac{3.34 + 11.94N + 6.34N^2}{N(1 + 0.586N)}, \quad \frac{v c_A'}{\tilde{d}} = \frac{4.49 + 22.48N + 70.16N^2}{N(1 + 5.66N)} \quad \text{(KS)} \tag{3.3}$$

where $\tilde{d}$ is the hard-core diameter reduced by the Kuhn statistical segment length $(\equiv d/2q)$.

Fig. 2 compares the Khokhlov–Semenov theory with the experimental phase boundary concentration c_I' for quasi-binary solutions of stiff-chain liquid-crystalline polymers. The agreement between the theory and experiments is much improved in comparison with Fig. 1. Especially for stiffer polymers with the persistence length $q \geq 37$ nm (i.e., schizophyllan, PBLG,

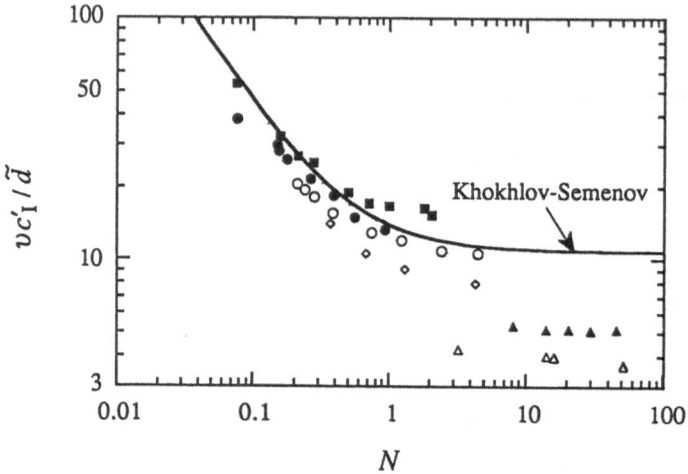

Fig. 2. Comparison of the Khokhlov–Semenov theory with experimental results for the phase boundary concentration c_I': the same symbols as those in Fig. 1.

and PHIC in toluene), the agreement is almost quantitative. However the data points for more flexible polymers are seen appreciably deviated from the theoretical curve of Khokhlov and Semenov.

3.2 Scaled Particle Theory for Wormlike Hard Spherocylinders

Since the phase boundary concentration for less stiff polymers is quite high (cf. Fig. 1), the second virial approximation Khokhlov and Semenov used may become worse for those solutions. To refine this deficiency, Sato and Teramoto [10] incorporated the higher virial terms into the Khokhlov–Semenov theory using the scaled particle theory (SPT). As mentioned in Sect. 2.1, the SPT considers a hypothetical scaled particle; in this case the scaled particle is a wormlike spherocylinder with a cylinder length λL_c and a hard-core diameter κd, where L_c and d are the cylinder length and hard-core diameter of the real particle, respectively, and λ and κ are the scale factors. The persistence length q of the scaled particle is assumed to be equal to that of the real particle.

Two important quantities in the SPT are the reversible work W of adding the scaled particle to a solution containing real solute particles at an arbitrary point under a constant osmotic pressure Π, and the probability $P(\mathbf{a};l)$ that the scaled particle with the tangent vector $\mathbf{a}$ at the contour point l may be inserted at some arbitrary point of a solution without overlapping any other particles. These two quantities can be exactly related to the Helmholtz free energy F by the equation,

$$F = \text{const.} + nW - TS_{\text{conf}} - \Pi V$$

$$= \text{const.} + \frac{nk_BT}{L} \int_0^{L_c} dl \int da f(\mathbf{a};l) \ln P(\mathbf{a};l) - TS_{\text{conf}} - \Pi V \tag{3.4}$$

where V is the volume of the solution.

When the scale factors λ and κ take sufficiently small values at the constant q, the scaled particle tends to a straight spherocylinder. Further, in the same condition, the scaled particle seldom overlaps simultaneously with two or more other real particles at its insertion into the solution. Taking advantage of these facts, we can obtain the following equation for $P(\mathbf{a};l)$:

$$P(\mathbf{a};l) = 1 - c'\Big[\lambda L_c d(1+\kappa) \int_0^{L_c} dl' \int da' \big|\sin \gamma(\mathbf{a},\mathbf{a}')\big| f(\mathbf{a}';l')$$

$$+ \tfrac{\pi}{4} d^2 L_c (1+\kappa)^2(1+\lambda) + \tfrac{\pi}{6} d^3 (1+\kappa)^3 \Big] \qquad (\lambda, \kappa \ll 1) \tag{3.5}$$

On the other hand, when λ and κ are much larger than unity, the scaled particle becomes much greater than the other particles in the solution and the latters can be regarded as a continuous medium. In this case, W is equal to the mechanical work necessary to produce a (macroscopic) void with the same size as the scaled particle against the osmotic pressure Π, so that it is given by

$$W = \Pi\Big[\tfrac{\pi}{4} \lambda L_c (\kappa d)^2 + \tfrac{\pi}{6}(\kappa d)^3 \Big] \qquad (\lambda, \kappa \gg 1) \tag{3.6}$$

Actually we need the expression of W (or $P(\mathbf{a};l)$) at $\lambda = \kappa = 1$. According to Cotter [4], we use the following interpolation formula:

$$W = \frac{k_BT}{L} \int_0^{L_c} dl \int da f(\mathbf{a};l) \ln P(\mathbf{a};l)$$

$$= C_{00} + C_{10}\lambda + C_{01}\kappa + C_{11}\lambda\kappa + C_{02}\kappa^2 + \Pi\Big[\big(\tfrac{\pi}{4}L_c d^2\big)\lambda\kappa^2 + \big(\tfrac{\pi}{6}d^3\big)\kappa^3\Big] \tag{3.7}$$

The coefficients C_{ij} are determined from the conditions that Eq.(3.7) reduces to Eq.(3.5) at small λ and κ limit, and W for the real system is obtained from Eq.(3.7) with $\lambda = \kappa = 1$. Using this result along with Eq.(3.4) and some thermodynamic relations, we finally obtain

$$S_{\text{tr}} = -nk_B\left[\text{const.} + \ln\left(\frac{c'}{1-vc'}\right) + \frac{Bc'}{2(1-vc')} + \frac{Cc'^2}{3(1-vc')^2} \right] \tag{3.8}$$

where

$$B = 6v + 2b\rho, \quad C = 4v'v'' + 2v'b\rho \tag{3.9}$$

with

$$b \equiv \tfrac{\pi}{4} L_c^2 d, \quad v \equiv \tfrac{\pi}{4}\Big(L_c d^2 + \tfrac{2}{3} d^3\Big), \quad v' \equiv v + \tfrac{\pi}{12} d^3, \quad v'' \equiv v - \tfrac{\pi}{24} d^3 \tag{3.10}$$

and

$$\rho \equiv \frac{4}{\pi} \iint |\sin \gamma| \bar{f}(a) \bar{f}(a') da da' \tag{3.11}$$

To obtain this result, the distribution function $f(a;l)$ was approximated by the distribution function $\bar{f}(a)$ averaged along the chain contour.

Assuming $\bar{f}(a)$ to be expressed by the Onsager trial function, we calculated numerically the isotropic–liquid crystal phase boundary concentrations using Eq.(3.2) with Eq.(3.8); for S_{conf} we used DuPré and Yang's interpolation formula [28]. The results are represented by the empirical equations given by

$$\frac{v \, c'_v}{\bar{d} a_v(\bar{d})} = \frac{A_v + N}{N} + S_v(N) \log\left(\bar{d}/\bar{d}_{0v}\right) \qquad (v = \text{I and A}) \qquad \text{(SPT)} \tag{3.12}$$

where $\bar{d} \equiv d/2q$ and $a_v(\bar{d})$ and $S_v(N)$ are functions of $\bar{d}$ and N, respectively, defined by

$$\begin{cases} a_v(\bar{d}) = a_{0v} + a_{1v} \log \bar{d} + a_{2v}\left(\log \bar{d}\right)^2 + a_{3v}\left(\log \bar{d}\right)^3 \\ S_v(N) = \left[S_{1v} + \left(S_{2v} + S_{1v} k_v^\circ\right)N\right] / N\left(1 + k_v^\circ N\right) \end{cases} \tag{3.13}$$

The numerical coefficients in these equations are given in Table 2. The deviation of Eq.(3.12) from the exact numerical calculation results is within 3% at $N \geq 0.15$ and $\bar{d} \leq 0.1$, and within 9% at $N \geq 0.05$ and $\bar{d} \leq 0.006$.

3.3 Comparison with Experiment

Fig. 3 shows the plot of $\left[v \, c'_I / \bar{d} a_I(\bar{d})\right] - S_I(N) \log\left(\bar{d}/\bar{d}_{0I}\right)$ against N for liquid crystalline polymers with different stiffness. Since the values of d and N for each polymer were estimated from the partial specific volume and the molecular weight, respectively, we have used no adjustable parameters in this figure. The data points for all the systems quite nicely follow the universal curve given by Eq.(3.12). When compared with Fig. 2, the agreement between theory and experiment is much improved for more flexible polymers (PHIC in DCM and the cellulose derivatives) in this figure.

Table 2. Numerical Parameters Appearing in Equations (3.12) and (3.13)

v	$\bar{d}_{0v}$	A_v	S_{1v}	S_{2v}	$k^\circ{}_v$	a_{0v}	a_{1v}	a_{2v}	a_{3v}
I	0.0343	0.176	−0.0797	0.0688	1.34	−3.603	−8.738	−0.8062	0.1854
A	0.0851	0.127	−0.0270	0.0743	1.26	−3.602	−8.318	0.0585	0.3733

The same plot for the another phase boundary concentration c_A' is shown in Fig. 4. The data points do not largely deviate from the theoretical curve drawn with Eq.(3.12), but the agreement seems not to be as good as in Fig. 3 especially for the solutions of PBLG and APC.

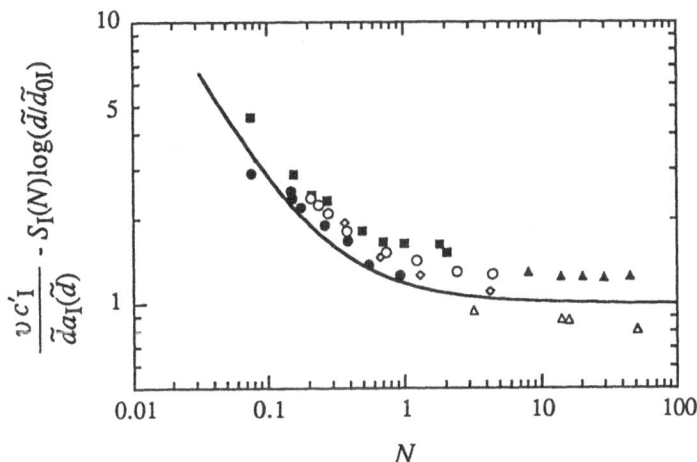

Fig. 3. Comparison of the scaled particle theory for wormlike hard spherocylinders with experimental results for the phase boundary concentration c_I': ($\bullet$) schizophyllan–water [12,13]; ($\blacksquare$) PBLG–DMF [14-19]; ($\bigcirc$)PHIC–toluene [20]; ($\diamondsuit$) PHIC–DCM [20]; ($\triangle$) HPC–water [21]; ($\blacktriangle$) APC–DBP [22].

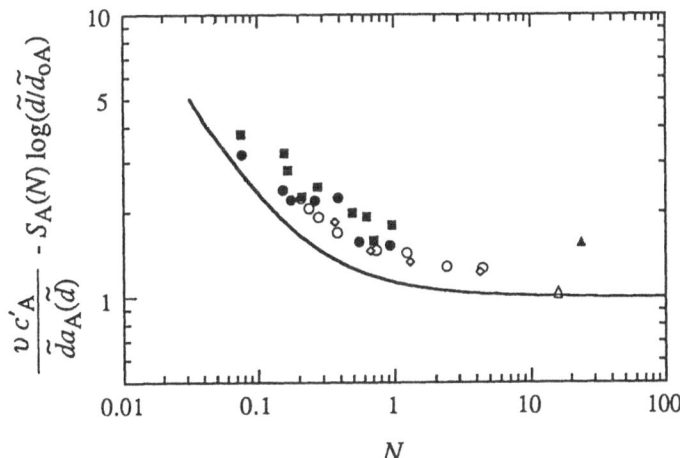

Fig. 4. The same comparison as in Fig. 3 for the phase boundary concentration c_A'.

The osmotic pressure Π and the orientational order parameter S are basic thermodynamic quantities for liquid crystalline polymer solutions, and can be used for a critical test of the scaled particle theory for wormlike hard spherocylinders. Both quantities can be calculated from F of this theory given by Eq.(3.2) along with Eqs.(3.1) and (3.8). The data points in Figs. 5 and 6 are the experimental results of Π and S for PBLG–DMF solutions obtained by Kubo and Ogino [17,18] and Abe and Yamazaki [29,30], respectively. If the values of d necessary to calculate S_{tr} in Eq.(3.8) are chosen to be 1.4 nm for Π and 1.5 nm for S, both experimental results can be successfully reproduced by the scaled particle theory for monodisperse solutes. Since the values of d used are close to that estimated from the partial specific volume (1.56 nm), the present theory describes consistently Π, S and the phase boundary concentrations for the PBLG–DMF system. Similar results has been obtained for other systems (e.g., PHIC and schizophyllan solutions [10]).

However the present scaled particle theory has its own limitations. For example, let us consider the typical flexible polymer, polystyrene. We know that the persistence length of this polymer is about 1 nm, and its diameter is 0.8 nm. If these parameters are input into the scaled particle theory for wormlike hard spherocylinders, the theory predicts for the polystyrene solution

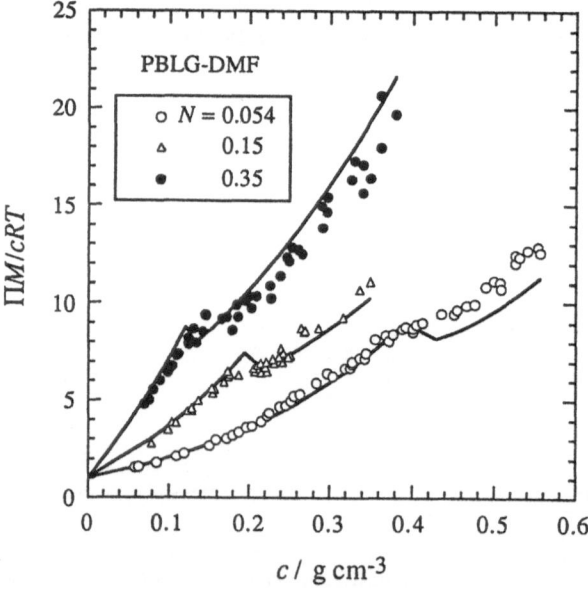

Fig. 5. Concentration dependence of the osmotic pressure for PBLG–DMF solutions; circles and triangles, data obtained by Kubo and Ogino [17,18]; curves, the theoretical results of the scaled particle theory for wormlike hard spherocylinders.

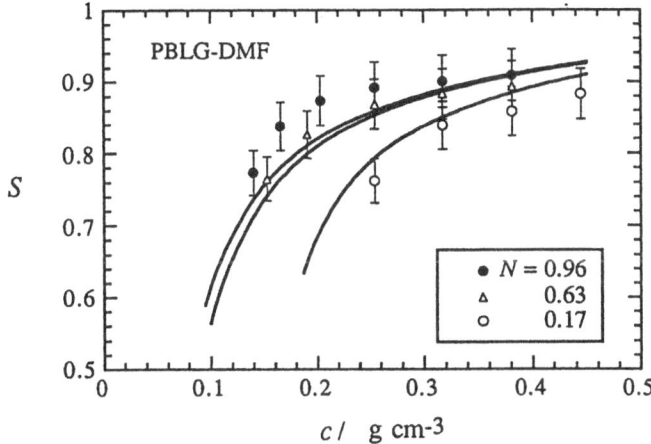

Fig. 6. Concentration dependence of the orientational order parameter for PBLG–DMF solutions; circles and triangles, data obtained by Abe and Yamazaki [29,30]; curves, the theoretical results of the scaled particle theory for wormlike hard spherocylinders.

to form a nematic phase at the polymer volume fraction above ca. 70 % for N larger than 20. Of course this prediction is wrong. Experimentally we have never observed the liquid crystalline state in any flexible polymer solutions. The wrong prediction comes from the interpolation formula used in the scaled particle theory. This formula is just the same as used by Cotter [4] for straight spherocylinders, and the final expression for S_{tr} has the same functional form as Cotter's one for the straight spherocylinder with the same contour length. However when q of the polymer is the same order as its d, we have to consider multiple contacts between pairs of different chains, which must change the expression of S_{tr} from that of the straight spherocylinder. Therefore the present scaled particle theory should be applied to the polymer with q much larger than d. From Fig. 3, it turns out that the minimum ratio of q to d where the scaled particle theory should be valid is about five (cf. Table 1).

4 POLYDISPERSITY AND ELECTROSTATIC INTERACTION EFFECTS

So far we have discussed quasi-binary solutions with one neutral stiff polymer sample and a good solvent. In this section we are concerned with more complex systems: quasi-ternary solutions consisting of two different molecular weight samples of a (neutral) polymer and a good solvent, and solutions with one stiff polyelectrolyte sample and aqueous salt.

The scaled particle theory described in the previous section can be straightforwardly extended to multicomponent systems with different wormlike hard spherocylinder species [31]. For a ternary solution containing two wormlike hard spherocylinder species 1 and 2 with the same d

and q, but different cylinder lengths ($L_{c,1}$ and $L_{c,2}$), the scaled particle theory gives the following expression of S_{tr}

$$S_{tr} = -nk_B \left[\text{const.} + \sum_{s=1}^{2} x_s \ln x_s + \ln\left(\frac{c'}{1 - \bar{v}c'}\right) + \frac{\bar{B}c'}{2(1 - \bar{v}c')} + \frac{\bar{C}c'^2}{3(1 - \bar{v}c')^2} \right] \tag{4.1}$$

with

$$\bar{B} = 6\bar{v} + 2b_1\bar{\rho}, \quad \bar{C} = 4\bar{v'}\bar{v''} + 2\bar{v'}b_1\bar{\rho}, \quad \bar{\rho} = x_1^2\rho_{11} + 2x_1x_2X'\rho_{12} + x_2^2X'^2\rho_{22} \tag{4.2}$$

where c' is the total polymer number density, x_s the mole fraction of species s, $b_1 \equiv \frac{\pi}{4}L_{c,1}^2 d$, $X' \equiv L_{c2}/L_{c1}$, and ρ_{ij} is calculated from Eq.(3.11) with $\bar{f}_i(\mathbf{a})$ and $\bar{f}_j(\mathbf{a'})$; the bar symbols attached to v, v', and v'' indicate the number averages of these volumes (cf. Eq.(3.10)).

Figure 7a compares an experimental ternary phase diagram for aqueous schizophyllan [13,32] with the theoretical phase diagram calculated from the scaled particle theory (Eq.(4.1)). Here c is the total polymer mass concentration and ξ the weight fraction of the shorter chain in the total polymer. With respect to the isotropic–liquid crystal (cholesteric) binodals and the tie lines, the agreement between the experiment and theory is almost quantitative for the hard-core diameter of 1.52 nm which is close to 1.68 nm estimated from the partial specific volume. However the scaled particle theory cannot predict the three phase coexistence of one isotropic and two liquid crystal phases as well as two liquid crystal phase coexistence, both of which were found experimentally (cf. the shadowed region and two squares in Figure 7a). Since the triphasic equilibrium must be determined by very delicate balances of the osmotic pressure and chemical potentials among the three coexisting phases, the failure to predict the three phase coexistence may be ascribed to a subtle difference in F between the real system and the theory. Indeed if we change the Kuhn segment number of the lower molecular weight sample from 0.0765 to 0.070, the scaled particle theory succeeds to predict the three phase separation as well as the two liquid crystal phase separation as shown in Figure 7b.

The present scaled particle theory can be also extended to charged stiff-chain polymer systems [33]. Let us consider polyelectrolytes interacting with each other by the pair potential u of the mean field

$$u = u_0 + w \tag{4.3}$$

where the u_0 is the hard-core potential and w is the electrostatic potential given by

$$u_0 = \begin{cases} \infty & \text{(the hard cores overlap)} \\ 0 & \text{(otherwise)} \end{cases} \quad w = \begin{cases} 0 & \text{(the hard cores overlap)} \\ w(r,\gamma) & \text{(otherwise)} \end{cases} \tag{4.4}$$

with r and γ being the distance and angle between the closest chain axes of two interacting polyelectrolytes. The electrostatic potential $w(r,\gamma)$ can be calculated from the Poisson–Boltzmann equation for charged cylinders [34,35].

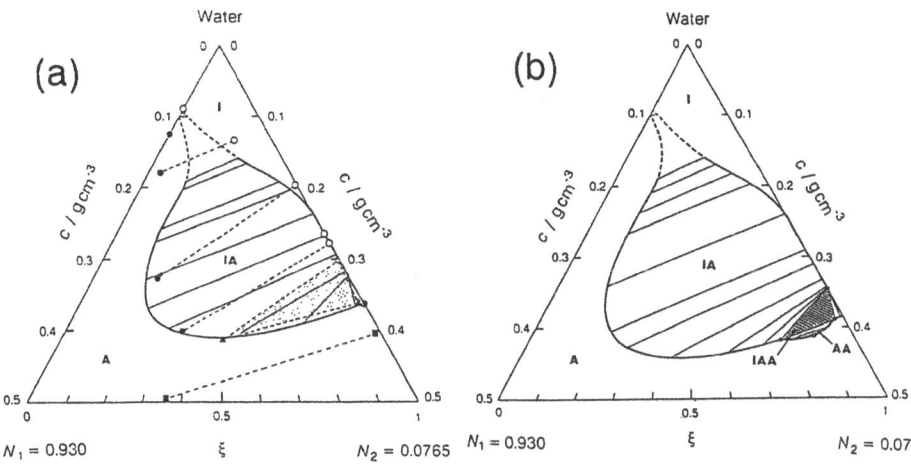

Fig. 7. (a) Ternary phase diagram for the aqueous solution containing two schizophyllan samples with $N = 0.93$ and 0.0765 at $25°C$; circles, the isotropic–cholesteric binodal points; shadowed triangular region, isotropic–cholesteric–cholesteric triphasic region; squares, cholesteric–cholesteric binodal points; the solid curves and thin segments, theoretical binodals and tie lines calculated from the scaled particle theory for wormlike hard spherocylinders using $d = 1.52$ nm. (b) Theoretical phase diagram for the ternary system with $N = 0.93$ and 0.070.

Now we choose the system without the potential w as the reference system, and regard the potential w as a thermodynamic perturbation. If the potential w is not very strong, and also the polymer concentration of the solution is not so high, the free energy F of the solution can be calculated in a perturbative way, and higher perturbation terms can be neglected. Considering only first perturbation term, we obtain F in the form [33]

$$F = -TS_{conf} - TS_{tr}|_{w=0} - \frac{1}{2}\langle\beta_{1,w}\rangle c' \tag{4.5}$$

Here $S_{tr}|_{w=0}$ is the translational entropy for the reference system which is given by Eq.(3.8), and $\langle\beta_{1,w}\rangle$ the binary cluster integral for the potential w given as an analytical function of the polymer charge density and the added salt concentration (or the Debye screening length) [33]. Equation (4.5) contains all the virial terms for the hard-core potential, but only the second virial term for the potential w.

Figure 8 shows the isotropic–liquid crystal phase diagram for aqueous solutions of a charged double helical polysaccharide xanthan with sodium chloride as an added salt [36,37]. The experimental phase boundary (mass) concentrations c_I and c_A of this system (circles) decrease with decreasing the added salt concentration. This demonstrates an important role of the electrostatic interaction in the isotropic–liquid crystal phase equilibrium in this system. The solid and dashed curves in the figure represent the theoretical results for the two phase boundary concentrations calculated from the above perturbation theory. The theory contains three molecular parameters: d, q, and the polymer charge density ν. The theoretical curves in the figure were calculated using these parameters estimated from dilute solution studies [38,39] and the molecular model of the xanthan double helix [40], and therefore they contain no fitting parameters. We obtain a good agreement between the theory and experiment except in the low molecular weight region at $C_s = 0.01$ M. We have also shown that the C_s dependence of c_I and c_A for a given sample in a wider range of C_s is successfully explained by the same theory [33].

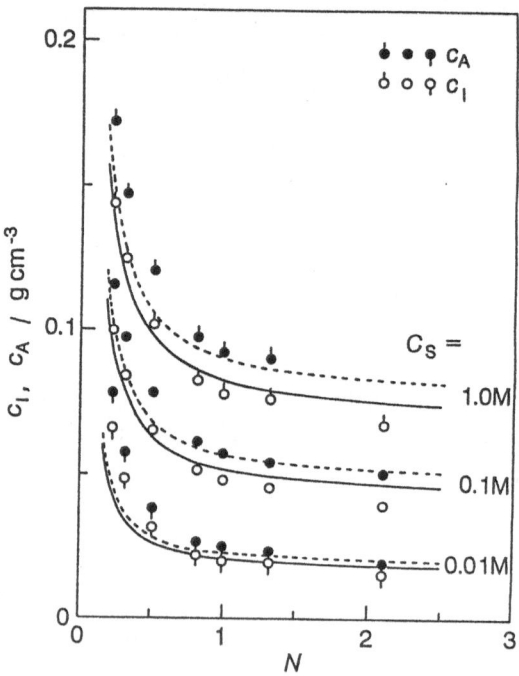

Fig. 8. Phase boundary concentrations for solutions of an ionic double helical polysaccharide xanthan and aqueous sodium chloride; circles, experimental data [36,37]; curves, the perturbation theory [33] with $d = 2.2$ nm $q = 120$ nm, and $\nu = 3.0$ (elementary charges)/nm.

REFERENCES

1 Onsager L (1949) Ann. N.Y. Acad. Sci. 51: 627
2 Flory PJ (1956) Proc. Roy. Soc. (London) A234: 73
3 Cotter MA (1974) Phys. Rev. A 10: 625
4 Cotter MA (1977) J. Chem. Phys. 66: 1098
5 Cotter MA, Martire DE (1970) J. Chem. Phys. 53: 4500

6 Cotter MA, Martire DE (1970) J. Chem. Phys. 52: 1909

7 Lasher G (1970) J. Chem. Phys. 53: 4141

8 Khokhlov AR, Semenov AN (1981) Physica 108A: 546

9 Khokhlov AR, Semenov AN (1982) Physica 112A: 605

10 Sato T, Teramoto A (1990) Mol. Cryst. Liq. Cryst. 178: 143

11 Flory PJ (1953) Principles of Polymer Chemistry, Cornell Univ. Press, Ithaca, New York

12 Itou T, Van K, Teramoto A (1985) J. Appl. Polym. Sci., Appl. Polym. Symp. 41: 35

13 Itou T, Teramoto A (1984) Polym. J. 16: 779

14 Wee EL, Miller WG (1971) J. Phys. Chem. 75: 1446

15 Miller WG, Rai JH, Wee EL (1974) Liquid Crystals and Ordered Fluids 2: 243

16 Miller WG, Wu CC, Wee EL, Santee GL, Rai JH, Goebel KG (1974) Pure Appl. Chem. 38: 37

17 Kubo K, Ogino K (1979) Mol. Cryst. Liq. Cryts. 53: 207

18 Kubo K (1981) Mol. Cryst. Liq. Cryst. 74: 71

19 Itou T, Funada S, Shibuya F, Teramoto A (1986) Kobunshi Ronbunshu 43: 191

20 Itou T, Teramoto A (1988) Macromolecules 21: 2225

21 Conio G, Bianchi E, Ciferri A, Tealdi A, Aden MA (1983) Macromolecules 16: 1264

22 Laivins GV, Gray GD (1985) Macromolecules 18: 1753

23 Yanaki T, Norisuye T, Hiroshi F (1980) Macromolecules 13: 1462

24 Itou S, Nishioka N, Norisuye T, Teramoto A (1981) Macromolecules 14: 904

25 Itou T, Chikiri H, Teramoto A, Aharoni SM (1988) Polym. J. 20: 143

26 Lifshitz IM (1969) Soviet Phys. JETP 28: 1280

27 Lifshitz IM, Grosberg AY, Khokhlov AR (1978) Rev. Mod. Phys. 50: 684

28 DuPré DB, Yang S (1991) J. Chem. Phys. 94: 7466

29 Yamazaki T, Abe A (1987) Polym. J. 19: 777

30 Abe A, Yamazaki T (1989) Macromolecules 22: 2145

31 Sato T, Shoda T, Teramoto A (1993) to be submitted to Macromolecules

32 Itou T, Teramoto A (1984) Macromolecules 17: 1419

33 Sato T, Teramoto A (1991) Physica A176: 72

34 Stroobants A, Lekkerkerker HNW, Odijk T (1986) Macromolecules 19: 2232

35 Philip JR, Wooding RA (1970) J. Chem. Phys. 52: 953

36 Inatomi S, Sato T, Teramoto A (1992) Macromolecules 25: 5013

37 Sato T, Kakihara T, Teramoto A (1990) Polymer 31: 824

38 Sato T, Norisuye T, Fujita H (1984) Macromolecules 17: 2696

39 Sho T, Sato T, Norisuye T (1986) Biophys. Chem. 25: 307

40 Okuyama K, Arnott S, Moorhouse R, Walkinshaw MD, Atkins EDT, Wolf-Ullish C (1980) In: French AD, Gardner KH (Eds) Fiber Diffraction Methods, American Chemical Society, Washington DC

Light Scattering Studies of a Nematic to Smectic - A Phase Transition in Rigid Rod Polymer Solutions

Jin-Hua Wang, Franklin Lonberg, Xiaolei Ao and Robert B. Meyer

Martin Fisher School of Physics, Brandeis University, Waltham, MA 02254

Section 1. Introduction

An important approach to understanding phase transitions in liquid crystals is to study systems in which the dominant interparticle interactions are hard repulsion, rather than the complicated interparticle potentials found in most liquid crystals. Computer simulations have shown [1] that a nematic to smectic A phase transition may exist in a system as simple as monodisperse hard rods. Experimentally, Xin Wen, Robert Meyer and Donald Caspar [2] have reported the observation of the smectic A ordering in tobacco mosaic virus (TMV) suspensions, in which the dominant interparticle interactions are screened electrostatic repulsion. In this paper we present the first quantitative measurements of the nature of the nematic to smectic A phase transition in TMV. In solutions in which the concentration is just below the transition concentration to the smectic phase, large pretransitional smectic fluctuations are observed. We will describe the methods of measuring the correlation lengths of these presmectic fluctuations by light scattering. In addition, the concentration difference between the smectic and the nematic phase close to the transition point is determined to be very small, implying that the transition is very close to second order. This result is consistent with the conclusion of numerous theoretical approaches [3], such as density functional theory and scale particle theory, which have concluded that a nematic to smectic A transition in a system of rigid rods should be second order or very close to second order.

Tobacco mosaic virus (TMV) is a rod-shaped macromolecule of length 3000 Å, diameter 180 Å [4]. The TMV virus, in a neutral or slightly alkaline buffer, has several thousand net negative charges on its surface. Positive ions in the solution form a screening layer whose

A. Teramoto, M. Kobayashi, T. Norisuje (Eds.)
Ordering in Macromolecular Systems
© Springer-Verlag Berlin Heidelberg 1994

thickness depends on the ionic strength of the solution. At the ionic strength of the solutions discussed in this paper, screened electrostatic repulsion is much stronger than the Van der Waals attractive force. Because the interparticle interaction is purely repulsive, the phase transitions are driven by a maximization of entropy alone. In phases of higher order the mobility of the particles is greater than the mobility in the phase of lower order, and thus the total entropy of the system is higher[2]. If one has two solutions of equal concentration, and one solution has a higher ionic strength, that solution will have a shorter Debye screening length and is experimentally observed to be less ordered. The lower order in the shorter screening length solution is seen as an evidence of the key role played by excluded volume effects determining the order of the phase.

In this experiment, all samples are TMV suspensions in the 50 mM sodium borate, PH 8.5 buffer. The virus particles used are monodisperse viruses purified from infected tobacco plants by Marguerite Cahoon. The samples are prepared by mixing 50 mM sodium borate buffer solution with 22 % TMV stock suspensions in the same buffer. The samples are then put in fused quartz capillaries of diameters about 0.7 mm. The capillaries are sealed using a high temperature oxygen-gas flame. After being aligned in a 10k guass magnetic field, the samples are stored vertically so that any gravitationally caused concentration gradient will be in the direction of the long axes of the capillaries. The TMV concentration of the samples is controlled to the accuracy of 5 % during sample preparation. The phase diagram of these TMV solutions is shown in Figure 1.

Isotropic $\xrightarrow{\hspace{1cm}}$ 90 mg/ml $\xrightarrow{\text{Nematic}}$ 172 mg/ml $\xrightarrow{\text{Smectic A}}$ 180 mg/ml $\xrightarrow{\text{Crystal}}$

Figure 1.

Phase diagram of TMV suspension in the 50 mM, pH 8.5 sodium borate buffer.

Section 2. Light Scattering Setup

Because the TMV smectic A layer spacing is about 3400 Å, laser light scattering is used as a probe of the smectic ordering. When the smectic phase of TMV is illuminated by white light, one sees strong iridescence caused by the Bragg scattering from the smectic layers. When an aligned smectic phase is illuminated with a laser, a very sharp Bragg scattering peak is observed. In the nematic phase close to the concentration at which the smectic phase will appear, the Bragg peak becomes a diffuse cone of scattered light. By quantitatively measuring the intensity distribution of this diffuse scattering, one can determine the dimensions of the presmectic fluctuations.

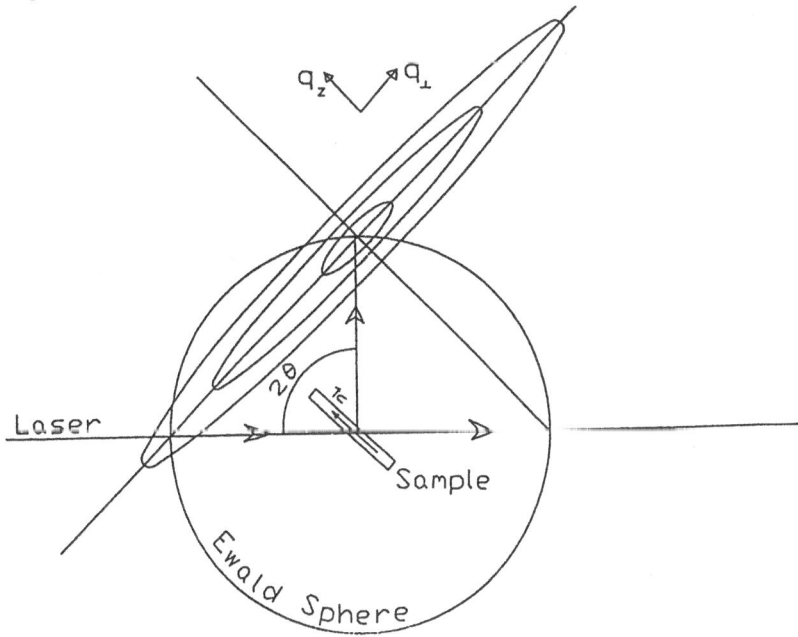

Figure 2.

A q-space representation of the light scattering geometry. The ellipses are constant intensity contours of the scattered light in q-space.

The crucial difference between the smectic A phase and the pretransitional nematic is that in the former one has quasi long rang order in the direction perpendicular to the smectic planes, and in the latter one has only finite size layered droplet-like fluctuations. The laser diffraction pattern to be expected from a nematic solution with presmectic fluctuations is understood by considering the Fourier transform of a periodic system of finite size. While the Fourier transform of an infinite periodic system is a delta function, the Fourier transform of a system with a finite fluctuation will be a diffuse peak. For a finite size elongated droplet composed of smectic layers, the Fourier transform will be a diffuse disc in reciprocal space. The intercept of this disc with the Ewald sphere gives a ring of scattered light. Measurement of the intensity profile along the ring gives the correlation lengths of presmectic fluctuations.

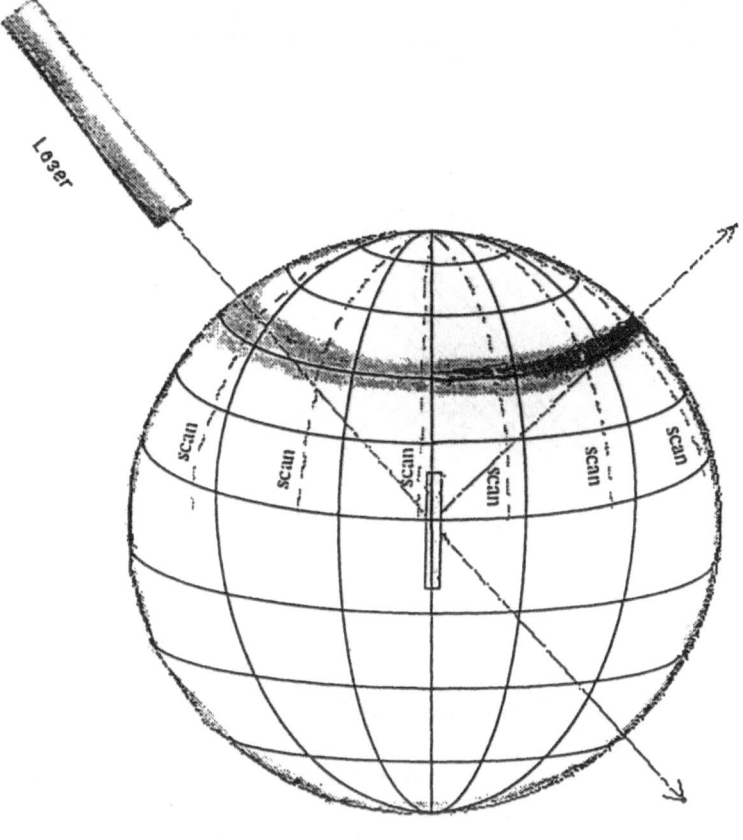

Figure 3.

The three dimensional light scattering geometry. Weakly iridescent samples produce diffusive scattering rings.

The motivation for the design of the actual experiment can be seen from Figure 3. If a smectic TMV sample were enclosed in a frosted glass sphere, and illuminated by a laser, one would see a sharp peak of the scattered light. If the sample were just below the smectic concentration, the sharp smectic peak would spread out into the diffuse "ring" shown in Figure 3. The experimental apparatus [5] is designed to measure this diffuse ring, which means that the light detector must move in two angular dimensions in order to cover the whole solid angle subtended by the ring of scattered light. The apparatus consists essentially of a detector which is mounted on an arm connected to two orthogonal rotation stages. Thus the detector can actually move over the surface of a sphere. In order to minimize the reflective distortion of the incident and scattered light, the sample and the detector are immersed in water whose index of refraction is close to that of the TMV solution.

Section 3. Experimental Results

In the geometry of Figure 3, the sample capillary points toward the pole of the sphere. By scanning along a set of longitudes, we have obtained the data shown in Figure 4. The zero longitude scan passes through the point of maximum intensity. In order to extract the physical content from the data, the intensity distribution along the ring is mapped into a two-dimensional q space. Figure 5 shows the same data now plotted as a function of the two components of the scattered wave vector. $q_\perp$ is the component perpendicular to the director of the liquid crystal, and $q_\parallel$ is the component parallel to the director. The contours of equal intensity are approximately elliptical, with small distortions due to misalignment of the light scattering system. The elliptical shapes of these contours, two of which are shown in Figure 4, reflect the ellipsoidal shapes of the presmectic droplets.

The scattering volumes viewed by the detector are different for different scattering angles. The raw data should, in principle, be corrected for different scattering volumes. However, such correction is found insignificant for the angle range presented here. The instrumental resolution function has been measured separately and is not broad enough to be significant for this data.

The determination of the correlation lengths is accomplished by fitting the raw intensity distribution data to the same empirical expression used to fit the x-ray scattering data to the correlation lengths in thermotropic liquid crystals [6].

$$I(q_{\parallel}, q_{\perp}) = \frac{I_0}{1 + \xi_{\parallel}^2(q_{\parallel} - q_0)^2 + \xi_{\perp}^2 q_{\perp}^2(1 + c\xi_{\perp}^2 q_{\perp}^2)}$$

This fitting process gives $\xi_{\parallel} = 10.6\ \mu m$, $\xi_{\perp} = 1.6\ \mu m$, $\xi_{\parallel}/\xi_{\perp} = 6.1$, $2\pi/q_0 = 3540$ Å and c is a small number (about 0.00015), which improves the fit; the profile is essentially lorenzian.

For any hard particle fluid, a first order phase transition is characterized by a finite discontinuity of fluid density and a second order phase transition by continuous change in density. Thus measuring the discontinuity in density is a direct way to determine the order of a phase transition. It is difficult to measure directly the concentration of a TMV solution with high precision. However, the birefringence of a well aligned TMV sample can be measured to an experimental accuracy about 1 %. For a well aligned TMV sample, the birefringence, Δn, is related to the concentration, c, by $\Delta n = \Delta n_0 \cdot c \cdot S(c)$, where S(c) is the orientational order parameter and Δn_0 is the specific birefringence constant which is about 1.94×10^{-5} ml/mg for a TMV nematic phase.

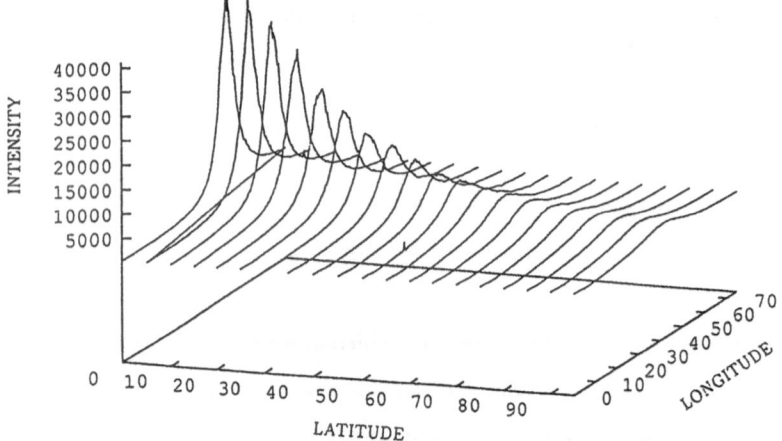

Figure 4.

The intensity distribution along the ring.

We have one sample which has a region of nematic phase in the top of the sample while the bottom fraction is in the smectic phase. There is no sharp boundary between the two phases, which suggests that the refraction index of the two phases are very close to each other. The measured birefringence of the two phases is the same to within 1 % experimental error. These two observations imply that the concentration difference between these two phase is very small. Thus the nematic to smectic phase transition in TMV suspension is very close to second order.

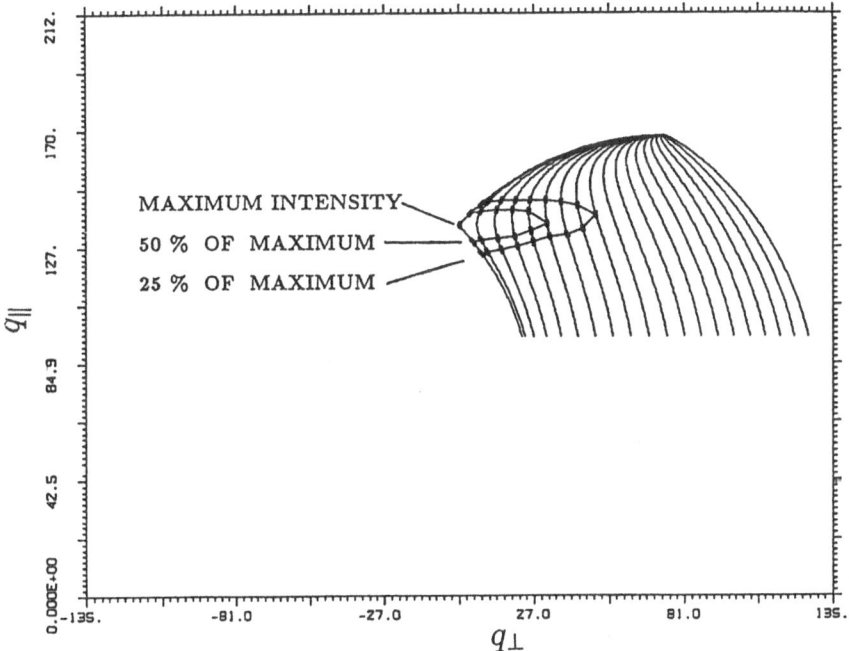

Figure 5.

The intensity distribution along the ring mapped into two dimensional q space. The 50% and the 25% intensity contours are close to ellipses, which is a characteristic feature of pre-smectic A fluctuations.

4. Conclusions and Further Experiments

1) The nematic to smectic A phase transition in rigid rod polymer solutions is very close to second order.

2) The pre-smectic A fluctuations in the TMV nematic phase of rigid rod macromolecules can be described by the same formalism used for the low molecular weight liquid crystals. This is an important result since the TMV particles are two orders of magnitude larger than the components of low molecular weight liquid crystals, and also the phase transition is driven by concentration instead of temperature.

3) Measurements of the dependence of the correlation lengths on concentrations by light scattering and measurements of the orientational order parameters by x-ray diffraction, are now in progress.

4) Because we use light scattering to investigate the nematic to smectic A phase transition, we can, in principle, and in fact are now measuring the *dynamics* of the pre-smectic fluctuations as well as the static correlation lengths.

Finally we wish to thank Seth Fraden and Donald Caspar for a series of useful discussions. This work is supported by a grant from the National Science Foundation through grant DMR-8803582, and by the Martin Fisher School of Physics of Brandeis University.

References

[1] Frenkel D, Lekkerkerker HNW, Stroobants A (1988) Nature 332:822

[2] Wen X, Meyer RB, Caspar DLD (1989) Phys Rev Lett 63:2760

[3] Vroege GJ, Lekkerkerker HNW (1992) Rep Prog Phys 55:1241

[4] Caspar DLD (1963) Adv Protein Chem 18:37

[5] Wen X, Schnitzer MJ, Meyer RB (1990) Rev Sci Intrum 61:2069

[6] Pershań PS (1988) Structure of liquid crystal phases. World Scientific Pub Co, Singapore, Teaneck NJ

Collective and Molecular Dynamics in Ferroelectric Liquid Crystals: From Low Molar to Polymeric and Elastomeric Compounds

F. Kremer[a], A. Schönfeld[b]

[a]Universität Leipzig, Fachbereich Physik, Linnéstraße 5, D-04103 Leipzig, FRG
[b] Max Planck Institut für Polymerforschung, Ackermannweg 10, D-55128 Mainz, FRG

Abstract

Broadband dielectric spectroscopy (10^{-2}Hz - 10^{10}Hz) is employed to analyse the collective and molecular dynamics in low molar weight and polymeric ferroelectric liquid crystals (FLC). Below 10^6Hz two collective dielectric relaxation processes are observed having the character of a soft - mode and that of a Goldstone - mode. The soft - mode is assigned to fluctuations of the amplitude of the helical superstructure, while the Goldstone - mode corresponds to fluctuations of the phase of the ferroelectric helix. In the frequency range between 10^6Hz and 10^{10}Hz <u>one</u> dielectric relaxation process is observed, the β - relaxation, which is assigned to the libration (hindered rotation) of the mesogenes around their long molecular axis. At the phase transition smectic A / smectic C* this process does not show any deviations from an Arrhenius - like temperature dependence. In comparing low molecular mass and the polymeric side group FLC, the collective and molecular dynamics is found to be qualitatively similar. This open new possible applications for FLC as sensors, actuators or (polymeric) flexible displays.

Introduction

Broadband dielectric spectroscopy (10^{-2}Hz - 10^{10}Hz) is a powerfull tool to study the collective and molecular dynamics in ferroelectric liquid crystals (FLC). In comparing low molecular weight and polymeric side - group systems qualitatively similar properties are observed. Hence it is possible to combine the electrical and dielectric properties of ferroelectric liquid crystals with that of side group polymers. This opens new fields of possible applications (e. g. flexible displays, piezoelectric sensors, etc.).

Experimental

To cover the entire frequency range from 10^{-2}Hz to 10^{10}Hz four different measurement systems were employed, (i) a frequency response analyser Solartron Schlumberger FRA1260 with a high impedance preamplifier of variable gain (10^{-2}Hz to 10^6Hz), (ii) a Hewlett Packard Impedance analyser HP4192A (10Hz to 10^7Hz), (iii) a Hewlett Packard coaxial line reflectometer HP4191A (10^6Hz to 10^9Hz) and (iv) a Hewlett

A. Teramoto, M. Kobayashi, T. Norisuje (Eds.)
Ordering in Macromolecular Systems
© Springer-Verlag Berlin Heidelberg 1994

Packard Network analysator HP8510 ($4.5 \cdot 10^7$Hz to 10^{10}Hz). Rubbed, polyimide coated metal electrodes ($\emptyset = 3$ - 5 mm, spacing: 20μm) were employed. They could be used over the whole frequency regime from 10^{-2}Hz to 10^{10}Hz. The orientation of the bookshelf geometry [1] was checked by measurements of the saturation polarization in the identical sample capacitor.

The data were quantitatively analyzed by fitting the experimental data using the generalized relaxation function according to Havriliak and Negami [2],

$$\epsilon^* = \epsilon_\infty + \frac{\Delta\epsilon}{(1 + (i\omega\tau)^\alpha)^\gamma}$$

where ϵ_∞ and ϵ_S describe the real part of the dielectric function for $\omega \gg \frac{1}{\tau}$ and $\omega \ll \frac{1}{\tau}$, respectively.

As low molecular weight FLC

k 322 S$_C^*$ 328 S$_A$ 338 is.

was used, as polymeric system the following sidechain polysiloxane was choosen.

g 0 S$_X$ 46 S$_C^*$ 98 S$_A$ 144 is.

The phase sequence was determined by polarizing light microscopy and by differential scanning calorimetry.

Results and Discussion

The dielectric spectrum splits up in two different dynamical regimes (fig.1) [3]. Below 1MHz two processes (Goldstone - and soft - mode) can be observed. Their huge dielectric strength indicate the collective character of these relaxations. The Goldstone - mode is assigned to the fluctuation of the phase of the helical superstructure. It is restricted to the ferroelectric Sc* - phase (fig.2). As it is possible to unwind the helical superstructure by applying a DC - bias field, the Goldstone - mode can be supressed continuously. In the unwound state the soft - mode becomes observable again (without DC - bias field it is covered by the much larger Goldstone - mode). The soft - mode is assigned to the fluctuation of the amplitude of the helical superstructure (tilt -

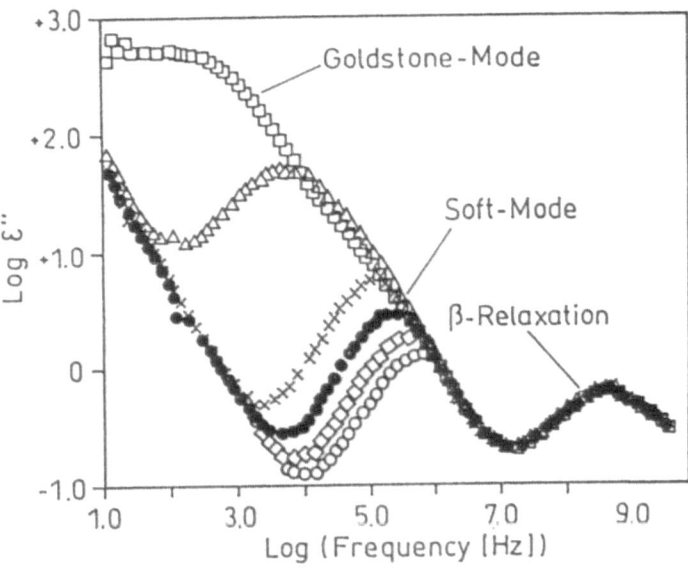

Fig. 1: *Goldstone - mode, soft - mode and β - relaxation near the phase transition S_A/S_C^* . (○) = 331.3K, (◇) = 330.4K, (•) = 329.4K, (×) = 328.5K, (△) = 327.5K, (□) = 326.5K ·*

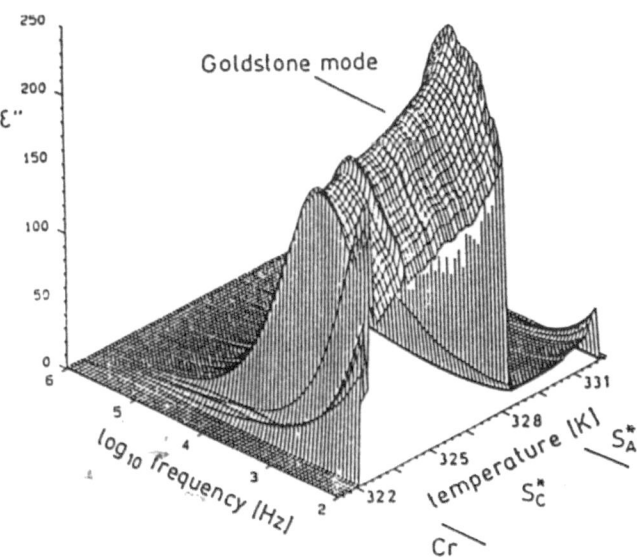

Fig. 2: *Dielectric loss ϵ'' vs. frequency and temperature; d.c. field strength = 0V/cm, a.c. field strength = 0.1kV/cm.*

184

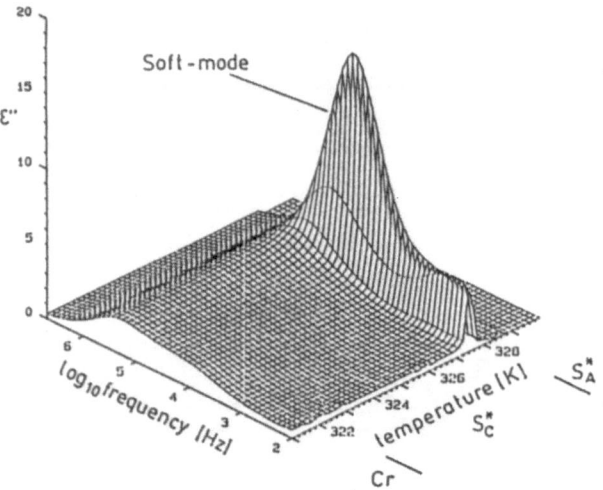

Fig. 3: *Dielectric loss ϵ'' vs. frequency and temperature; d.c. field strength = 8kV/cm, a.c. field strength = 0.1kV/cm.*

fluctuation). It occurs both in the S_A - phase and in the S_C^* - phase (fig.3). Cooling down in the S_A - phase the soft - mode increases in its dielectric strength and shifts to lower frequencies. At the phase transition S_A/S_C^* the soft mode has a maximum in the dielectric strength and a minimum in the frequency position. Near the phase transition the soft mode shows a linear temperature dependence of the inverse dielectric strength and in the frequency position (fig.4). These findings confirm well with a theoretical

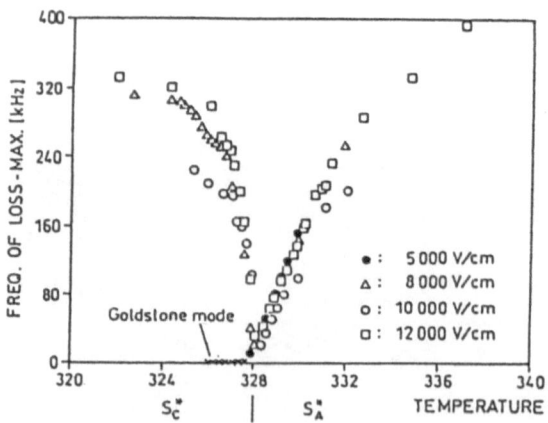

Fig. 4: *Frequency at maximum loss vs. temperature for different exter- nal electric d. c. bias fields.*

description which is based on an extended Landau expansion [4].

In the frequency regime from 10^6 Hz to 10^{10}Hz <u>one</u> dielectric loss process, the β - relaxation, is observed (fig. 5). It is assigned to the hindered rotation (libration) of the

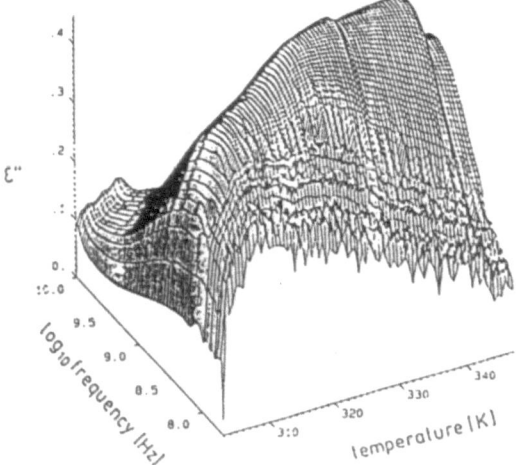

Fig. 5: *Frequency and temperature dependence of the dielectric loss ϵ''.*

mesogene around its long molecular axis [3]. The experimental data were fitted with the generalized relaxation function according to Havriliak and Negami. The temperature dependence of the dielectric strength $\Delta\epsilon$ and the relaxation time τ are shown in fig.6. Cooling down from the isotropic phase $\Delta\epsilon$ increases at the phase transition is./S_A

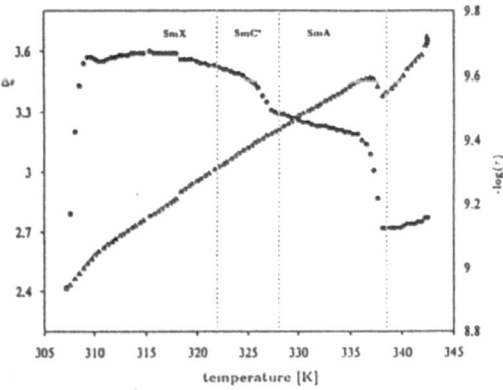

Fig. 6: *Temperature dependence of the relaxation rate $1/\tau$ ($\triangle$) and the dielectric strength $\Delta\epsilon$ ($\bullet$) of the β - relaxation.*

because the mesogenes orient in the bookshelf geometry which leads to a change of the dipole moment interacting with the outer electric field. For the same argument a step occurs at the phase transition S_A/S_C^*. The relaxation time τ shows an Arrhenius - type temperature dependence which is not influenced at the phase transition S_A/S_C^*. This leads to a modified picture for the molecular origin of ferroelectricity in FLC. In the common explanation [5,6] this is based on a "free"rotation of the mesogenes around their long molecular axis in the nonferroelectric S_A -phase and its strong hinderance in the S_C^* - phase. But this should result in a strong decline in $\Delta\epsilon$ and an increase of the relaxation time in contrast to the experimental results. Instead the β - relaxation has the character of a librational motion. While in the nonferroelectric S_A - phase the distribution of lateral dipole moments is isotropic, it becomes strongly anisotropic in the S_C^* - phase. This anisotropy is caused by the chirality and the tilt [7].

In the polymeric sample also three dielectric loss processes, Goldstone - mode, soft - mode and β - relaxation, can be observed (fig.7). So there are no qualitative

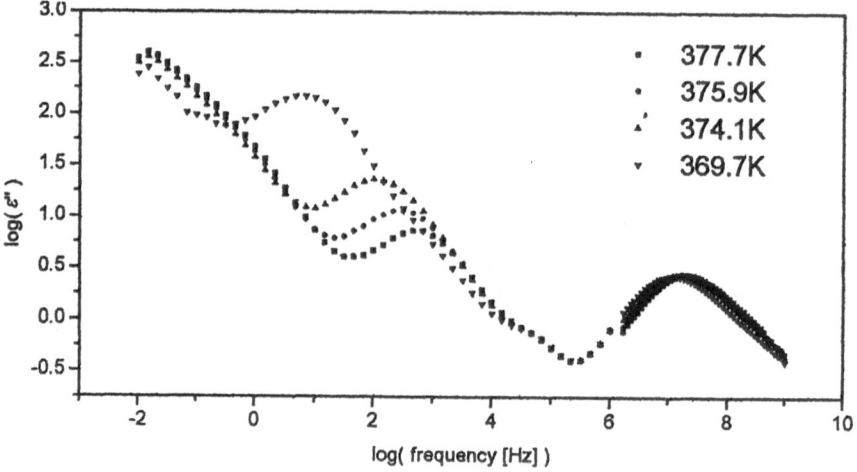

Fig. 7: *Goldstone - mode, soft - mode and β - relaxation for the polymeric sample.*

differences with respect to the dynamics of low molecular weight FLCs. At a more closer look some typical polymer properties influence the dynamics: While the Goldstone - mode in low molecular weight FLCs is nearly temperature independent, a pronounced temperature dependence is observed in the polymeric sample (fig8). Comparing the relaxation frequencies the Goldstone - mode in polymers is shifted nearly one decade to lower frequencies. The helical superstructure can be unwound using a DC - bias field of similar strength. In the unwound state the soft - mode can be observed (fig.9). Fitting the experimental data with the Havriliak - Negami function shows a similar behaviour of the inverse dielectric strength (fig.10). For the frequency position of the maximum

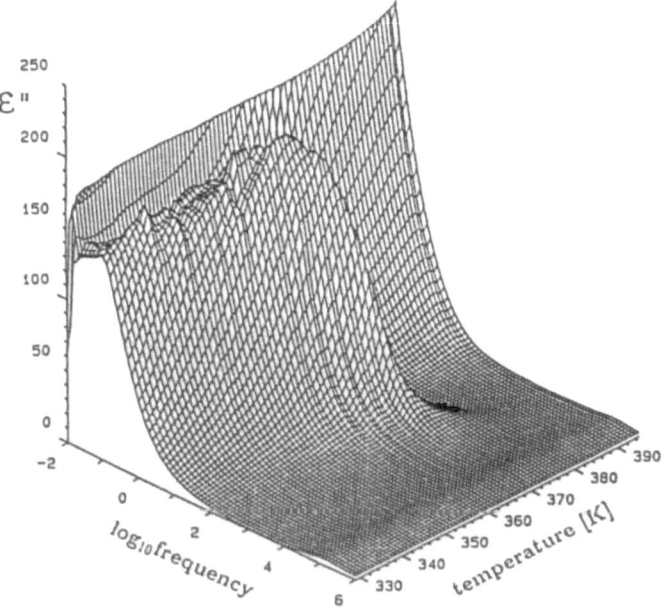

Fig. 8: *Dielectric loss ϵ'' vs. frequency and temperature; d.c. field strength = 0V/cm.*

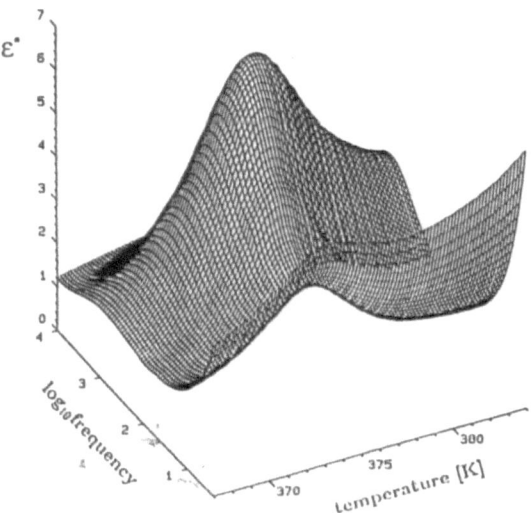

Fig. 9: *Dielectric loss ϵ'' vs. frequency and temperature; d.c. field strength = 15kV/cm.*

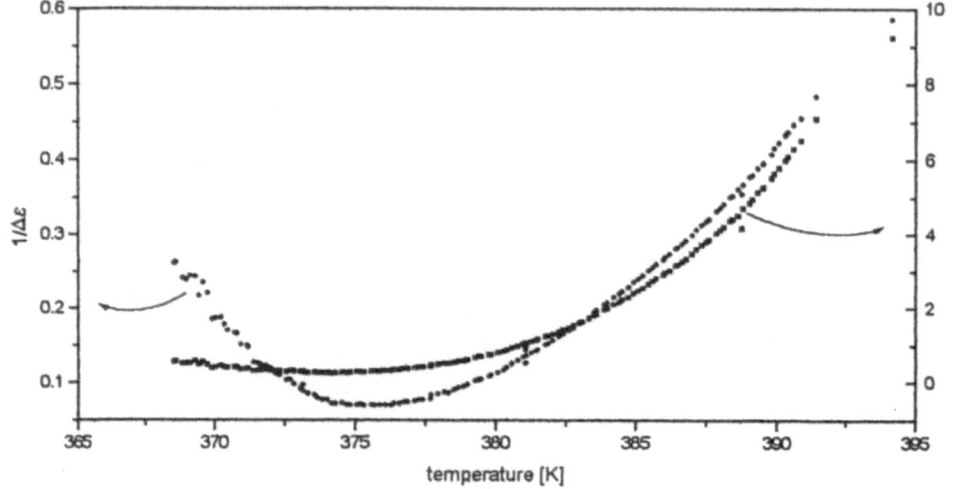

Fig. 10: *Frequency at maximum loss and invers dielectric strength vs. temperature.*

ϵ'' only in the S_A - phase a slowing down can be observed (fig.10). While in the S_C^* - phase the process is expected to become faster again (see fig.4), the soft mode remains at nearly the same frequency position. This may be due to the polymer viscosity. Further more the phase transition S_A/S_C^* is not as sharp as for the low molecular weight FLC which is also typical for polymers. In the high frequency regime a fully comparable dynamic behaviour is found. There is <u>one</u> relaxation process (fig.11), the

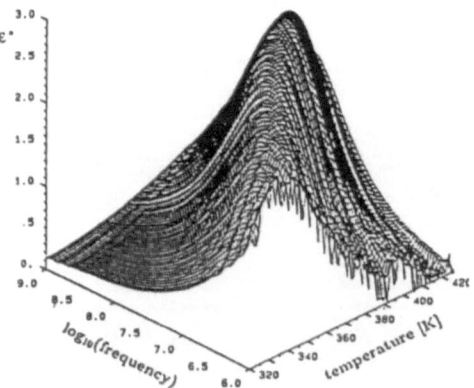

Fig. 11: *Frequency and temperature dependence of the dielectric loss ϵ'' for the polymeric sample.*

β - relaxation, which has also an Arrhenius - type temperature dependence (fig.12).

Only the steps in the dielectric strength that occur at the phase transitions are not

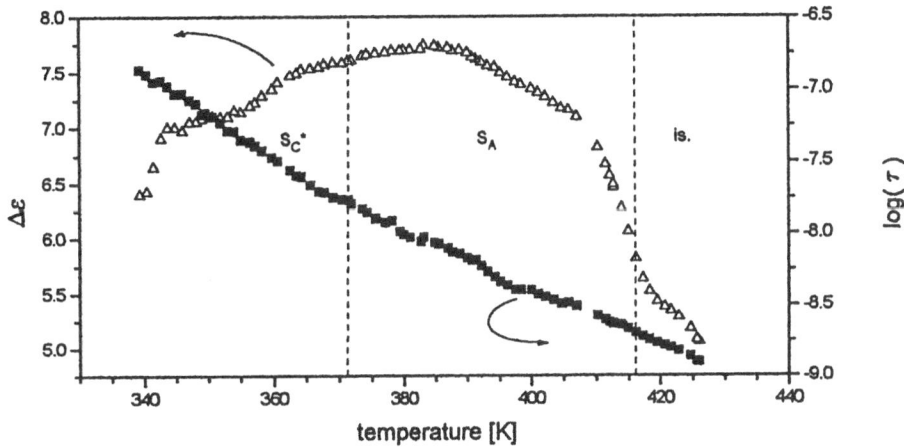

Fig. 12: *Temperature dependence of the relaxation rate $1/\tau$ ($\triangle$) and the dielectric strength $\Delta\epsilon$ ($\bullet$) of the β - relaxation.*

as pronounced as for the low molecular weight samples, because the orientation of the polymers in the bookshelf geometry is only partially achieved (fig.12).

References

[1] Clark N. A., Lagerwall S. T.; (1980) Appl. Phys. Lett. **36**, 899

[2] Havriliak S., Negami S.; (1966) J. Polym. Sci. **C14**, 89

[3] Kremer F., Schönfeld A., Hofmann A., Zentel R., Poths H.; (1992) **3**, 249

[4] Carlsson T., Zeks B., Filipic C., Levstik A.; (1990) Phys. Rev A **42**, 877

[5] Lalanne J. R., Buchert J., Destrade C., Nguyen H. T., Marcerou J. P.; (1989) Phys. Rev. Lett. **62**, 3046

[6] Beresnev L. A., Blinov L. M., Osipov M. A., Pikin S. A.; (1988) **158A**, 43

[7] Osipov M. A., Meister R., Stegemeyer H.; Liq. Cryst., submitted

The Rheology of PBLG Liquid Crystalline Polymers at High Concentrations

R.G. Larson[*], J. Promislow[*†], S.-G. Baek[‡], and J.J. Magda[‡]

* AT&T Bell Laboratories, Murray Hill, NJ, 07016
† currently at Dept Chem Eng, Stanford Univ, Stanford, CA. 94305
‡ Dept Chem Eng, Univ of Utah, Salt Lake City, Utah 84112

INTRODUCTION

The rheological behavior of most liquid crystalline polymers (LCP's), especially thermotropic ones, is not yet well understood. This is true, despite notable progress towards a molecular-level understanding of the rheology of some lyotropic LCP's, especially modestly concentrated solutions of poly(γ-benzylglutamate), or PBG, in the solvent metacresol. Many aspects of the rheology of PBG solutions seem to be well described by the Doi molecular theory,[1] particularly the behavior of the first normal stress difference N_1, which is negative in sign over a range of shear rates. In fact, semi-quantitative agreement between the Doi theory and measurements of N_1 has been achieved for PBG solutions of modest concentration.[2] However, the success of the Doi theory in describing the rheological behavior of some PBG solutions does not extend to many other LCP's.

One of the most peculiar differences between the Doi theory and the rheology of many LCP's is the existence of pronounced shear thinning at low shear rates — i.e. the so-called "Region I" behavior.[3-7] Fig. 1 shows an idealized three-region flow curve of shear viscosity vs. shear rate $\dot{\gamma}$ proposed by Onogi and Asada;[3] it shows shear thinning at high and low rates of shear, and a "Newtonian"-like constant-viscosity plateau at intermediate shear rates. For some polymer solutions, it has even been found that as $\dot{\gamma}$ is decreased to very low rates, Region I shear thinning gives way to a second Newtonian viscosity plateau at the lowest $\dot{\gamma}$.[8,9] However, the Doi theory for rod-like polymers predicts only one Newtonian plateau and one shear thinning region; these can be identified as Regions II and III of Fig. 1. The Doi theory does not predict a low-shear-rate shear-thinning regime that could be identified as Region I. Thus the origin of Region I behavior is an important mystery, whose explanation might lead to a much better understanding of the rheology of both lyotropic and thermotropic LCP's.

We emphasize, however, that the viscosity curves measured for many LCP's often do not establish the existence or nonexistence of "Region I" flow behavior. One reason is that many LCP's do not show a clear Region II plateau within the range of accessible shear rates, but instead show either a single shear-thinning region[10] or two shear-thinning regions separated only by an inflection point, and not by a definite plateau region. If there is but a single shear thinning regime, it is obviously not possible to tell from the viscosity data alone whether that regime should be labeled Region I or Region III, or whether it encompasses both Region I and Region III, with Region II absent. Fortunately, the Doi theory predicts that as one enters Region III by increasing the shear rate, the sign of N_1 should change sign from positive to negative. From this sign change in N_1, Region-III shear thinning can be identified if the

A. Teramoto, M. Kobayashi, T. Norisuje (Eds.)
Ordering in Macromolecular Systems
© Springer-Verlag Berlin Heidelberg 1994

viscosity data are supplemented by N_1 data. N_1 data for all solutions to be studied here are published elsewhere.[11,12]

There is a second reason that the existence of Region I is often not known with certainty from viscosity data alone. Some fluids show a Region-II plateau at the lowest shear rates, followed by shear thinning at higher shear rates; from this it is easy to identify Regions II and III with the Newtonian and shear thinning regions, respectively. But the absence of shear thinning at the lowest experimental shear rates does not prove the nonexistence of Region I; it only proves that Region I was not accessed at the lowest *measurable* shear rates. The existence of Region I in many PBG solutions is therefore uncertain. Only two research groups, Asada et al.,[4] and Moldenaers et al.,[12] have measured viscosity curves for PBG solutions at shearing stresses as low as 1 dyn/cm^2, where Region I might most readily be found. These groups have obtained conflicting results. Asada et al. found region I behavior at both moderate (11%) and high (40%) concentrations, while Moldenaers et al. found no region I for a concentration of 12% and only a possible hint of a region I viscosity upturn at a higher concentration of 25%. Thus, at present, the existence of Region I behavior in PBG solutions is controversial.

In addition to the possible existence of Region I behavior, limited literature data indicate that PBG solutions begin to deviate in other ways from the Doi theory when they become more concentrated. Kiss and Porter investigated the rheology of racemic PBG solutions with average molecular weight 335,000, and concentrations up to 25%. They found that the low-shear-rate viscosity showed a *minimum* at a concentration of around 22%; a 25% solution of this polymer had a higher viscosity than a 22% solution, in disagreement with the Doi theory. Similarly, Asada et al. found that for PBLG with molecular weight 260,000, a 40% solution had a higher low-shear-rate viscosity than a 20% solution.

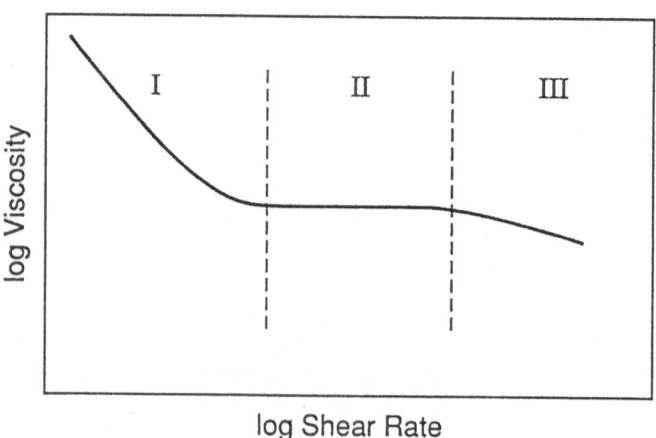

Figure 1.   Onogi and Asada's[3] conception of a three-region flow curve for liquid crystalline polymers.

Here we examine the shear viscosity of PBG solutions with systematically increasing concentration, starting from modest concentrations, where the Doi theory applies, and raising it toward high concentrations, around 40%, where the molecular packing density is high enough that such solutions might acquire some of the characteristics of thermotropics. Measurements of the first and second normal stress difference $N_1(\dot{\gamma})$ and $N_2(\dot{\gamma})$ show[8,11] that there is a local maximum in N_1 versus shear rate, as well as a region of negative N_1. Neither the maximum in N_1 nor the region of negative N_1 disappear as the concentration C increases, but both shift to higher $\dot{\gamma}$, in qualitative agreement with the Doi molecular theory. At least one aspect of the behavior of the second normal stress difference $N_2(\dot{\gamma})$, namely the increasingly negative slope of N_2 versus $\dot{\gamma}$ at low and modest shear rates, is also explained by the Doi theory. Thus, for the most part, $N_1(\dot{\gamma})$ and $N_2(\dot{\gamma})$ do not show dramatic changes in qualitative behavior at high concentrations of PBLG, other than a continuous shifting to higher shear rates of both the local maximum in N_1 and the range of negative N_1; and this shifting is by and large predicted by the Doi molecular theory. Since the Doi theory predicts that as the shear rate is increased the beginning of Region-III shear thinning should be accompanied by a change in sign in N_1, the measurements of N_1 help us distinguish Region I from Region III, and to identify other features of the shear viscosity data that lie outside the predictions of the Doi theory.

The viscosity/shear-rate data presented here for PBLG solutions cover a wide enough range of concentrations, 3%-40%, to encompass both solutions whose viscosities and normal stress differences are well described by the Doi theory, and those that deviate substantially from it. We shall also present linear viscoelastic data G' and G'', for some of these solutions.

EXPERIMENTAL

The PBLG solutions

($M_{vis} = 238,000$; $M_w/M_n \approx 1.7$) were described elsewhere; [13] we shall refer to this particular polymer as PBLG238. These solutions were studied using two rheometers with cone-and-plate fixtures, one at AT&T Bell Labs and the other at the University of Utah. At Bell Labs, the Rheometrics System 4 rheometer was used, with a cone-and-plate radius, R, of 25 mm, and a cone angle, α, of 0.04 radians at the temperature $T = 27.5 \pm 1.0°$ C. At the University of Utah, studies were made with a Weissenberg rheogoniometer at $T = 25 \pm 0.1°$ C. Data obtained at the University of Utah confirmed those obtained at Bell Labs and are not reported here.

RESULTS

Viscosities

Fig. 2a shows the shear viscosity at shear rates $\dot{\gamma}$ of 2,7,20, and 70 sec^{-1} for concentrations ranging from 3% to 40%; for comparison, the predictions of the Doi theory[14] are shown in Fig. 2b. In Fig. 2b, ν is the number concentration of rod-like molecules and ν^* is the the critical concentration at which the isotropic phase becomes unstable to the nematic. η^* and D_r^* are the viscosity and rotary diffusivity in the isotropic phase at concentration ν^*. The theoretical results plotted in Fig. 2b are taken from the original, approximate, Doi theory, which gives reasonable predictions of the shear viscosity for a wide range of shear rates. The value of the rotational diffusivity parameter D_r^* in the Doi theory is estimated from the shear rate at which N_1 changes sign from positive to negative for a 12.5% solution of PBLG238; an estimate $D_r^* \approx 6 \text{ sec}^{-1}$ is thereby obtained.[11]

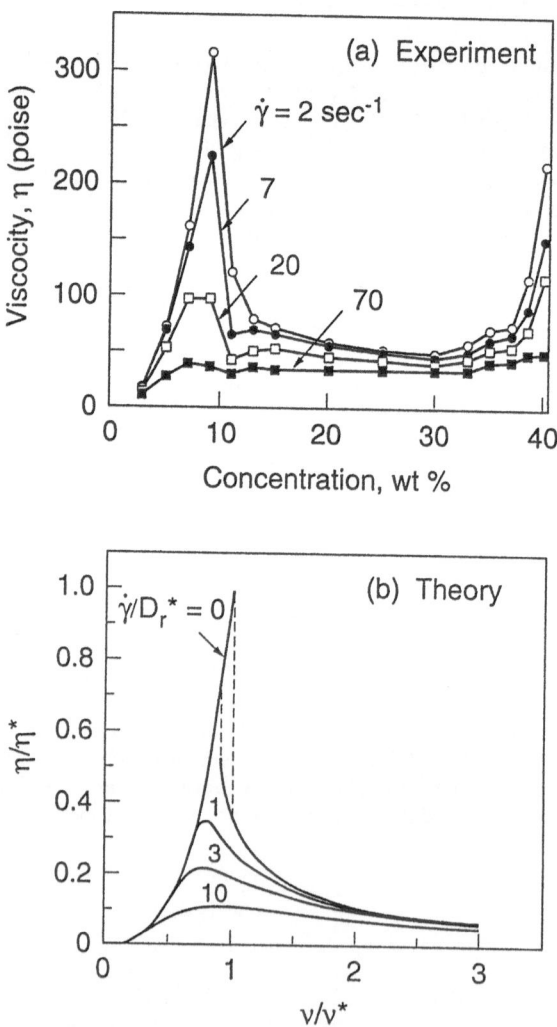

Figure 2.   a) Measured viscosity as a function of concentration of PBLG238 for various shear rates $\dot{\gamma}$. b) Viscosity predicted by the Doi theory.[14]

For concentrations of PBLG that are less than 30%, our experimental results are in excellent qualitative agreement with the Doi theory; see Fig 2a&b. In both theory and experiment, there is a maximum in viscosity at a concentration near the isotropic-anisotropic transition. As $\dot{\gamma}$ increases, the height of the maximum decreases greatly and shifts toward slightly lower concentrations. At concentrations above the isotropic-anisotropic transition, the theory shows a monotonic decrease in η as C increases; the rate of this decrease steepens as $\dot{\gamma}$ decreases; this too compares favorably with the experimental data. Similar favorable comparisons between theory and the early experiments of Kiss and Porter[8] were discussed in the Introduction. Our results, however, show a small discrepancy not reported earlier; there is a slight dip in viscosity at a concentration near the anisotropic side of the biphasic region; the theory does not show this dip.

But the greatest deviations from the Doi theory occur for C $\geq$ 30%; at these high concentrations, the experiments show at each $\dot{\gamma}$ an increase in η with increasing C. The rate of this increase becomes steep for concentrations above 38.5%. This increase in viscosity with increasing concentration is not predicted by the Doi theory.

Figs. 3-5 show the viscosity plotted against shear rate for the concentration ranges 3—9%, 13—30%, and 30—40%. The first of these ranges is the isotropic regime where the low-shear-rate viscosity increases with increasing concentration. The second range of concentration, covered in Fig. 4, is the liquid crystalline regime where the low-shear-rate viscosity decreases with increasing shear rate. The shear-rate and concentration dependencies in Figs. 3-4 are in qualitative agreement with the Doi theory. At concentrations C above 30%, however, the viscosity η increases with increasing concentration, Fig. 5, and this is *not* predicted by the Doi theory. Note that the $\eta(\dot{\gamma})$ curves in Fig. 5 abruptly change shape between the concentrations of 37% and 40%. A solution with a concentration of 38.5% showed behavior qualitatively similar to that of the 40% solution. Thus for C $\leq$ 37%, the viscosity tends toward a Newtonian plateau at the lowest shear rates accessed, while for C $\leq$ 38.5%, there is instead an inflection point, and $\eta(\dot{\gamma})$ follows an approximate power law at the lowest shear rates, indicative of "Region I" behavior. The power-law exponent for the 40% solution is around 0.46.

At a still higher concentration, 45%, a solution of PBLG238 is a "quasi-solid" in that it does not flow under gravity to the bottom of its container even after months of storage. When attempting to shear this "gel-like" material reliable steady state viscosities could not be obtained, as the sample edge became highly distorted by the flow. Lee and Meyer[15] have observed that a racemic mixture of PBLG and PBDG in a mixed solvent dioxane/dichloromethane undergoes a transition from a nematic to what appears to be a columnar phase when the volume fraction of polymer reaches 0.36. For PBLG in metacresol, a volume fraction of 0.36 would correspond to a weight percentage of 42% PBLG. Since a columnar phase posseses positional order in the plane perpendicular to the molecular alignment direction, the "solid-like" rheology we observed for 45% PBLG238 is consistent with the formation of such a phase at high volume fraction of PBLG.

The shear-viscosity data for the 40% solutions reported in Figs. 5 were obtained after the samples had been pre-sheared for many strain units at rates of 10 sec^{-1} or above. If a freshly loaded sample is sheared at low rates of shear *without pre-shear*, lower

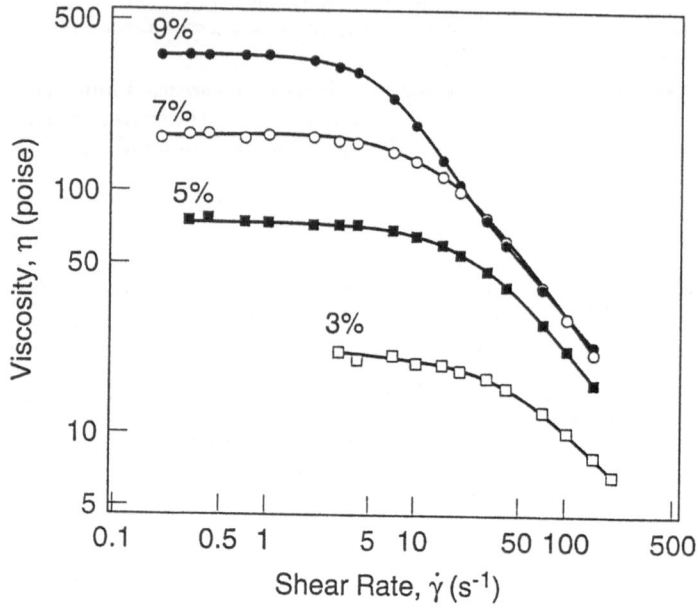

Figure 3. Viscosity as a function of shear rate for PBLG238 at concentrations in the isotropic range.

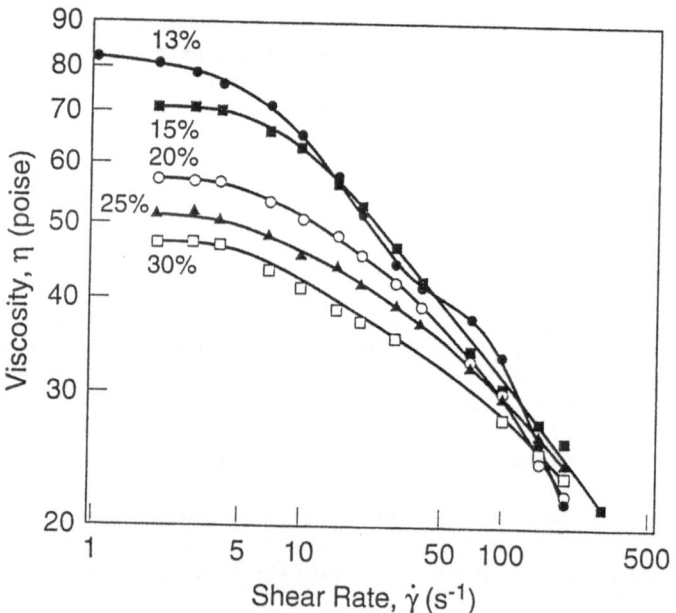

Figure 4. Viscosity as a function of shear rate for PBLG238 at moderate concentrations in the liquid crystalline regime.

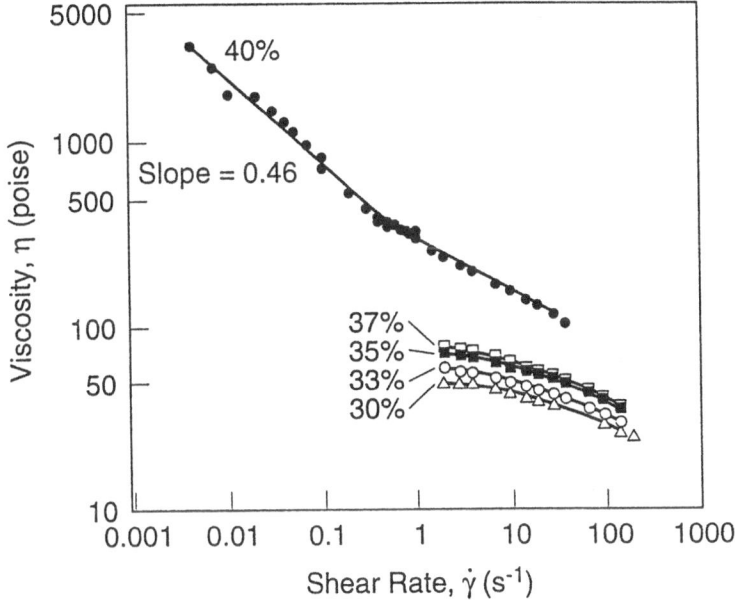

Figure 5. Viscosity as a function of shear rate for PBLG238 at high concentrations in the liquid crystalline regime.

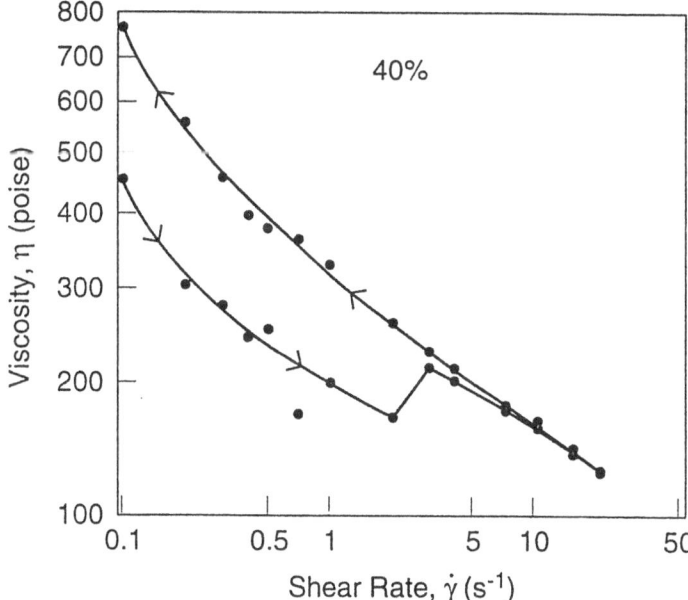

Figure 6. Viscosity as a function of shear rate for 40% PBLG238.

viscosities are obtained. Fig. 6 shows the hysteretic behavior of the 40% solution that was sheared at increasing, and then decreasing, shear rates following sample loading. Note that on the branch of increasing shear rate, there is a viscosity jump at $\dot{\gamma} = \dot{\gamma}_c \approx 3\ \text{sec}^{-1}$. There is little or no hysteresis at shear rates above $\dot{\gamma}_c$. If, after $\dot{\gamma}$ has been decreased along the upper branch of Fig. 6, $\dot{\gamma}$ is again increased, the upper curve is again traversed. By traversing the hysteresis loop at different rates, we investigated the possibility that these hysteretic effects could have been an artifact induced by solvent evaporation. But we found that the same hysteretic behavior was obtained no matter how fast the loop was traversed; in one case the low-shear-rate viscosity was changed from a low to a high value in only three minutes. Thus evaporation effects can be ruled out as an explanation for the phenomenon. Furthermore, if the sample is sheared at $\dot{\gamma} > \dot{\gamma}_c$, and then is left in a quiescent state in the rheometer for a day or more before the viscosity is remeasured, the low-shear-rate viscosity after the rest period is then found to be somewhere between the values on the upper and the lower branches in Fig. 6. Similar behavior was observed with the 38.5% solution. Hence, *shearing at* $\dot{\gamma} \geq \dot{\gamma}_c$ *induces a structural change in the material that raises its viscosity.* The viscosity decreases again only if the solution is allowed to rest for more than a day. This structural change has little effect on the first normal stress difference N_1; in traversing the hysteresis loop of Fig. 6, the value of N_1 increases by only about 15%, compared to a 100% increase in η.

Dynamic Moduli

G' and G'' were measured on the System 4 at a strain amplitude of 5%, within the linear viscoelastic regime. The data were measured after loading each sample, with no pre-shearing other than that produced by sample loading. In Fig. 7, G' and G'' data for six samples with concentrations ranging from 13% to 37% are all plotted on the same "master curve," obtained by shifting both along the horizontal and along the vertical axes. While there is no theoretical basis for it, such horizontal and vertical shifting does allow us to present all these data on a single figure; the characteristic relaxation times and moduli are given in Table 1. These characteristic parameters are defined by the cross-over point of the extrapolated straight lines that define the low frequency behavior of G' and G''; see Fig. 7. Data for all concentrations have been shifted to superpose onto the curves for the 13% solution. For example, from Table 1, data for the 37% solution was shifted horizontally to the left by the factor $2.94/0.15 = 19.6$, and vertically downward by factor $18.9/1.35 = 14$. At low frequency, the power laws $G'' \propto \omega^{0.92}$ and $G' \propto \omega^{1.4}$ are followed. Interestingly, the shifting required to bring G' and G'' into superposition also helps one correlate the shear-rate regimes where the first normal stress difference N_1 is negative. Fig. 7 shows that for each solution, the negative N_1 regime occurs in a range of shear rates just below the cross-over frequency $\omega = 1/\tau$. Thus, the second sign change in N_1, where N_1 changes from negative to positive, occurs roughly at the shear rate $\dot{\gamma} \sim 1/\tau$. This result is not too surprising. The cross-over frequency $\omega = 1/\tau$ is roughly the rate of molecular relaxation; so that when $\dot{\gamma} = 1/\tau$, the shearing rate is fast enough to overcome molecular relaxation processes, and N_1 changes sign. If this correlation of G' and G'' with N_1 behavior is general, it could be used to estimate the range of shear rates over which one should expect N_1 to be negative for other LCP's.

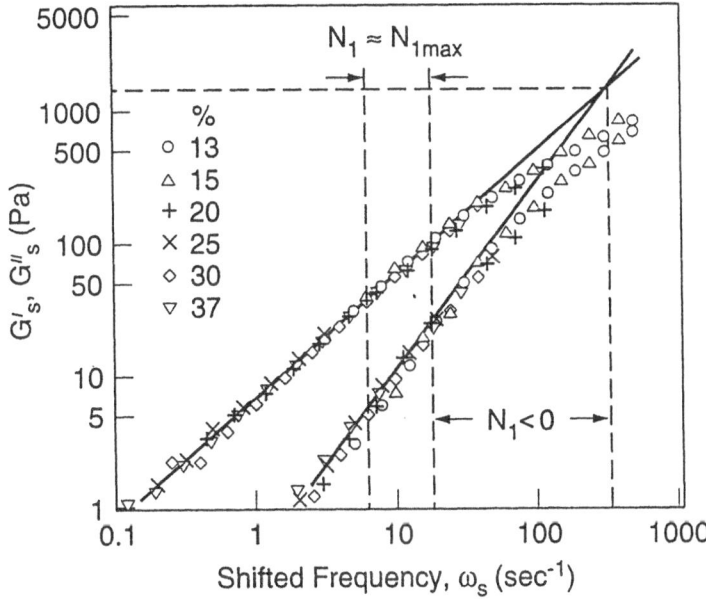

Figure 7. "Master curve" plotting shifted values of G' and G'' as functions of a shifted frequency for concentrations of PBLG238 from 13% to 37%. The concentration-dependent shift factors can be computed from Table 1, as discussed in the text. The regions of shifted shear rates where N_1 is near its positive maximum, and where N_1 is negative, are shown.

Table 1: Relaxation Times and Moduli From Dynamic Data

Concentration	Relaxation Time, τ (msec)	Modulus G (kPa)
13%	2.94	1.35
15%	2.35	1.49
20%	0.68	2.70
25%	0.29	6.75
30%	0.20	10.80
37%	0.15	18.90

When the concentration of PBG is raised to 40%, there is a drastic change in the low-frequency behavior of G' and G''. Fig. 8 plots the storage modulus G' versus frequency for the 37% and the 40% solutions. For the 40% solution, G' seems to approach an elastic plateau as the frequency is reduced, indicating that at this concentration some kind of elastic structure has formed in the solution. Recall that we also observed the appearance of region I in the shear viscosity $\eta(\dot{\gamma})$ for C > 37%. The behavior of $\eta(\dot{\gamma})$ and $G'(\omega)$ are remarkably similar to some thermotropic LCP's.

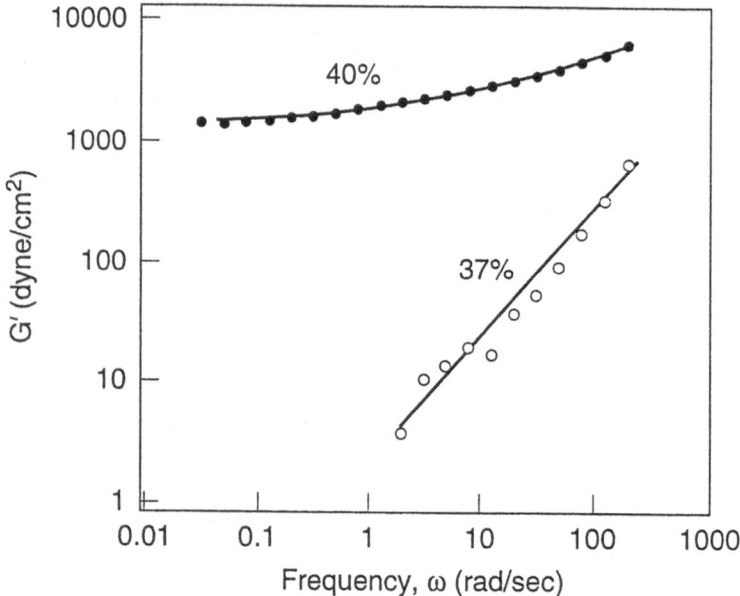

Figure 8.   Plot of G' versus frequency for 37% and 40% solutions of PBLG238.

DISCUSSION

At concentrations greater than 30%, the first normal stress difference N_1 follows Doi-theory predictions in that the range of negative N_1 continues to shift to higher shear rates as the concentration increases; however the shear viscosity begins to deviate significantly from the Doi theory in two ways. First, the shear viscosity increases with increasing concentration for $C \geq 30\%$, even at low shear rates. And secondly, when $C \geq 38.5\%$, the viscosity no longer approaches a Newtonian plateau as the shear rate becomes small, but instead follows a power-law in shear rate, consistent with so-called "Region I" behavior of Onogi and Asada. Also, for $C \geq 38.5\%$, when a freshly loaded sample is subjected to shear starting at low rates that are then systematically increased, there is a change in fluid structure at a critical shear rate $\dot{\gamma}_c$ at which the viscosity jumps discontinuously. Thereafter, if the shear rate is reduced, higher viscosity values of η are obtained than were measured for the freshly loaded sample. Asada et al. also presented shear-viscosity data for PBLG solutions that also showed a "jump" in shear viscosity at a shear rate near the transition from region I to region II. Their viscosity-shear rate curves show no hysteresis; but this could be accounted for if the viscosities obtained at the lowest shear rates were measured on a freshly loaded sample without having been sheared previously at high rates.

The solutions displaying "Region I" behavior also have a low-frequency elastic plateau in the storage modulus measured in small-amplitude oscillatory shear. These characteristics of highly concentrated PBLG solutions are analogous to the rheological characteristics of some thermotropic liquid crystalline polymers.

According to the only available theories for Region I shear thinning, namely the theory of Marrucci,[16] and a related theory of Wissbrun,[17] Region I behavior occurs because of the effect of shear on the polydomain texture of liquid crystalline materials. While these theories show some promise, at present they do not seem able to explain why some materials show Region I shear thinning only at extraordinarily low shear rates or not at all, while others show Region I behavior over a wide range of shear rates. Thus, the mechanism for "Region I" shear thinning is still unresolved.

References

1. Kuzuu N, Doi M (1983) J. Phys. Soc., Japan, 52:3486; (1984) 53:1031
2. Magda JJ, Baek SG, DeVries KL, Larson RG (1991) Macromolecules 24:4460
3. Onogi S, Asada T (1980) In: Astarita G, Marrucci G (eds) Rheology. Plenum, New York
4. Asada T, Tanaka T, and Onogi S (1985) J Appl Polym Sci: Appl Polym Symp 41:229
5. Wissbrun KF (1980) Brit Polym J 12:163
6. Wissbrun KF (1981) J Rheol 25:619
7. Wissbrun KF, Kiss G, Cogswell FN (1987) Chem Eng Comm 53:149
8. Kiss G Porter RS (1978) J Polym Sci: Polym Symp 65:193
9. Grizutti N private communication
10. Kalika DS, Giles DW, Denn MM (1990) J Rheol 34:139
11. Baek S-G, Magda JJ, Larson RG (1993) J Rheol, submitted
12. Moldenaers P, Yanase H, Mewis J (1990) In: Weiss RA, Ober CK (eds) Liquid Crystalline Polymers. ACS Symposium Series 435
13. Larson RG, Mead DW (1991) J Polym Sci: Polym Phys Ed 29:1271
14. Doi M, Edwards SF (1986) The Theory of Polymer Dynamics. Oxford Press, London
15. Lee S-D, Meyer RB (1990) Liq Cryst 7:451
16. Marrucci C (1984) proceedings of the Ninth International Congress on Rheology, Acapulco, Mexico, Oct. 8-13
17. Wissbrun KF (1985) Faraday Disc Chem Soc 79:161

Microdomain Structures of Triblock Copolymers of the ABC Type

I. Noda, Y. Matsushita and Y. Mogi

Department of Applied Chemistry, Nagoya University
Furo-cho, Chikusa-ku, Nagoya 464-01 Japan

Abstract: To understand microdomain structures of triblock copolymers of the ABC type at the molecular level, we studied the variation of morphology of microdomain with composition, the molecular weight dependence of lamellar domain spacing and the chain conformation of middle-block polymer in lamellae by transmission electron microscopy(TEM), small-angle X-ray scattering(SAXS) and small-angle neutron scattering(SANS), in comparison with microdomain structures of diblock copolymers of the AB type, because the middle-block chain of the triblock copolymers has a bridge conformation, while the one end of block chain of diblock copolymers is always free.

Samples used here were isoprene(I)-styrene(S)-2-vinylpyridine(P) triblock copolymers of which the two end-block polymers have the same chain lengths and the corresponding triblock copolymers with deuterated middle-block polymers, prepared by a succesive anionic polymerization technique. Film specimens for TEM, SAXS and SANS were prepared by solvent-casting and annealing.

TEM and SAXS studies revealed that the morphologies of microdomains change in the order of spherical, cylindrical, ordered tricontinuous double-diamond and lameller structures as the S block polymer content decreases. The order is similar to that of diblock copolymers, but the composition ranges and the types of packing in lattice are different. SAXS studies revealed that the lamellar domain spacings of the triblock copolymers are larger than those of SP diblock copolymers at least for high molecular weight samples in accordance with the theories of microphase separation in the strong segregation limit. Moreover, SANS studies suggested that the middle-block polymer of the triblok copolymers is contracted along the direction parallel to lamellae at the almost same magnitude as the block chain of the diblock copolymers.

INTRODUCTION

It is well-known that block copolymers with mutually imcompatible components form microdomain structures in bulk. Studies on the microdomain

A. Teramoto, M. Kobayashi, T. Norisuje (Eds.)
Ordering in Macromolecular Systems
© Springer-Verlag Berlin Heidelberg 1994

structures have been carried out extensively for AB diblock copolymers, so that the variation of morphology with composition[1], the molecular weight dependence of lamellar domain spacing[1,2] and the conformation of block chains in lamellae[1,3,4] have been elucidated. In clarifying the morphology of microdomains of block copolymers at the molecular level, studies on the morphology of triblock copolymers of the ABC type in comparison with diblock copolymers of the AB type are of interest, because the triblock copolymers have the middle-block chain of which both ends are held at the different domain boundaries, or the middle-block chain has a "bridge" conformation, while the one end of block chain of diblock copolymers is always free. However, the systematic study on the morphology of triblock copolymers of the ABC type has not been carried out, not only because the preparation of well-defined triblock copolymer samples is difficult but also because the morphologies are various and complicated. It is to be noted that the chain conformation of triblock copolymers of the ABC type is simpler than that of triblock copolymers of the ABA type since the former has only the bridge conformation for the middle-block polymer, while the latter can have a "loop" conformation as well as the bridge conformation, though the morphologies of the former microdomains might be more complicated than those of the latter.

Since morphologies of diblock copolymers are determined by volume fractions of block components in the strong segregation limit, the equilibrium morphologies of microdomains formed by the A and B block polymers and by the B and C block polymers in triblock coplymers of the ABC type are assumed to be geometrically symmetric to each other if the volume fractions of the A and C block polymers are equal. In this work, therefore, we prepared triblock copolymers of the ABC type having the two end-block polymers with the same chain lengths, but having the middle-block polymers with various chain lengths, and studied 1) the variation of morphology of microdomains of the triblock copolymers with composition by transmission electron microscopy (TEM) and small-angle X-ray scattering (SAXS)[5,6,7], 2) the molecular weight dependence of domain spacings of the lamellar structures by SAXS[8] and 3) the chain conformation of the middle-block polymer in the lamellar structures by small-angle neutron scattering (SANS)[8], in comparison with those of diblock copolymers.

SAMPLES AND METHODS

Samples used here were isoprene(I)-styrene(S)-2-vinylpyridine(P) triblock copolymers of which the I and P block polymers have the same chain lengths. To study the variation of morphology with composition the samples with various S block polymer contents were prepared, while the samples with the same compositions but various molecular weights, and the corresponding samples with deuterated styrene (D) block polymers were prepared to study the lamellar domain spacings and the chain conformation of the middle-block polymers in the lamellar structures, respectively. They were prepared by a successive anionic polymerization technique in tetrahydrofuran (THF) with cumylpotassium as an initiator at $-78°C$ under high vacuum[5].

The number-average molecular weights, M_n, and the indexes of molecular weight distribution, M_w/M_n, of triblock copolymers and their precursors were determined with a Hewlett-Packard membrane osmometer Type 502 and a Tosoh gel permeation chromatograph, respectively. The compositions of triblock copolymers were determined with a Perkin-Elmer elemental analyser Type 240B and by a pyrolysis-gas chromatography, and the microstructures of I block polymers were determined by ^1H NMR with a Varian Gemini 200-MHz FT. These characterizations indicated that the volume fractions of I and P block polymers are nearly equal, the molecular weight and composition distributions are fairly narrow and the microstructures of I block polymers consist of 40% of 1,2- vinyl and 60% of 3,4-vinyl configurations[5,8].

Film specimens for TEM, SAXS and SANS were prepared by solvent-casting from dilute THF solutions, followed by drying at room temperatures *in vacuo* for a week and annealing at 120°C *in vacuo* for 10 days. The film specimens stained by OsO_4 or phosphotangustic acid (PTA) were observed with a transmission electron microscope of JEOL Type 2000FX[5,6,7]. SAXS measurements were carried out at room temperatures with a Kratky U-slit camera of Anton Paar Co., by using the wave length $\lambda = 0.154$ nm. The scattering intensities were measured at the edge-view where X-ray incidents along the direction parallel to the film surface. Since there is no difference between q values at diffraction peaks, q_{max} in the measured and desmeared data, where q ($= (4/\pi\lambda)\sin\theta$) is the magnitude of scattering vector and 2θ is the scattering angle, we determined q_{max} values from the measured data[7,8].

SANS measurements were carried out at room temperatures with a SANS spectrometer equipped with a two-dimensional position sensitive detector at the National Institute of Standards and Technology, by using a neutron beam with $\lambda = 0.90$ nm and the width of the distribution $\Delta\lambda/\lambda = 0.25$. The scattering intensities were measured at the through-view where a neutron beam incidents along the direction perpendicular to the film surface, and they were averaged circularly at the same q values on the two-dimensional position sensitive detector[8].

VARIATION OF MORPHOLOGY OF MICRODOMAINS WITH COMPOSITION

Figure 1 shows typical examples of transmission electron micrographs of microdomains of ISP triblock copolymers with various compositions[5]. In the micrographs of film specimens stained by OsO₄, the black, white and gray images denote I, S and P domains, respectively. Four types of morphology were observed for the microdomains; a) lamellar structures, b) complicated three-dimensional structures called the "ordered tricontinuous double-diamond (OTDD) structures", as discussed later, c) cylindrical and d) spherical structures.

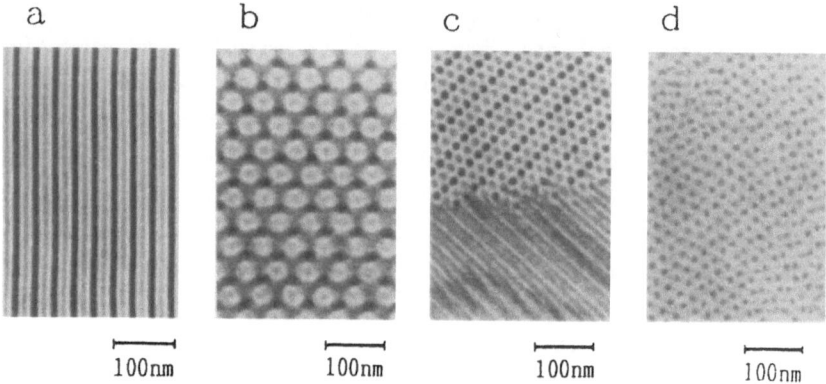

Figure 1. Four types of morphology of microdomain structures of ISP triblock copolymers: a) lamellar, b) OTDD, c) cylindrical, d) spherical structures. (Reproduced from ref. 5).

Figure 2 shows a triangle diagram for the variation of morphology with volume fraction of middle-block polymer (S) in the triblock copolymers having the two end-block polymers (I and P) with the same chain lengths[5,7]. These morphologies were not changed by annealing, so that they are considered to be equilibrium ones. To determine these structures more in detail we will discuss TEM and SAXS data.

Lamellar Structures

As shown in Figure 1a the repeating unit of lamellar structures consists of four layers of three phases such as I, S, P and another S layers (I-S-P-S). Thus, we call this structure the "three-phase four-layer lamellar structure"[5,7]. This structure can be considered to be a one-dimensional super-lattice composed of two sub-lattices (I-S) and (P-S). Figure 3 shows a SAXS intensity profile of the lamellar structure where the diffraction peaks of integer-order are observed in contrast to that of diblock copolymers reflecting the difference in electron density profiles of both block copolymers[8].

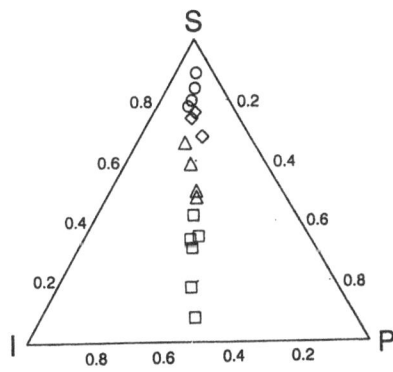

Figure 2. Triangle diagram for the variation of morphology with volume fraction of middle-block polymer (S) in ISP triblock copolymers: lamellar (□), OTDD (△), cylindrical (◇) and spherical (○) structures.

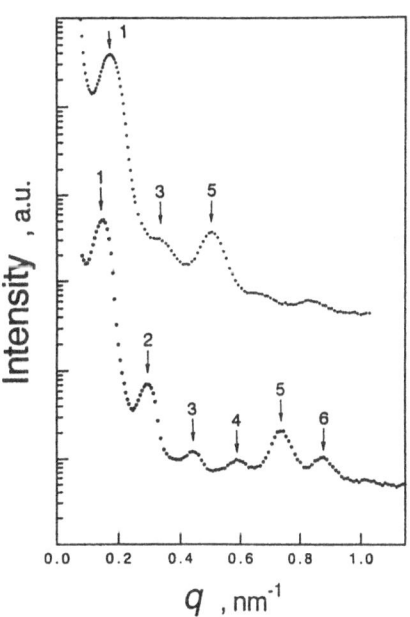

Figure 3. SAXS intensity profiles of lamellar structures of SP diblock (upper) and ISP triblock (lower) copolymers. The numbers denote the orders of diffraction.

Ordered Tricontinuous Double-Diamond Structures

As shown in Figure 1b the micrograph of this structure is complicated. Therefore, it is difficult to correlate directly the two-dimensional micrograph to a three-dimensional structure in contrast to the lamellar structure. Unfortunately, however, no distinct peak is observed in the SAXS intensity profile from this structure[7]. Thus we will discuss the micrographs more in detail. Figure 4 shows our model for this structure, where the end-block polymers (I and P) constitute two kinds of diamond framework mutually interpenetrated, while the middle-block polymers (S) constitute a matrix domain[6]. Figure 5 shows the arrangment of two diamond frameworks. where we assume tetrapot structures formed by the identical four cylindrical struts as the basic units of the diamond framework for simplicity, though the actual surfaces may have a constant mean curvature[6]. In the micrograph of film specimens stained by OsO4, the I and P domains give primarily dark and gray images, respectively, while the S domain gives no image. However, the shade of image in the micrograph also depends on the intensity of electron beams transmitted through the structures.

208

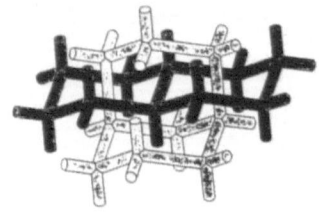

Figure 4. Schematic view of double-diamond frameworks represented by cylindrical struts. (Reproduced from ref. 6).

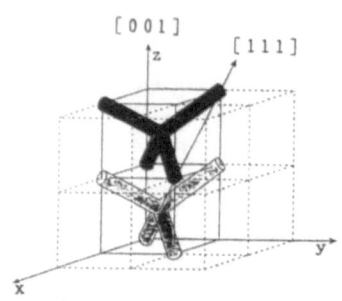

Figure 5. Unit cell of the double-diamond frameworks in the coordinate. (Reproduced from ref. 6).

Figure 6 shows the comparison between the micrograph and the images of the model at the [111] direction[6]. Figure 6a shows the image only for a single layer of diamond framework of I domain with 3-fold symmetry. The I strut, which is thick along the direction, gives a dark image, whereas the remaining three struts give gray images, because their dimensions along the direction are small. Figure 6b shows the image only for a single layer of diamond framework of the P domain. One of the P struts gives a gray image, whereas the other struts give almost no image by the same reason as the case of I domain. Superimposing the two images in Figures 6a and b, considering the relationship between their positions shown in Figure 5, we have the overall image as shown in Figure 6c, which is in good agreement with the electron micrograph shown in Figure 6d. In the micrograph of film specimens stained by PTA, the P domain

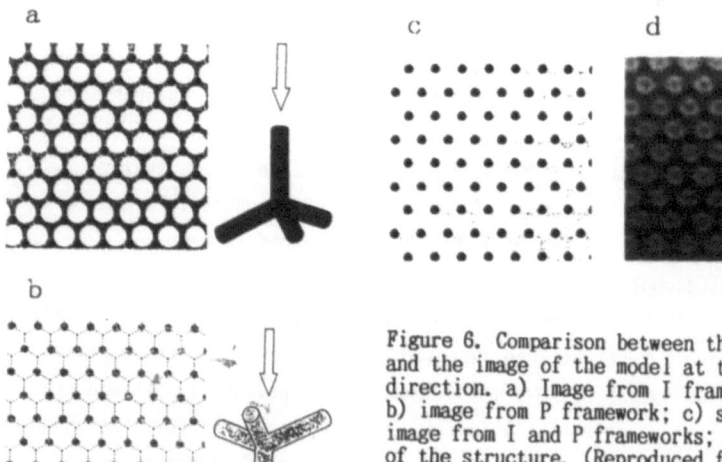

Figure 6. Comparison between the micrograph and the image of the model at the [111] direction. a) Image from I framework; b) image from P framework; c) superimposed image from I and P frameworks; d) micrograph of the structure. (Reproduced from ref. 6).

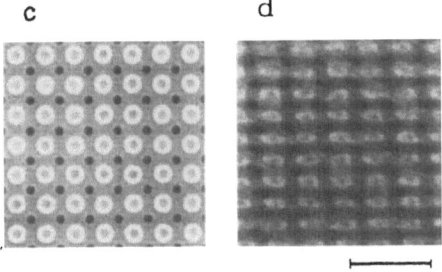

c   d

Figure 7. Comparison between the micrograph and the image of the model at the [001] direction. c) Superimposed image from I and P frameworks; d) micrograph of the structure. (Reproduced from ref. 6).

100nm

gives a dark image, while the I and S domains give no image. If the model is correct. therefore, the micrograph is expected to agree with the image at the [111] direction, where the I domains in Figure 6a are replaced by the P domains. We confirmed that this is the case[7].

Figure 7 shows the comparison between the micrograph and the overall image of the model at the [001] direction obtained in the same way as in Figure 6[6]. Apparently, they well agree with each other. Moreover, tilting the specimen on the microscope stage by 54.7° which is the rotational angle between the two directions [001] and [111], we confirmed that the micrograph in Figure 7 changes to that in Figure 6. From the above disscusion we conclude that this morphology is represented by the proposed model, which consists of the two kinds of diamond framework domains mutually·interpenetrated as sub-lattices. Thus, we name this morphology the "ordered tricontinuous double-diamond (OTDD) structure" after the ordered bicontinuous double-diamond (OBDD) structure for diblock copolymers[6].

Cylindrical and Spherical Structures

In the SAXS intensity profile of cylindrical structures two diffraction peaks appear to be observed at the q_{max}'s where $q_{max1}:q_{max2} = 1:\sqrt{5}$[7]. The profile is clearly diffrent from that of hexagonal close-packing for the cylindrical structures of diblock copolymers, of which the diffraction peaks are observed at the q_{max}'s where $q_{max1}:q_{max2}:q_{max3} = 1:\sqrt{3}:\sqrt{7}$. In the SAXS intensity profile of spherical structures three diffraction peaks appear to be observed at the q_{max}'s where $q_{max1}:q_{max2}:q_{max3} = 1:\sqrt{3}:\sqrt{5}$[7]. The profile is also clearly different from that of simple cubic lattice for the spherical structures of diblock copolymers, of which the diffraction peaks are observed at the q_{max}'s where $q_{max1}:q_{max2}:q_{max3} = 1:\sqrt{2}:\sqrt{3}$. The micrographs of cylindrical and spherical structures in Figure 1 show that these structures

210

consist of the two kinds (I and P) of cylinder and sphere, respectively. Since the arrangements of two kinds of cylinder or sphere are symmetric, in other words, these structures consist of two sub-lattices having the same type of packing, we may speculate that the cylindrical structures have the tetragonal packing of two kinds of cylinder and the spherical structures have the packing of the CsCl type, where the element of a simple cubic lattice is located at the body-center of another simple cubic lattice. We need a detail analysis on the SAXS data to have a definite conclusion on these structures, but we may conclude that all the morphologies of triblock copolymers of the ABC type having the end-block polymers with the same lengths bear the characteristics of super-lattice structures in the strong segragation limit.

Figure 8 shows the variation of morphology with volume fraction of polystyrene block in the triblock copolymers (T), together with that in styrene-isoprene diblock copolymers(D)[5,7]. Comparison between T and D reveals that the composition ranges of double-diamond and lamellar structures are entirely different for both block copolymers, though those of spherical and cylindlical structures are almost the same. Thus we conclude that the variation of morphology with composition for triblock copolymers of the ABC type having the end-block polymers with the same chain lengths is similar to that of diblock copolymers, but their structures and composition ranges are different.

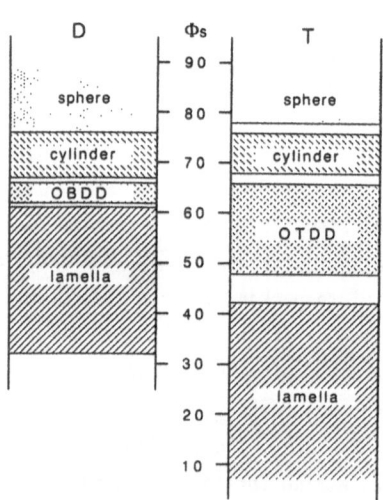

Figure 8. Variation of morphology with volume fractions of S block polymer (Φ_s) of ISP triblock copolymers (T) in comparison with that of SI diblock copolymers (D).

MOLECULAR WEIGHT DEPENDENCE OF LAMELLAR DOMAIN SPACING

Studies on the molecular weight dependence of domain spacing of microdomain structures and the chain conformation of block polymer in the microdomains are important not only to understand the microdomain structures at the molecular level, but also to clarify the chain conformations of polymers in anisotropic and confined spaces.

The molecular weight dependence of domain spacing of AB diblock copolymers have been extensively studied for lamellar structures[1,2], because they can be safely assumed to be in equilibrium in contrast to other morphologies. As shown in Figure 2 the triblock copolymers have lamellar structures around the composition of I:S:P=1:1:1. In the present work, therefore, we prepared the triblock copolymers having the same composition of 1:1:1, but various molecular weights to study the moleculalar weight dependence of the lamellar domain spacing by SAXS[8]. The lamellar domain spacings D of the triblock copolymers were evaluated by using the Bragg equation, $D=2n\pi/q_{max}$ where n is the order of diffraction. Figure 9 shows double logarithmic plots of the experimental D values against the number-average molecular weight of ISP triblock copolymers. Using the least-squares method we obtained the following empirical equation over the molecular weight range from 40k to 280k[8].

$$D = 0.085_5 M_n^{0.77} \ (nm) \tag{1}$$

The exponent 0.77 is much larger than that in the following empirical equation for the molecular weight dependence of lamellar domain spacing of styrene-2-vinylpyridine (SP) diblock copolymers shown in Figure 10[2].

$$D = 0.033_7 M_n^{0.64} \ (nm) \tag{2}$$

The molecular weight dependence of domain spacing of block copolymers are determined by the following three factors: 1) Repulsive interactions between different block polymers, 2) Loss of conformational entropy to maintain the uniform segment density and 3) Loss of placement entropy to confine junction

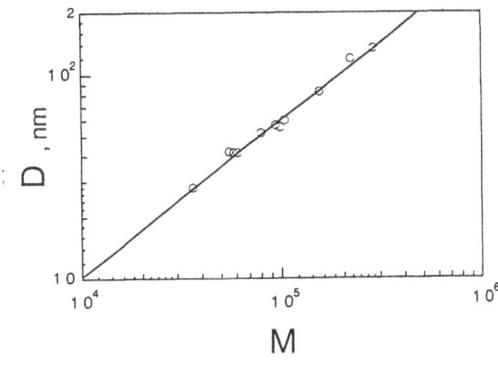

Figure 9. Double logarithmic plots of experimental lamellar domain spacings against number-average molecular weights of ISP triblock copolymers. (Reproduced from ref. 8).

points in the interfacial region. According to Helfand-Wasserman (H-W)[2,9], the lamellar domain spacing of symmetrical diblock copolymers in the strong segregation limit is given by

$$-0.816 \chi^{1/2}bZ + 0.297D^{3.5}/(Z^{1/2}b)^{2.5} + D = 0 \tag{3a}$$

where Z is the number of segment of diblock copolymer, b is the segment length and χ is the Flory-Huggins interaction parameter. The first, second and third terms in eq 3a correspond to the factors 1), 2) and 3), respectively. When Z $\gg$ 1, that is, the third term is neglected, moreover, we have

$$D = 1.34 \chi^{1/7}bZ^{9/14} \tag{3b}$$

Assuming that the free-end of a block chain of diblock copolymers is located in the middle of the domain, Semenov derived the following equation for high molecular weight polymers on the basis of Gaussian statistics[2,10].

$$D = (4/6^{1/2})(3/\pi^2)^{1/3} \chi^{1/6}bZ^{2/3} \tag{4}$$

Using a different theoretical method Ohta-Kawasaki (O-K) also presented a theoretical equation for the diblock copolymers in the strong seglegation limit[2,11].

$$D = (2/3^{5/6}) \chi^{1/6}bZ^{2/3} \tag{5}$$

Figure 10 shows the comparison between the theories and experimental data of lamellar domain spacings of SP diblock copolymers which are practically symmetric, because the molecular parameters of both block components are almost the same[2]. Apparently, the theories of H-W and Semenov are in good agreement with the data.

Since the chain conformation of the block polymer of diblock copolymers, of which the free-end is located in the middle of domain[4], is similar to the bridge conformation of middle block chain of triblock copolymers, the same procedure in the theory of Semenov may be applied to the calculation of conformational entropy for the lamellar structure of ABC triblock copolymers. For a "symmetrical triblock copolymer" that the numbers of segment, segment lengths and densities of A, B and C block polymers of triblock copolymers are the same, we have[8]

$$D = (4/6^{1/2})(6/\pi^2)^{1/3} \chi^{1/6}bZ^{2/3} \tag{6}$$

Here, we assume that χ values between A and B block polymers, and between B and C block polymers are the same. Figure 11 shows the comparison between the experimental D values of ISP triblock and SP diblock copolymers[8]. In this figure the molecular volumes V_m are employed instead of molecular weights for the comparison. Apparently, the lamellar domain spacings of ISP triblock copolymers are larger than those of SP diblock copolymers at least for the high molecular weight samples.

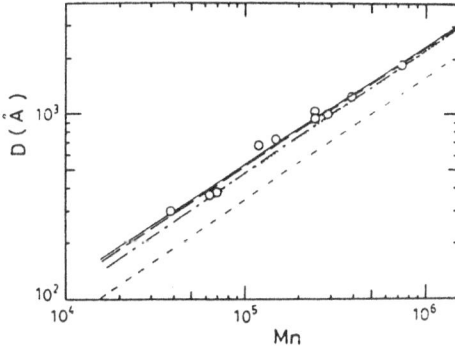

Figure 10. Comparison between the theories and experimental lamellar domain spacings of SP diblock copolymers. (—) eq 2; (···) H-W (eq 3a); (– –) H-W (eq 3b); (–·–·) Semenov (eq 4); (– – –) O-K (eq 5). (Reproduced from ref. 2).

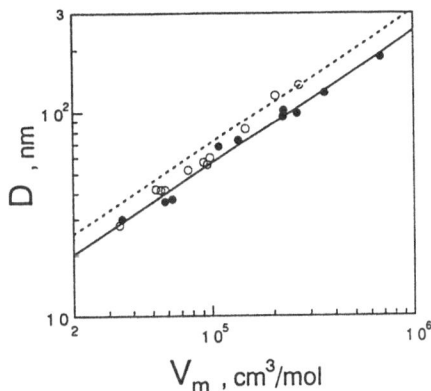

Figure 11. Comparison between lamellar domain spacings of SP diblock (●) and ISP triblock (○) copolymers, The full and broken lines denote eq 2 and the calculated values which are $2^{1/3}$ times larger than the full line, respectively. (Reproduced from ref. 8).

Comparing eq 6 with eq 4, we can predict that the domain spacing of triblock copoymers is $2^{1/3}$ (1.26) times larger than that of diblock copolymers. The difference results from the fact that the number of boundary in the repeating structure of triblock copolymers is twice larger than that of diblock copolymers, because the domain spacing of ABC triblock copolymers consists of A-B-CC-B-A, while that of AB diblock copolymer is A-BB-A. As shown in Figure 11 the experimental results are in good agreement with the theoretical prediction at least for the high molecular weight samples.

With decreasing molecular weight the contribution of factor 3), which is neglected in the theory of Semenov, increases. Assuming that the conformational entropy of B chain is the same as that of A or C chain with the free-end in the theory of Helfand-Wasserman, we have the following equation for lamellar structures of triblock copolymers of the ABC type with different molecular parameters[8].

$$
-2(\gamma_{AB} + \gamma_{BC})\{(Z_A/\rho_{0A}) + (Z_B/\rho_{0B}) + (Z_C/\rho_{0C})\}/kT +
$$

$$
\frac{0.353\{(Z_A^{1/2}/\rho_{0A}b_A)^{2.5} + (Z_B^{1/2}/\rho_{0B}b_B)^{2.5} + (Z_C^{1/2}/\rho_{0C}b_C)^{2.5}\} D^{3.5}}{\{(Z_A/\rho_{0A}) + (Z_B/\rho_{0B}) + (Z_C/\rho_{0C})\}^{2.5}}
$$

$$
+ 2D = 0 \qquad (7)
$$

with $\gamma_{KJ} = kT\alpha_{KJ}^{1/2}\{(\beta_K + \beta_J)/2 + (\beta_K - \beta_J)^2/6(\beta_K + \beta_J)]$

$\alpha_{KJ} = \chi_{KJ}(\rho_{0K}\rho_{0J})^{1/2}$

$\beta_K = (\rho_{0K}b_K^2/6)^{1/2}$

where Z_K is the number of segment in the K-block chain, ρ_{0K} is the number density of K-segment, b_K is the statistical segment length of K-chain and χ_{KJ} is the Flory-Huggins interaction parameter between K-and J-chains. For the symmetrical triblock copolymers eq 7 corresponds to eq 6, if the third term is neglected. Using the molecular parameters of I, S and P block polymers and assuming that $\chi_{IS} = \chi_{SP} = 0.1$ where the subscripts I, S and P denote I, S and P block polymers, respectively, we evaluated theoretical values for ISP triblock copolymers from eq 7 with and without the third term.

Figure 12 shows the comparison between the theories and the experimental data of ISP triblock copolymers[8]. As expected the theoretical line from eq 7 without the third term agree with the experimental data at the high molecular weights. With decreasing the molecular weight the data deviate from the line and follow the theoretical line with the third term.

CHAIN CONFORMATION OF BLOCK POLYMERS IN LAMELLAR MICRODOMAINS

The chain conformation of block polymer of AB diblock copolymers in lamellar microdomains have been extensively studied, not only because lamellar microdomains can be safely assumed to be in equilibrium in contrast to other morphologies as mentioned above, but also because we can determine the dimensions of block chain along the directions parallel and perpendicular to lamellae if the lamellae are oriented along the direction paralell to the surface of film specimens. Recently, we reported that the molecular weight dependences of radii of gyration of S block polymer of SP diblock coplymers along direction perpendicular to lamellae, defined as y-axis, $R_{g, y}$ and along the direction parallel to lamellae, defined as x-and z-axes, $R_{g, x}$ and $R_{g, z}$ are given by[3]

$$R_{g, y} = 0.0045_3 M^{0.64} \text{ (nm)} \tag{8}$$

$$R_{g, x} = R_{g, z} = 0.028_9 M^{0.43} \text{ (nm)} \tag{9}$$

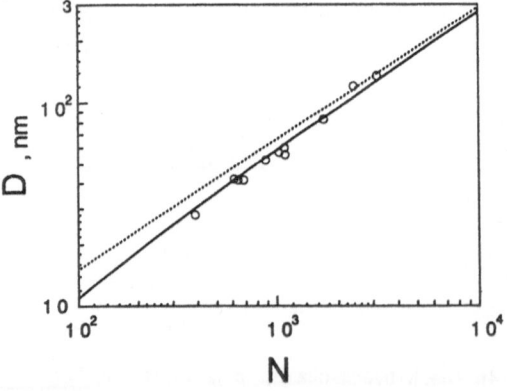

Figure 12. Comparison between the experimental and theoretical lamellar domain spacings of ISP triblock copolymers, where N is the degree of polymerization of the triblock copolymers.
(—) eq 7; ($\cdots$) eq 7 without the third term.

These equations indicate that the block chain of SP diblock coplymers is
extended along the direction perpendicular to lamellae in accordance with the
theories of microphase separation in the strong segregation limit, but it is
contracted along the direction parallel to lamellae to keep the volume occupied
by the block polymer coil constant when the microphase separation occurs. It is
to be noted that the contraction is not predicted by the theories[3].

To study the chain conformation of middle-block polymers in the lamellar
structures of triblock copolymers by SANS, we prepared three triblock
copolymers with deuterated styrene (D) blocks and the same composition of
I:D:P=1:1:1, together with the corresponding ISP triblock copolymers[8].
If lamellae are predominantly oriented along the direction parallel to the film
surface, no diffraction is observed so that a single-chain scattering is
measured at the through-view by SANS without contrast matching as reported for
SP diblock copolymers[3]. For the present triblock copolymers, however,
diffraction peaks were observed even at the through-view[8]. This implies that
the orientaion was not predominant. To extract the single-chain scattering from
the data at the through-view, therefore, we subtracted the circular-averaged
scattering intensity at the edge-view of the same specimens from the scattering
data at the through-view so as to minimize the diffraction peaks by adjusting
the scattering intensity at the edge-view, because the shapes of diffraction
peaks at the through-and edge-views are very similar and the diffraction is
much stronger than the single-chain scattering at the edge-view.

The radii of gyration $R_{g,x}$ evaluated from the Guinier plot thus obtained are
plotted against molecular weight of middle block polymers, together with the
data of diblock copolymers given by eq 9 in Figure 13[8]. This figure suggests

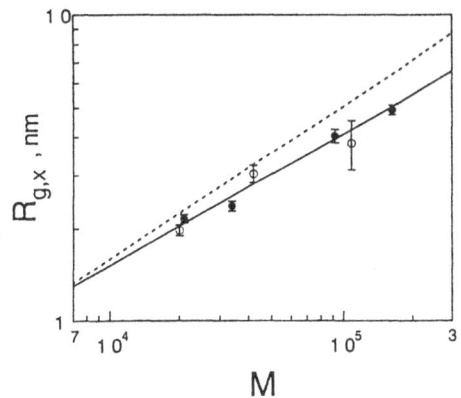

Figure 13. Double logarithmic plots
of radius of gyration of the middle
block polymer along the direction
parallel to lamellae against molecular
weights for ISP triblock copolymers
(O) in comparison with those for SP
diblock copolymers (●). The full and
broken lines denote eq 9 and the
molecular weight dependence of
unperturbed radius of gyration for
polystyrene, respectively.
(Reproduced from ref. 8)

that the middle-block chain is contracted along the direction parallel to lamellae at the almost same magunitude as the block chain of diblock copolymers. As mentioned above, the block chain of diblock copolymers is contracted along the direction parallel to lamellae to keep the volume occupied by the block polymer coil constant. This may be the case for the triblock copolymers, because the triblock copolymers in the molecular weight range in Figure 13 may be extended along the direction perpendicular to lamellae at the almost same magnitude as the diblock copolymer as infered from the domain spacing data in Figure 11. However, we cannot have a definite conclusion, since the experimental errors are fairly large due to the imperfect orientation of lamellae.

Acknowledgment: The authors wish to thank Dr. Charles C. Han of National Institute of Standards and Technology for his cooperation in SANS measurements.

References
1 Bates FS (1990) Ann Rev Phys Chem 41:525
2 Matsushita Y, Mori K, Saguchi R, Nakao Y, Noda I, Nagasawa M (1990) Macromolecules 23:4313
3 Matsushita Y, Mori K, Mogi Y, Saguchi R, Noda I, Nagasawa M, Chang T, Glinka CJ, Han CC (1990) Macromolecules 23:4317
4 Matsushita Y, Mori K, Saguchi R, Noda I, Nagasawa M, Chang T, Glinka CJ, Han CC (1990) Macromolecules 23:4387
5 Mogi Y, Kotsuji H, Kaneko Y, Mori K, Matsushita Y, Noda I (1992) Macromolecules 25:5408
6 Mogi Y, Mori K, Matsushita Y, Noda I (1992) Macromolecules 25:5412
7 Mogi Y, Kotsuji H, Nomura M, Ohnishi K, Matsushita Y, Noda I to be published
8 Mogi Y, Mori K, Kotsuji H, Mori K, Matsushita Y, Noda I, Han CC (1992) Macromolecules in press
9 Helfand E, Wasserman ZR (1976) Macromolecules 9:879
10 Semenov AV (1985) Sov Phys-JETP(Engl Transl) 61:733
11 Ohta T, Kawasaki K (1986) Macromolecules 19:2621

Evolution of Ordering in Thin Films of Symmetric Diblock Copolymers

T.P. Russell

IBM Almaden Research Center
San Jose, California 95120

A.M. Mayes

Massachusetts Institute of Technology
Cambridge, MA 02139

M.S. Kunz

Ciba-Geigy
Basel, Switzerland

Abstract: The manner in which thin films of asymmetric diblock copolymer of polystyrene and poly(methyl methacrylate) form a multilayered structure has been investigated by transmission electron microscopy and neutron reflectivity. It is shown that spin coated films of the diblock copolymer are initially microphase separated with an average repeat period that is much smaller than the equilibrium period. Upon heating, two different relaxations were found. One involves a rapid, local relaxation of the copolymer chains in which the phase size increases to a size scale comparable to the equilibrium period. However, the microstructure is a bicontinuous network of polystyrene and poly(methyl methacrylate). Upon further heating the copolymer chains are transported within the bicontinuous network to form the multilayered structure.

INTRODUCTION

Diblock copolymers are comprised of two chemically distinct homopolymers covalently joined at one end. In general, specific interactions of the two blocks at an interface forces a preferential segregation of the components to the interface. Herein, thin films of symmetric diblock copolymers of polystyrene, PS, and poly(methyl methacrylate), PMMA, denoted P(S-b-MMA), on either a Si or Au substrate will be discussed, though the physics can be extended to copolymers in general. In the case of P(S-b-MMA), the PS segments have a lower surface energy than that of the PMMA segments and, consequently, PS will be located preferentially at the air surface. On a Si substrate, the more polar PMMA segments interact strongly with the native oxide layer on the Si, PMMA, thus, segregates, to the substrate[1]. On Au, however, PS is found preferentially at the substrate[2]. Due to these specific interactions and the fact that the PS chain is covalently linked to the PMMA chain, at equilibrium a multilayered structure comprised of alternating layers of PMMA and PS is found[1-3]. This multilayered structure mandates that the film thickness at any point on the specimen be defined in terms of the repeat period, L. On a Si substrate the film thickness is given by $(n + 1/2)L$ and on Au by nL, where n is an integer. If the initial film thickness does not conform to this constraint, then either islands or holes are found on the surface with a step height of L.[5-7]. This multilayered structure has led to quantitative descriptions of many aspects of the lamellar microdomain morphology of diblock copolymers[4-8].

While the driving forces for the multilayer formation are understood, the mechanism by which the ordering occurs has received only limited attention[9]. It is important to characterize the morphology of the initial film prior to annealing. Are the copolymer chains phase mixed or are they microphase separated? Does the ordering occur via a

A. Teramoto, M. Kobayashi, T. Norisuje (Eds.)
Ordering in Macromolecular Systems
© Springer-Verlag Berlin Heidelberg 1994

diffusion of single chains or is there a coordinated motion of many chains? In this article neutron reflectivity and transmission electron microscopy studies have been used to shed some light on theses issues. It is shown that the initial films are in a nonequilibrium, microphase separated state. Upon annealing, a rapid local relaxation of the copolymer chains occurs with the formation of a bicontinuous network of PS and PMMA. This bicontinuous network provides a path by which copolymer chains can laterally diffuse along the interface and be delivered to the multilayer front propagating into the sample from either the air or substrate interface.

EXPERIMENTAL

Neutron reflectivity measurements were performed on the SPEAR reflectometer at the Manuel P. Lujan, Jr. Neutron Scattering Center at the Los Alamos National Laboratory. The diblock copolymer used in these studies contained a deuterium labelled PS block, denoted P(d-S-b-MMA), with $M_w = 8.0 \times 10^4$ and $M_w/M_n = 1.05$. The fraction of the PS segments in the copolymer was 0.49. Solutions of the copolymer in toluene were deposited on polished 5 cm diameter Si substrates (5 mm thick) and spun at 2000 rpm to remove excess polymer and solvent. The films were then dried at 80°C to remove the solvent. Subsequent heating of the films was performed at 170°C under vacuum. Specimens were annealed for a predetermined time and quenched to room temperature for the reflectivity measurements. Samples for the transmission electron microscopy studies were prepared from P(S-b-MMA) having $M_w = 8.6 \times 10^4$ with $M_w/M_n = 1.03$ and a PS fraction of 0.47. Films of the P(S-b-MMA) were spin coated onto substrates that consisted of 2 mm thick polyimide onto which a ~100 nm layer of Au was evaporated. The films were annealed for a given period of time at 170°C under vacuum and quenched to room temperature. The specimens were subsequently microtomed to reveal an edge view of the copolymer morphology. This sample preparation technique, developed by Kunz and Shull[10], provides an easy means of examining the structure of thin films. To enhance contrast, the thin sections were exposed to RuO_4 vapors that preferentially stained the PS portion of the copolymer. Electron microscopy studies were performed using a Philips 400T electron microscope operated at 100KeV.

RESULTS AND DISCUSSION

Shown in Figure 1 is the neutron reflectivity profile for an 800Å P(d-S-b-MMA) film on a Si substrate after spin coating and drying below the glass transition temperature of the copolymer. The reflectivity is shown as a function of $k_{z,0}$, the neutron momentum in vacuum normal to the film surface. $k_{z,0} = (2\pi/\lambda) \sin\theta$ where λ is the neutron wavelength and θ is the grazing angle of incidence. At low $k_{z,0}$ a region of total external reflection is observed after which the reflectivity decreases rapidly with increasing $k_{z,0}$. Evident in the reflectivity profile are the Kiessig fringes arising from the total thickness of the specimen. No other distinct interferences are immediately evident in the profile. Shown as the dashed line in the figure is the reflectivity profile calculated assuming that the scattering length density was invariant with depth in the sample. A homogeneous scattering length density profile would describe either a phase mixed copolymer, where the PS and PMMA segments were intimately mixed, or a microphase separated copolymer where the microphases were randomly oriented in the specimen. The disagreement between the measured and calculated reflectivity profiles clearly demonstrates that neither of these cases describes the as prepared specimen. The best fit to the reflectivity profile was obtained from the scattering length density profile shown in the inset. Here one has effectively two damped cosine functions propagating into the specimen from both the substrate and air interfaces. Since the PS segments of the copolymer are labelled, the increased scattering length density at the air surface shows that PS is preferentially located at the air surface. Whereas, adjacent to the native oxide layer on the Si, the scattering length density is close to that of pure PMMA. From the connectivity of the PS and PMMA blocks, the periodic variation in the scattering length density of the two components can be understood. From these data it is evident that there is a strong tendency, even during the initial film casting process, for the P(d-S-b-MMA) copolymer to form a multilayered structure. However, from these data alone, one can not conclude that the P(d-S-b-MMA) is microphase separated. In fact,

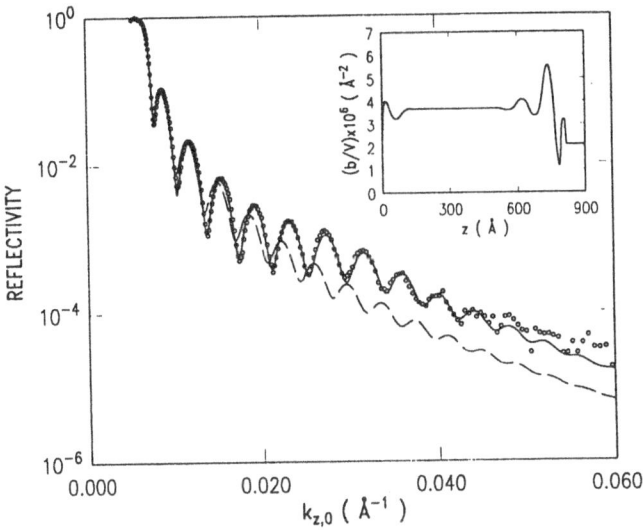

Figure 1. Neutron reflectivity as a function of the neutron momentum normal to the surface, $k_{z,0}$, for P(d-S-b-MMA) after spin coating from a toluene solution onto a Si substrate (used with permission, Ref. 9).

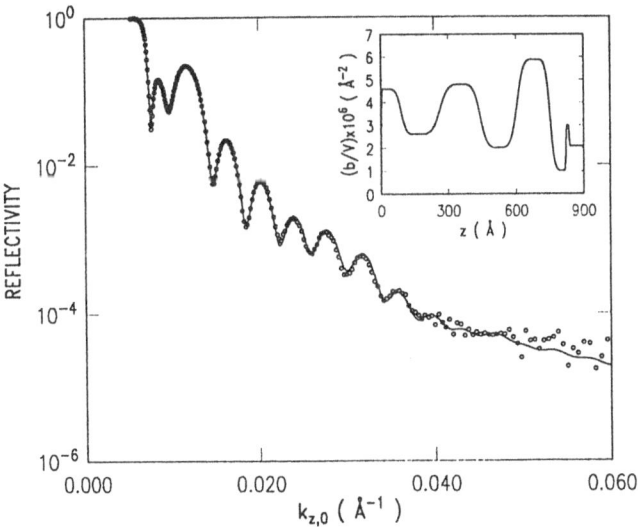

Figure 2. Neutron reflectivity profile of P(d-S-b-MMA) after annealing at 170°C for six minutes (used with permission, Ref. 9).

scattering length density profiles similar to that shown in Figure 1 have been observed for phase mixed copolymers [11]. One must keep in mind that the reflectivity yields the scattering length density laterally averaged over distances of $\sim 1\mu m$, the coherence length of the neutron.

One rather surprising feature of these data is the observed period. From the scattering length density profile a period of ~ 13 nm is found. This is nearly 1/3 the period of the lamellar microdomain structure in the bulk at equilibrium. Even if the copolymer is phase mixed, a period only slightly smaller than the equilibrium period would be expected. Consequently, these results demonstrate that the spin coating process has trapped the copolymer chains in a highly non-equilibrium structure. If the copolymer chains are microphase separated, then these results could only be explained by having the blocks from two different copolymer chains interdigitated and tilted with respect to plane of the lamellae.

Upon heating the films to 170°C for only six minutes, the reflectivity profile shown in Figure 2 was obtained. Relative to the data shown in Figure 1, dramatic changes have occurred. In addition to the Kiessig fringes from the total film thickness, one sees the appearance of a strong interference maximum at $k_{z,0} \sim 0.013 \text{Å}^{-1}$. The solid line is the best fit to the reflectivity profile, calculated using the scattering length density profile shown in the inset. Immediately evident is the fact that the period has nearly tripled and is now identical to the period of the equilibrium lamellar microdomain structure in the bulk, $L = 320\text{Å}$ From the substrate interface one sees a half layer of pure PMMA followed by a full layer of PS. Proceeding further towards the air surface, a diminishment in the ordering is observed. The strong interference maximum in the reflectivity profile is a direct result of this partially ordered, multilayer structure. It is clear, though, that for these thin films, the ordering from the substrate interface appears to dominate since the purity of the layers is much higher at the substrate. This can be explained by the relative strengths of the interactions of the copolymer blocks with the two interfaces.

Further annealing of the specimen at 170°C causes the ordering of the copolymer chains with respect to the surface to improve. After several hours a near perfect multilayered structure is formed. The characteristics of such a multilayered structure has been discussed in great detail previously[4] and will not be dealt with here.

These reflectivity studies prompted an investigation of the ordering process using transmission electron microscopy, TEM. While it would be most desirable to perform the TEM studies on the identical specimens used in the reflectivity studies, the thickness of the Si substrate used for reflectivity prohibited microtoming. Attempts were made to perform TEM studies on an Ultem (polyimide) substrate coated with silicon oxide. However, the brittleness of this silicon oxide layer, coupled with the differences in the thermal expansion coefficients of the coating and the Ultem resulted in a severe cracking of the oxide layer. As an alternative, Au was evaporated onto the Ultem sheet and the copolymer was spin coated onto the surface of the Au. In this case, previous studies have shown that PS is located at the Au substrate[2]. The interactions of the PS with the Au are much weaker than the interactions of the PMMA with silicon oxide. Nonetheless, the interactions are sufficiently strong to induce the formation of a multilayered structure. With these differences in mind, it was hoped that TEM would provide some information to augment the reflectivity results.

Shown in Figure 3 is a TEM micrograph obtained from a thin section of P(S-b-MMA) microtomed normal to the surface of the film. The Ultem substrate and the evaporated Au layer are indicated in the figure. What is immediately evident in the micrograph is the the copolymer is microphase separated and that the microdomains appear to be randomly arranged in the sample. The approximate center to center distance between the PS or PMMA microdomains is ~ 13 nm, close to the period observed in the reflectivity studies. Thus, the TEM results, coupled with the reflectivity results, demonstrate that the copolymer is microphase separated and that, in the center of the films, the microdomains are randomly arranged in space. However, in the vicinity of the substrate and air

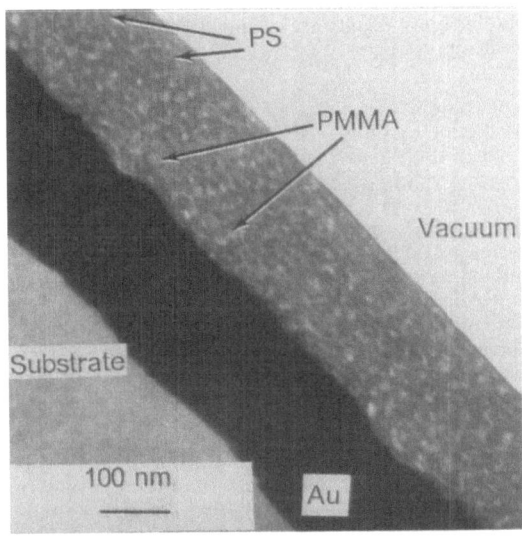

Figure 3. Transmission electron micrograph of an edge view of P(S-b-MMA) after spin coating from a toluene solution.

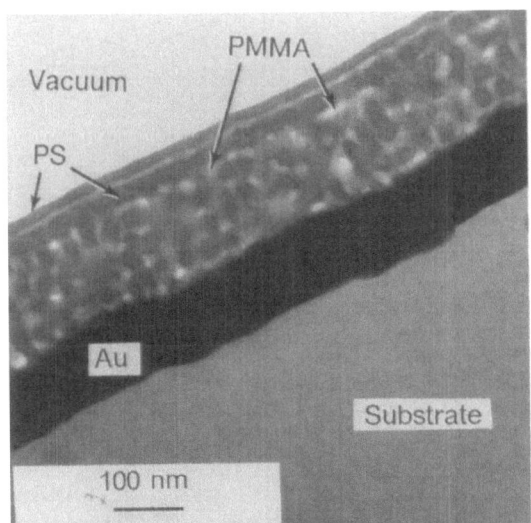

Figure 4. Transmission electron micrograph of an edge view of a thin film of P(S-b-MMA) on an Au coated polyimide substrate after annealing for two hours at 175°C under vacuum.

interfaces, the interactions of the blocks with the interface causes and orientation of the microdomains. The reflectivity results are an indication of the distance over which the influence of the interface propagates.

A TEM image of a copolymer annealed at 175°C for 2hrs. is shown in Figure 4. It should be noted that the interactions of PS and PMMA with the Au or air interface are much weaker than that of the PMMA with the native oxide layer on Si. Consequently, the time scale for the ordering will be extended over much longer times. In this micrograph it is seen that the average center to center distance has increased substantially over that observed initially. In fact, it is now close to the microdomain spacing in the bulk at equilibrium. This increased period requires a rapid, local relaxation of the chains. Since the microphase separated structure is retained, this relaxation could involves a simple disentanglement of interdigitated chains. This would occur very quickly and would be irreversible. In the micrograph, it is also evident that there is an orientation of the lamellar multilayer parallel to the air interface. No such orientation is seen from the Au interface. As expected the PS layer adjacent to the air interface is one-half the period in the bulk. This results from the bilayer nature of the lamellar microdomain morphology.

A careful examination of Figure 4 reveals that from the second PS layer adjacent to the surface through the remainder of the sample, the PS phases are not one single shade of gray. There are areas, roughly circular in shape, that are darker than the remaining PS phase. This is a strong indication that the PS and the PMMA are forming a bicontinuous network. This observation is critical since it provides the mechanism by which the copolymer can be delivered to the growing multilayered structure. In a bicontinuous network it is very easy for the copolymer chains to diffuse laterally along the network, keeping the junction points at the interface between the PS and PMMA microphases.

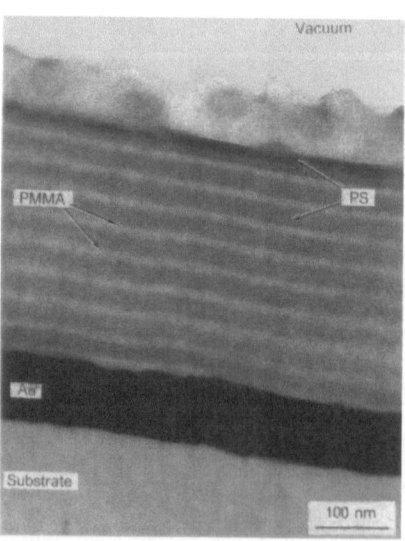

Figure 5. Transmission electron micrograph of an edge view of a thin film of P(S-b-MMA) on an Au coated polyimide substrate after annealing for twenty-four hours at 175°C under vacuum.

Finally, after annealing the sample at 170°C for 24 hours, the micrograph obtained in Figure 5 was obtained. This micrograph shows the near perfect orientation of the lamellar microdomains parallel to the substrate surface. In fact, within the image, no defects are evident even though the interactions between the PS and PMMA blocks of the copolymer and the two interfaces are rather weak. The orientation is quite strong as evidenced by these data.

CONCLUSION

Herein it has been shown that the self-assembly of symmetric diblock copolymer into a multilayered structure in thin films, occurs in a rather interesting manner. First, a non-equilibrium microphase separated structure was observed whereby the copolymer chains appeared to be interdigitated. Rapidly, this interdigitation is relaxed with the formation of a bicontinuous network of PS and PPMA microdomains. The average period in this network was nearly identical to that of the of the lamellar microdomain period in the bulk. This network provides a convenient route by which copolymer chains can be transported to the growing multilayer front with a diffusion coefficient that is not significantly different from the bulk diffusion coefficient. As more copolymer chains are fed to the interface, the multilayered structure grows into the bulk of the sample.

Acknowledgements·

The authors wish to acknowledge Dr. Gregory Smith of the Los Alamos National Laboratory for assistance with the neutron reflectivity studies. This work was partially supported by the U.S. Department of Energy, Office of Basic Energy Sciences under contract DE-FG03-88ER45375. This work benefited from the use of SPEAR at the Los Alamos National Laboratory which is supported by the U.S. Department of Energy, Office of Basic Energy Sciences and other Department of Energy programs under contract W-7405-ENG-36 to the University of California.

REFERENCES

1 Coulon G, Russell TP, Deline VR, Green PF (1989) Macromol 22:2581
2 Russell TP, Coulon G, Deline VR, Miller DC (1989) Macromol 22:4600
3 Anastasiadis SH, Russell TP, Satija SK, Majkrzak CF (1989) Phys Rev Lett 62:1852
4 Anastasiadis SH, Russell TP, Satija SK, Majkrzak CF (1990) J Chem Phys 92:5677
5 Coulon G, Collin B, Ausserre D, Chatenay D, Russell TP (1990) J Phys France 51:2801
6 Ausserre D, Chatenay D, Coulon G, Collin B (1990) J Phys France 51:2571
7 Collin B, Chatenay D, Coulon G, Ausserre D, Gallot Y (1992) Macromol 25:1621
8 Mayes AM, Johnson RD, Russell TP, Smith SD, Satija SK, Majkrzak CF, (1993) Macromol 26:1407
9 Russell TP, Mayes AM, Bassereau P (1993) Physica A in press
10 Kunz MS, Shull KR (1993) Polymer 34:2427
11 Menelle A, Anastasiadis SH, Russell TP, Satija SK, Majkrzak CF (1992) Phys Rev Lett 68:67

Rheology of Polymeric Materials with Mesoscopic Domain Structure

Masao Doi, James Harden and Takao Ohta*

Department of Applied Physics, Faculty of Engineering
Nagoya University, Chikusa, Nagoya Japan 464
* Department of Physics, Faculty of Science,
Ochanomizu University, Bunkyo-ku, Tokyo 112 Japan

Abstract

Computer simulation is carried out for rheological properties of a hexagonally ordered mesophase of block copolymers. When the applied strain is small, the stress response is that of usual viscoelastic solid. On the other hand, when the applied strain is large, anomalous rheological response is observed:(i) for an oscillatory shear of low frequency, the stress-strain curve becomes lozenge shape, and (ii) for a step shear, the stress relaxes in two steps. A simple model describing such behavior is proposed based on the observation that the anomalous behavior appears when the slippage of lattice planes takes place. The model reproduces many characteristic features found by the computer simulation. The possible relevance of this model to more general cases of ordered liquids is discussed.

1. Introduction

Block copolymers form ordered phase by microphase separation [1, 2]. When an ordered phase is formed, the rheological properties change drastically[3, 4]. In the disordered state, the rheological responses are those of typical viscoelastic liquids. There is a clear regime of linear viscoelasticity: the viscosity approaches a constant value at low shear rate , and the storage modulus $G'(\omega)$ and the loss modulus $G''(\omega)$ behave as $G'(\omega) \sim \omega$ and $G''(\omega) \sim \omega^2$ at low frequency. In the ordered phase, on the other hand, the steady state viscosity increases indefinitely as the shear rate is lowered (yield behavior) [5, 6], and the frequency dependence of the complex modulus becomes fractional at low frequency, typically $G'(\omega) \sim G''(\omega) \sim \omega^\alpha$ with $\alpha \simeq 0.5$ [7, 8].

Theoretical study of the block polymers has been mostly concerned with equilibrium structures [9]-[11], but increasing attention has recently been paid to dynamics [12]-[14]. The dynamics of the ordered phase of block copolymers has two aspects. One is the problem of individual chains: how a copolymer chain can move in a nonuniform environment. The other is the problem of the structural change of mesophases: how the domains change their shape and relative position in response to externally applied perturbations. The former problem has been studied by several groups from the point of view of the relaxation of inter-domain chain entanglements [12, 13]; while the latter problem has not been tackled so far, with a few exceptions [14].

In this paper, we shall focus our attention to the latter problem. We consider a microphase-separated state of AB- diblock copolymers, and study its dynamical response to shear deformation. We consider the case that the mesophase is a defect-free hexagonal array of cylindrical domains, and restrict our attention to shear deformations applied perpendicularly to the axes of the cylinders. Thus the problem we consider is effectively two-dimensional.

A. Teramoto, M. Kobayashi, T. Norisuje (Eds.)
Ordering in Macromolecular Systems
© Springer-Verlag Berlin Heidelberg 1994

First we attack the problem by computer simulation[15]. By using the cell dynamic approach, we study how a shear strain deforms the domains and lattice structure and how such deformation is reflected in the rheological properties. Our simulation indicates that the rheological properties are quite different from both those of homogeneous copolymer phases and of linear viscoelastic crystalline solids. We shall show that this anomalous behavior is correlated with the relative slipping of lattice planes.

Next, we develop a simple model to explain the anomalous behavior[16]. We consider a multi-layer of system of macrolattice planes which interact with each other via periodic potential, and calculate the system response when the top layer is sheared relative to the bottom. We will show that this simple model explains the characteristic features of the rheology of the ordered phases in the simulations, and also qualitatively reproduces some of the rheological behavior observed in experiments on microphase-separated block copolymer solutions in selective solvents.

2. Computer Simulation
2.1 Model

We consider a mesophase of AB diblock copolymers. Since our purpose here is to study the effect of the deformation of domains, we do not consider the motion of individual chains. Since the characteristic time of chain motion is much shorter than the time scale of the structural change of domains, chain conformations may be assumed to be in equilibrium for a given domain conformation. Thus we describe the state of the system by the local volume fraction of A and B segments, $\phi_A(r)$ and $\phi_B(r)$. We assume that the system is incompressible ($\phi_A(r) + \phi_B(r) = 1$), so that the state of the system is described by a single order parameter

$$\psi(r) = \phi_A(r) - \phi_B(r) \tag{1}$$

Due to mass conservation of A and B species, the spatial average of $\psi(r, t)$ is determined solely by the block ratio $f \equiv N_A/(N_A + N_B)$ via

$$\bar{\psi} \equiv \frac{1}{V} \int dr \psi(r) = 2f - 1 \tag{2}$$

where N_A and N_B are, respectively, the degree of polymerization of A and B blocks.

For the dynamics of $\psi(r, t)$, we assume the standard time dependent Ginzburg Landau equation. If there is no macroscopic flow, this is written as

$$\frac{\partial \psi}{\partial t} = M \nabla^2 \frac{\delta H\{\psi\}}{\delta \psi} \tag{3}$$

Here M is a transport coefficient, and $H\{\psi\}$ is the free energy functional of the order parameter field $\psi(r)$ in units of the thermal energy $k_B T$.

Ohta and Kawasaki [11] calculated $H\{\psi\}$ using a mean-field approximation and proposed that $H\{\psi\}$ for block copolymers can essentially be written as a sum of two terms: a short-range interaction term $H_S\{\psi\}$ and a long-range interaction term $H_L\{\psi\}$,

$$H\{\psi\} = H_S\{\psi\} + H_L\{\psi\} \tag{4}$$

$H_S\{\psi\}$ has the same form as that for polymer blends [17]:

$$H_S\{\psi\} = \int d\mathbf{r}[W(\psi) + \frac{1}{2}D(\nabla\psi)^2] \tag{5}$$

where $W(\psi)$ is a function of ψ with two local minima, whose explicit form will be given later; and where the term $D(\nabla\psi)^2$ represents the free energy cost for spatial composition inhomogeneity. On the other hand, $H_L\{\psi\}$ is obtained from the long range interaction potential $G(\mathbf{r})$ which is the solution of $\nabla^2 G(\mathbf{r} - \mathbf{r}') = -\delta(\mathbf{r} - \mathbf{r}')$ as

$$H_L\{\psi\} = \frac{\alpha}{2} \int d\mathbf{r} \int d\mathbf{r}' G(\mathbf{r} - \mathbf{r}')[\psi(\mathbf{r}) - \bar{\psi}][\psi(\mathbf{r}') - \bar{\psi}] \tag{6}$$

where $\alpha \propto [Nbf(1 - f)]^{-2}$ with a constant of proportionality that is of order unity, and $N = N_A + N_B$ is the total copolymer degree of polymerization.

The above form of the free energy functional is an approximation for the exact power series expansion of the free energy with respect to $\psi(\mathbf{r})$. The exact form contains higher order, long range couplings of $\psi(\mathbf{r})$, and a much more involved form of $G(\mathbf{r})$. However, as it was demonstrated by Ohta and Kawasaki [11], this form of the free energy reproduces many essential features of block copolymer mesophase behavior, including (i) the two thirds power law for the domain size R in the strong segregation limit ($R \propto N^{2/3}$), and (ii) the appropriate structural transitions of the mesophases when the block ratio f is varied. Furthermore, Bahiana and Oono [18] conducted a computer simulation based on this free energy and showed that the microphase-separated structures are actually formed. Therefore, although the above expression for the free energy is approximate, we believe that it describes the essential physics of the ordered phase.

In the present paper we consider the case that there is a macroscopic shear flow, the velocity field of which is given by

$$v_x(\mathbf{r}) = \dot{\gamma}(t)y, \qquad v_y = v_z = 0 \tag{7}$$

where the shear rate $\dot{\gamma}(t)$ is a given function of time. Under such a flow field, the time evolution equation for $\psi(\mathbf{r}, t)$ given in eq.(3) acquires a convective term and becomes

$$\frac{\partial \psi}{\partial t} + \nabla \cdot (v\psi) = M\nabla^2 \frac{\delta H\{\psi\}}{\delta \psi} \tag{8}$$

In the simulation, we will choose the coordinate system so that the x-direction is parallel to a basis vector of the ordered hexagonal lattice formed by the equilibrium mesophase.

Substituting eqs.(5), (6) and (7) into eq.(8), we have

$$\frac{\partial \psi}{\partial t} = -\dot{\gamma}y\frac{\partial \psi}{\partial x} + M\nabla^2[-D\nabla^2\psi + F(\psi)] - M\alpha(\psi - \bar{\psi}) \tag{9}$$

where $F(\psi) \equiv dW/d\psi$. Notice that due to the Laplacian in eq.(3), the long range term (the term proportional to α) becomes local in the equation of motion. If we set $\alpha = 0$, eq.(9) becomes the same as that used previously in the study of the macrophase separation of binary blends under shear flow [19].

Given the structure of the deformed mesophase, the stress can be calculated by the following formula derived in Ref.[20].

$$\sigma_{\alpha\beta} = -k_B T \frac{D}{V} \int d\boldsymbol{r} \frac{\partial \psi}{\partial r_\alpha} \frac{\partial \psi}{\partial r_\beta} + k_B T \frac{\alpha}{V} \int d\boldsymbol{r} \int d\boldsymbol{r'} r_\beta \frac{\partial G(\boldsymbol{r})}{\partial r_\alpha} \psi\left(\boldsymbol{r'} + \frac{\boldsymbol{r}}{2}\right) \psi\left(\boldsymbol{r'} - \frac{\boldsymbol{r}}{2}\right) \quad (10)$$

To solve the equation we used follow the cell dynamic approach introduced by Oono and Puri [21]. We consider a $L \times L$ square lattice of cell size a, and define the concentration $\psi(\boldsymbol{n}, t)$ for each cell at the lattice point $\boldsymbol{n} = (n_x, n_y)$, where n_x and n_y are integers between 1 and L. The Laplacian ∇^2 is then transformed as

$$\nabla^2 X \rightarrow \frac{1}{a^2}(<< X >> -X) \quad (11)$$

where $<< X >>$ stands for the average over cells neighboring the cell $\boldsymbol{n}$. For the present system, we used the following scheme:

$$<< X(\boldsymbol{n}, t) >> = \frac{1}{6} \sum_{i \in \{N\}} X(i, t) + \frac{1}{12} \sum_{i \in \{NN\}} X(i, t) \quad (12)$$

where $\{N\}$ and $\{NN\}$ represent the nearest neighbor cells and the next nearest neighbor cells, respectively. Equation (9) is then transformed to the following difference equation:

$$\psi(\boldsymbol{n}, t+1) = \psi(\boldsymbol{n}, t) - \dot{\gamma} n_y \frac{1}{2} [\psi(n_x + 1, n_y, t) - \psi(n_x - 1, n_y, t)]$$

$$+ << I(\boldsymbol{n}, t) >> -I(\boldsymbol{n}, t) - \alpha[\psi(\boldsymbol{n}, t) - \bar{\psi}] \quad (13)$$

where $I(\boldsymbol{n}, t)$ is the discrete version of the thermodynamic force $\delta H_S\{\psi\}/\delta\psi$ arising from the short range part of the free energy:

$$I(\boldsymbol{n}, t) = F(\psi(\boldsymbol{n}, t)) - D[<< \psi(\boldsymbol{n}, t) >> -\psi(\boldsymbol{n}, t)] \quad (14)$$

In eq.(13), the units of energy, length and time are chosen in such a way that the lattice constant a and the transport coefficient M can be eliminated. An example of the equilibrium configuration generated by eq.(13) with $\dot{\gamma} = 0$ is shown in Fig. 1.

2.2 Results of Computer Simulation
2.2.1 Oscillatory Strain

When an oscillatory strain

$$\gamma(t) = \Gamma \sin(\omega t), \quad (15)$$

is applied, the shear stress becomes oscillatory as well. Fig. 2 shows plots of the shear stress $\sigma_{xy}(t)$ against the shear strain $\gamma(t)$. We shall call such plots Lissajous patterns. Due to the

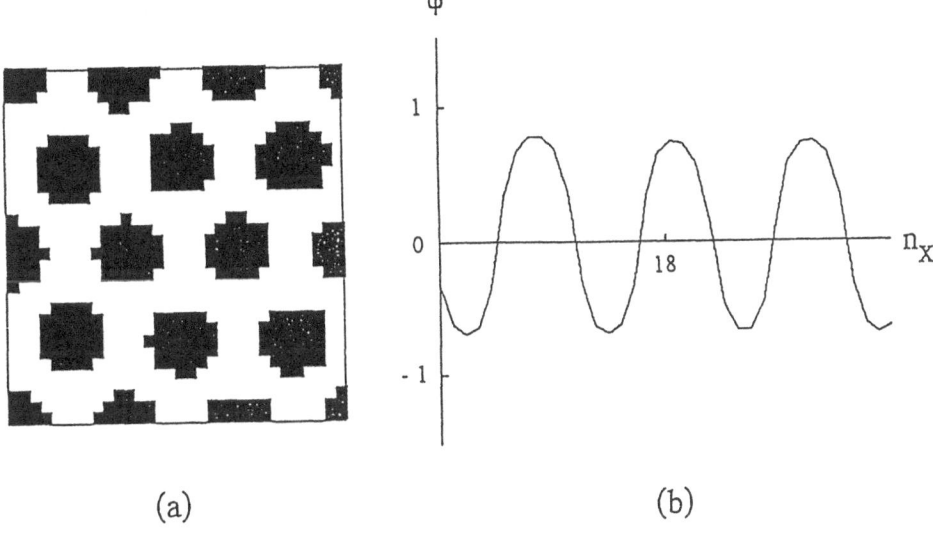

(a)   (b)

Figure 1: (a) Equilibrium domain pattern for the system of L=32. The dark area denotes the region of $\psi > \bar{\psi}$. (b) The concentration profile along the line of $n_y = 10$.

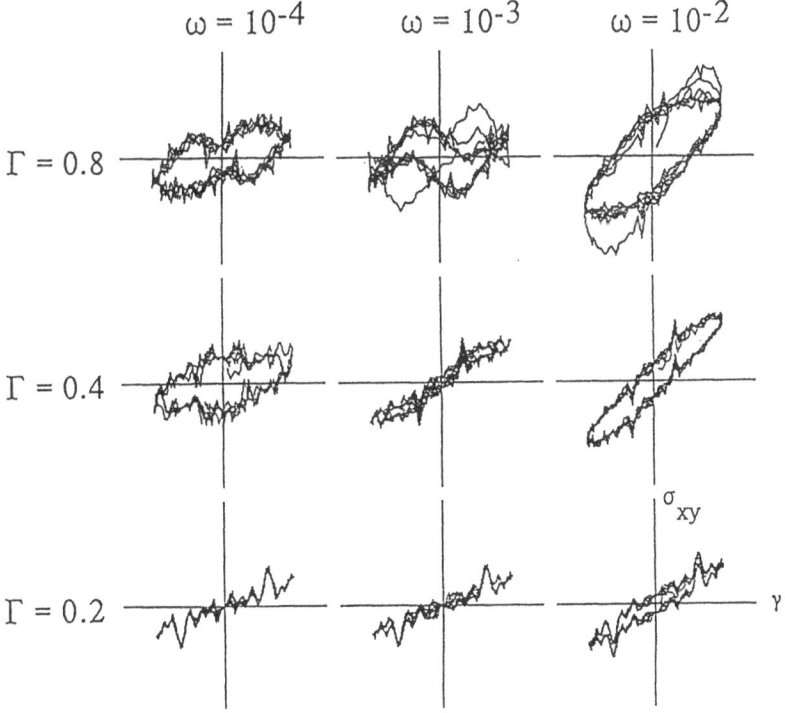

Figure 2: Lissajous pattern of the stress-strain relation. The scale of the abscissa is adjusted for different values of Γ.

small size of the system, there are considerable fluctuations in the stress. However, one can see several general trends:

(1) For the small amplitude, $\Gamma = 0.2$, the curves are typical of the viscoelasticity of soft crystalline solids. At low frequencies $\omega = 10^{-4}$ and 10^{-3}, the Lissajous patterns are essentially lines, indicating that the system responds as a pure elastic solid. On the other hand, the curve becomes elliptic for $\omega = 10^{-2}$, indicating an appearance of viscous component in the stress at higher frequency.

(2) For a larger amplitude, $\Gamma = 0.4$, some anomalous behavior appears. The Lissajous patterns for $\omega = 10^{-2}$ and 10^{-3} can be still interpreted in terms of the linear viscoelastic model. At high frequency, $\omega = 10^{-2}$, the pattern is elliptic; while at lower frequency, $\omega = 10^{-3}$, the pattern becomes a line, which in this case is slightly bent at the extremities indicating that there is a small non-linear effect in the equilibrium stress-strain relationship. The unusual behavior is seen at the lowest frequency, $\omega = 10^{-4}$. Here the Lissajous pattern adopts a rather wide lozenge shape.

(3) At an even larger amplitude, $\Gamma = 0.8$, the nonlinear effect becomes more apparent. The pattern is deformed even at the highest frequency, $\omega = 10^{-2}$. Furthermore, for lower frequencies, there even appears a region of negative slope near the center of the Lissajous pattern.

2.2.2 Step Strain

To understand the unusual behavior of the Lissajous patterns, we carried out computer simulation for an applied step-shear.

$$\gamma(t) = \begin{cases} 0 & \text{for } t < 0 \\ \gamma & \text{for } t > 0 \end{cases} \tag{16}$$

Fig. 3 shows the relaxation of the shear stress σ_{xy} for several values of the strain. When the applied strain is small ($\gamma = 0.1, 0.3$), the stress relaxes to the final stationary value with a relaxation time of about $\tau = 10^2$. On the other hand, when the applied strain is large ($\gamma = 0.4$), unusual behavior is observed. After the initial relaxation at $t \simeq 10^2$, the stress level remains stationary for a significant period of time, and then starts to decrease again at $t \simeq 1.2 \times 10^4$. Fig. 4 shows the mesophase structural evolution during the course of the stress relaxation for $\gamma = 0.4$. Just after the strain is imposed, the domains are deformed, and the lattice is deformed from a hexagonal to a rectangular one (Fig. 4(a)). The domain deformation quickly relaxes in a time scale of the order of 10^2. However, each domain is still distorted (slightly elongated along the x direction) even at 5000 time steps(Fig. 4(b)). By watching the time evolution of the domain patterns, one can see that a layer slippage takes place as the configuration evolves from that shown in Fig. 4(c) to that in Fig. 4(d): the layer indicated by A has moved to the right while the layer B has moved to the left. The time at which the slippage takes place approximately corresponds to the time at which the stress shows the second relaxation.

3. Theoretical Analysis
3.1 Layer Slippage Model

To understand the anomalous results obtained by the simulation, we consider a simple model. In this model, we do not consider the deformation of the domains explicitly, and

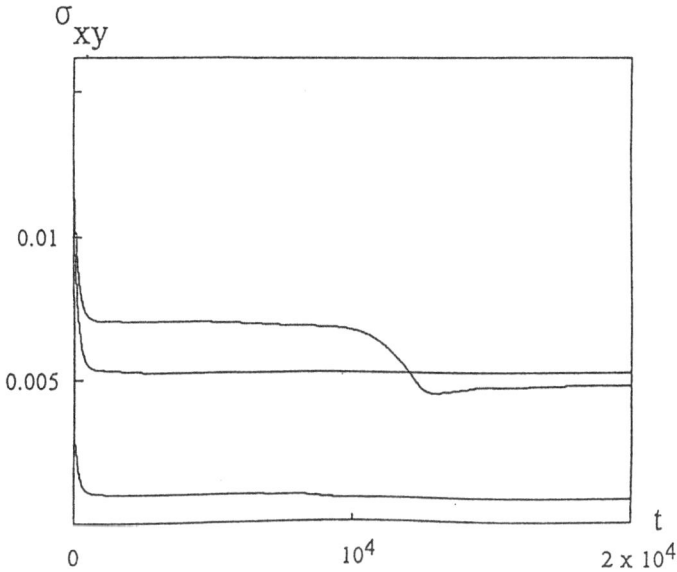

Figure 3: Time evolution of the shear stress σ_{xy} after the application of a step-strain. The applied strains are $\gamma = 0.4, 0.3$, and 0.1 from top to bottom.

describe the state of the system by the horizontal displacement of the layer x_i (see Fig. 5). Thus, the state of the system is entirely characterized by the following dimensionless parameters

$$\tilde{\gamma}_i = 2\pi\frac{x_{i+1} - x_i}{a} \tag{17}$$

where a is the periodicity of the lattice structure and i is the layer index. Notice that $\tilde{\gamma}_i$ is related to the local shear strain $\gamma_i \equiv (x_{i+1} - x_i)/h$ h being the distance between the layers, by

$$\tilde{\gamma}_i = 2\pi\frac{h}{a}\gamma_i \tag{18}$$

Since $\tilde{\gamma}_i$ and γ_i are strictly proportional, we shall write $\tilde{\gamma}_i$ as γ_i in the following.

We now consider the shear stress σ_i acting between the $i+1$ and i th layers. In equilibrium, $\sigma_i(t)$ must be a periodic function of $\gamma_i(t)$

$$\sigma_i(t) = F(\gamma_i(t)) \qquad \text{with} \qquad F(\gamma_i + 2\pi) = F(\gamma_i) \tag{19}$$

However, if the system is not in equilibrium, viscoelastic effects are important. Therefore, we assume the following relation between stress and strain for each layer,

$$\sigma_i(t) = F(\gamma_i(t)) + \int_{-\infty}^{t} dt' G(t - t')\dot{\gamma}_i(t') \tag{20}$$

Figure 4: Evolution of the domain patterns in the relaxation process after the application of a step-shear: (a) $t = 1$ (the pattern immediately after the shear is applied), (b) $t = 5000$, (c) $t = 10000$, and (d) $t = 15000$. In order to aid in the identification of the domains under the sheared boundary condition, two extra cells are added at the boundaries.

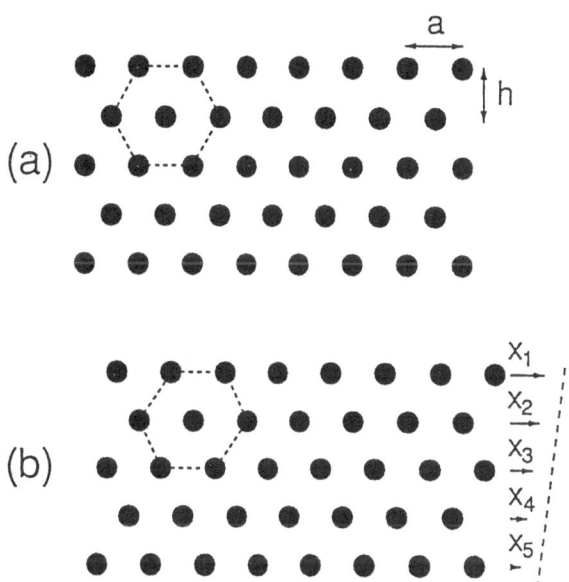

Figure 5: Sketch of a cross section of a cylindrical mesophase in the equilibrium and deformed states. The dashed lines in (a) and (b) indicate, respectively, the equilibrium and sheared hexagonal unit cells.

where $G(t)$ is a relaxation function which vanishes asymptotically as $t \to \infty$.

To explicitly carry out the analysis, we assume the simplest form for $F(\gamma)$ and $G(t)$:

$$F(\gamma) = G_1 \sin \gamma \tag{21}$$

$$G(t) = G_2 \exp(-t/\tau) \tag{22}$$

Equations(20)-(22) describe the local stress-strain relation for each layer. The macroscopic stress-strain relation for the multilayer system can be obtained from these equations. Since inertial effects are negligible, all σ_i must be equal. We thus replace σ_i by the macroscopic shear stress σ, to obtain

$$\sigma(t) = G_1 \sin(\gamma_i(t)) + \int_{-\infty}^{t} dt' G_2 \exp(-(t - t')/\tau) \dot{\gamma}_i(t') \tag{23}$$

The macroscopic strain $\bar{\gamma}$ is simply the average of the local layer strains:

$$\bar{\gamma} = \frac{1}{N} \sum_{i=1}^{N} \gamma_i \tag{24}$$

where N is the number of layers in the system. Equations(23) and (24) define the relation between the macroscopic stress σ and macroscopic strain $\bar{\gamma}$.

In the following, we shall develop equations in a dimensionless form. We choose G_1 as the unit of stress (or elastic constant), and τ as the unit of time. Thus eq.(23) is written as

$$\sigma(t) = \sin(\gamma_i(t)) + \int_{-\infty}^{t} dt' G \exp[-(t - t')]\dot{\gamma}_i(t') \tag{25}$$

where $G = G_2/G_1$. Equation(25) can also be expressed in a differential form:

$$(G + \cos\gamma_i)\dot{\gamma}_i + \sin\gamma_i = \sigma + \dot{\sigma} \tag{26}$$

We now show that this model reproduces the characteristic features of the reults of the sumulation.

3.2 Stress Relaxation after Step Strain

Consider a step-strain experiment. Suppose a step-strain $\bar{\gamma}$ is applied at time $t = 0$. If all γ_i are strictly equal to $\bar{\gamma}$ at $t = 0_+$, then they remain so for $t > 0$. In practice, however, there are some fluctuations. Let us consider how the fluctuations $\delta\gamma_i = \gamma_i(t) - \bar{\gamma}$ propagate in time. Linearizing eqs.(24) and (26) with respect to $\delta\gamma_i$ and $\delta\sigma = \sigma - \bar{\sigma}$, we have

$$\sum_{i=1}^{N} \delta\gamma_i = 0 \tag{27}$$

$$(G + \cos\bar{\gamma})\delta\dot{\gamma}_i + \cos(\bar{\gamma})\delta\gamma_i = \delta\sigma + \delta\dot{\sigma} \tag{28}$$

From eq.(27), it follows that the right hand side of eq.(28) is equal to zero, and hence

$$\delta\dot{\gamma}_i = -\frac{\cos\bar{\gamma}}{G + \cos\bar{\gamma}}\delta\gamma_i \tag{29}$$

If $\bar{\gamma} > \pi/2$, the coefficient of $\delta\gamma_i$ on the right hand side of eq.(29) is positive. Therefore, the fluctuation grows exponentially in time, and a bifurcation of the solution takes place. We solved the time evolution of eqs.(24) and (26) numerically, and the results are shown in Fig. 6.

When the strain is small, the stress relaxes with a single relaxation time $\tau = 1$ associated with viscoelastic domain relaxation, approaching the equilibrium value, while , the double step relaxation is observed when the applied strain is large. The double relaxation is due to the slippage. This is clearly demonstrated in Fig. 7 where the trajectories of $\gamma_i(t)$ are shown.

3.3 Oscillatory Shear

Next we consider the oscillatory shear:

$$\bar{\gamma}(t) = A \sin\omega t \tag{30}$$

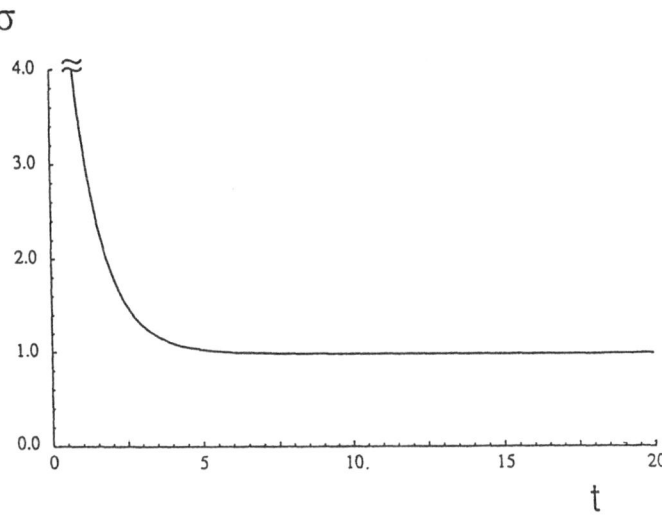

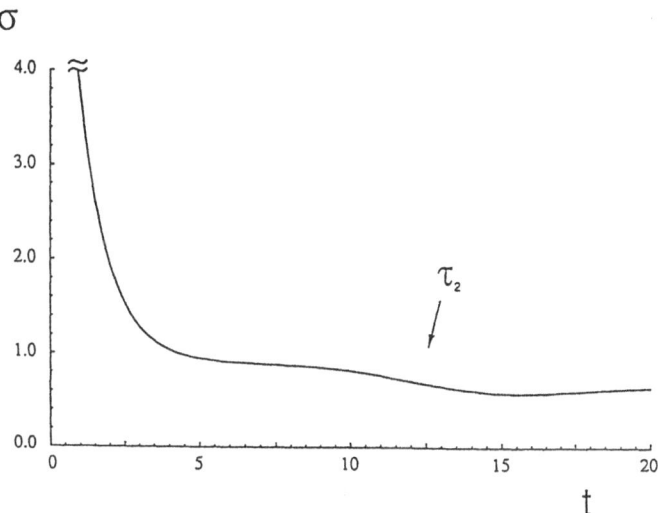

Figure 6: Top: Plot of shear stress $\sigma(t)$ vs t, after an applied step-shear strain of $\bar{\gamma} = 1.4$. Bottom:the same plot for $\bar{\gamma} = 2.0$

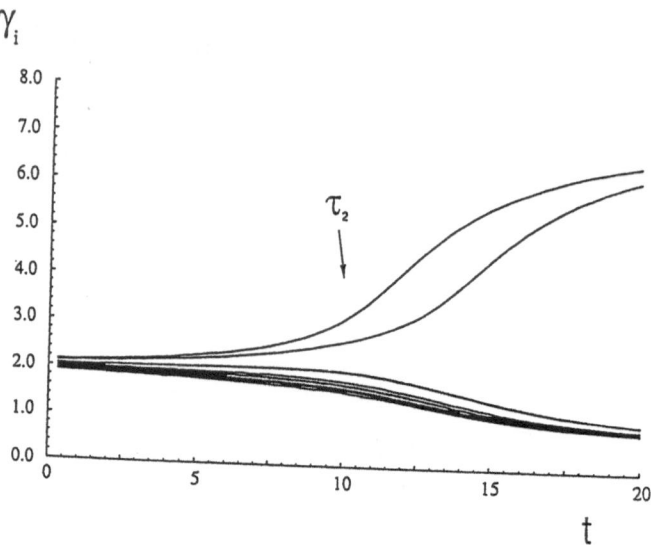

Figure 7: Plot of lattice plane shear strains γ_i vs t corresponding to the stress relaxation plot shown in Fig. 6.

In this case, the linearized equation of eq.(26) becomes

$$\delta\dot{\gamma}_i = -\frac{\cos\bar{\gamma} - \dot{\bar{\gamma}}\sin\bar{\gamma}}{G + \cos\bar{\gamma}}\delta\gamma_i \qquad (31)$$

where we have assumed $G \gg 1$. Analysis of eq.(31) indicates that for $\omega \to \infty$, the solution becomes unstable at the critical amplitude $A_{c_1} = 2.405$, which is the smallest solution of $J_0(x) = 0$, where J_0 is the Bessel function of order 0. On the other hand, for $\omega \to 0$, the solution becomes unstable at $A_{c_2} = \pi/2 \simeq 1.56$.

In order to ascertain the the frequency dependence of the stability limits and the resulting stress-strain patterns in the slipping and non-slipping regimes, we have made numerical calculations of the time evolution of eqs.(24) and (26) under imposed periodic shear strains. The results of numerical integration of eqs.(24) and (26) are summarized in Fig. 8, while specific examples of stress-strain patterns for $G = 2$ at low and high ω and for large and small A are shown in Fig. 9. The numerical analysis indicates that in the region where the homogeneous solution becomes unstable, various solutions are possible. The characteristic features of the solutions are the following:

(i) $\omega \to 0$: For $A > \pi/2$, the Lissajous pattern adopts a lozenge shape like that shown in Fig. 9(b). This pattern can be understood as follows. Consider the idealized sketch given in Fig. 10. Initially, all layers start with approximately zero displacement, and follow the homogeneous solution. Accordingly the stress increases as $\sigma = \sin\gamma + G\gamma$ (c.f. the interior line in Fig. 10. However, as $\bar{\gamma}$ approaches $\pi/2$, the homogeneous solution becomes unstable, and slip planes develop. Thus for $\bar{\gamma} > \pi/2$, further macroscopic strain is accomplished by slippage up to $\bar{\gamma} = A$. During this period, most layers have their local strain frozen around

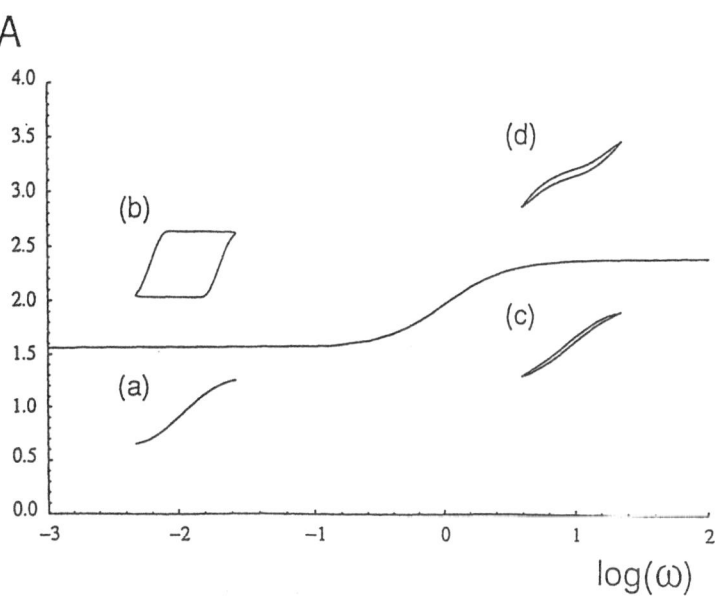

Figure 8: Sketch of the various regimes of oscillatory stress-strain patterns for $G = 2$ as a function of frequency ω and shear strain amplitude A.

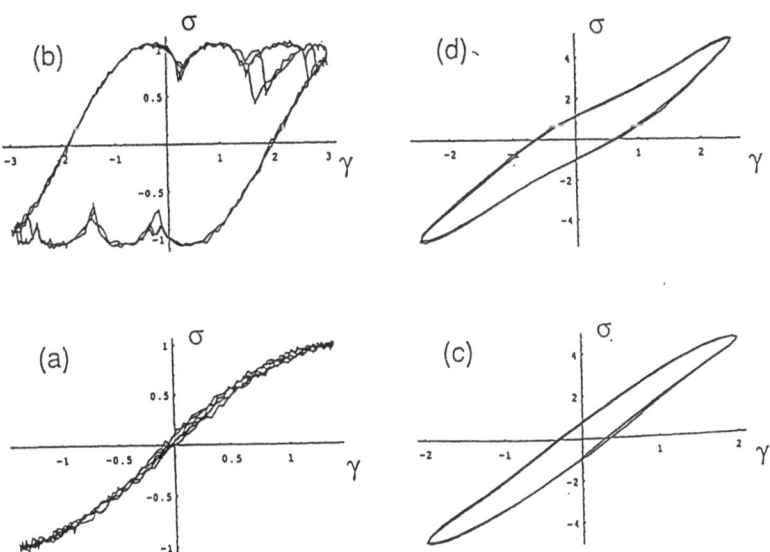

Figure 9: Examples of the Lissajous patterns obtained at low and high ω and large and small A through numerical integration of the multi-layer model.

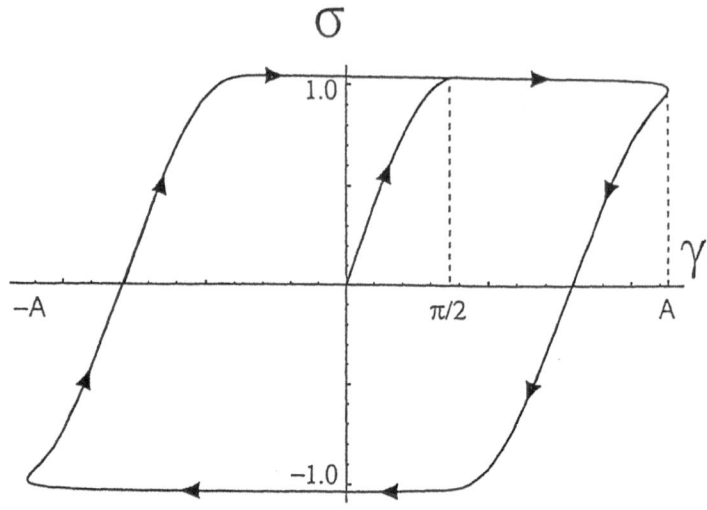

Figure 10: Sketch of the lozenge stress-strain pattern

$\gamma_i = \pi/2$, resulting in a constant stress level $\sigma = 1$ from the initiation of slippage at $\bar{\gamma} \simeq \pi/2$ up to $\bar{\gamma} = A$ (c.f. the top edge of Fig. 10.). When $\bar{\gamma} = A$, the applied strain rate changes sign, and the layers once again follow the homogeneous solution until $\bar{\gamma}$ decreases by roughly π (c.f. the right edge of Fig. 10). At this point, $\bar{\gamma} \simeq A - \pi$ and slip planes develop once again. Subsequent macroscopic strain up to $\bar{\gamma} = -A$ occurs through slippage with $\sigma = -1$ (c.f. the bottom edge of Fig. 10). Analogous processes produce the left and top edges of Fig. 10, thus completing one cycle of deformation.

(2)$\omega \rightarrow \infty$: If $A < A_{c_1}$, the homogeneous solution is stable, so that the shear stress is given by

$$\sigma = \sin\bar{\gamma} + G\bar{\gamma} \tag{32}$$

Thus the Lissajous pattern is that of strain-softening, i.e. the stress decreases with increasing strain near the extremum of the stress-strain curve. On the other hand, $A > A_{c_1}$, the numerical calculations indicate the layers separate into two groups: one oscillating around $\gamma = \pi$, and another oscillating around $\gamma = -\pi$, i.e. half of the layers follow $\gamma_a(t) = \bar{\gamma} + \pi$, and the rest follow $\gamma_r(t) = \bar{\gamma} - \pi$. Thus, the stress in this case is given by

$$\sigma = -\sin\bar{\gamma} + G\bar{\gamma} \tag{33}$$

Notice that in this case, the Lissajous pattern is that of strain-hardening.

4. Conclusion

The present study indicates that there are at least two relaxation processes in the microphase-separated state of diblock copolymers. One is the shape relaxation of each domain, and the other is a lattice relaxation via slippage. In real systems, there is of course a third relaxation mechanism due to chain entanglement as in the case of any polymer liquid.

We have presented a simple phenomenological model which describes the first two processes. The model is different from the conventional picture based on the creation and propagation of defects. In the present model, the plasticity is caused by the slipping between the lattice planes. This model is motivated by the result of the computer simulation. Whether such picture is true or not in real system should be checked by experiments. At this stage we point out that there is no reason to believe that the plasticity of the mesophase of block copolymers is caused by the same mechanism as in metals. Block copolymer system is much softer than metals, and there is a strong viscous dissipation effect. This is the key feature which distinguishes the non-linear rheological properties of these materials from the visco-plastic behavior of ordinary metals. The theory presented here may be relevant to other ordered phase in liquids such as concentrated emulsions, colloid crystals, regular foams etc [22, 23]. Further study is needed to clarify this point.

References

[1] *Block Copolymers: Science and Technology*; Meier, D.J., Ed.; MMI Press/Harwood Academic Pub: New York, 1983.

[2] Hashimoto, T. in *Thermoplastic Elastomers*, Legge, N.R.,Ed.; Hanser: New York, 1987.

[3] Watanabe, H.; Kotaka, T., in *Current Topics in Polymer Science, Vol.II*; Ottenbrite, R.M.; Utracki, L.A.; Inoue, S., Eds.; Hanser Publishers: New York, 1987, pp. 61-96; a review of experimental results.

[4] Bates, F.S.; Fredrickson, G. H. *Ann. Rev. Phys. Chem.*, 1990, *41*, 525.

[5] Watanabe, H.; Kotaka, T.; Hashimoto, T.; Shibayama, M.; Kawai, H. *J.Rheology*, 26, *1982*, 153.

[6] Hashimoto, T.; Shibayama, M.; Kawai, H.; Watanabe, H.; Kotaka, T. *Macromolecules*, 16, *1983*, 361.

[7] Bates, F.S. *Macromolecules*, 1984, *17*, 2607.

[8] Rosedale, J.H.; Bates, F.S. *Macromolecules*, 1990, *23*, 2329.

[9] Helfand, E.; Wasserman Z.R.; *Macromolecules*, 1980, *13*, 994.

[10] Leibler, L.; *Macromolecules*, 1980, *13*, 1602.

[11] Ohta, T.; Kawasaki, K. *Macromolecules*, 1986, *19*, 2621; ibid 1990, *23*, 2413.

[12] Witten, T.A.; Leibler, L.; Pincus, P. *Macromolecules*, 1990, *23*, 824.

[13] Rubinstein, M.; Obukhov, S.P. *Macromolecules*, in press.

[14] Kawasaki, K.; Onuki, A. *Phys.Rev.A*, 1990, *42*, 3664.

[15] T. Ohta, Y. Enomoto, J.L. Harden and M. Doi, *Macromolecules*, to be published.

[16] M. Doi, J.L. Harden and T. Ohta, *Macromolecules*, to be published.

[17] DeGennes, P.G.; *J. Chem. Phys.*, 1980, *72*, 4756.

[18] Bahiana, M.; Oono, Y. *Phys. Rev. A*, 1990, *41*, 6763.

[19] Ohta, T. Nozaki, H.; Doi, M. *J. Chem. Phys.*, 1990, *93*, 2664.

[20] Kasawaki, K.; Ohta, T. *Physica A*, 1986, *139*, 223.

[21] Oono, Y.; Puri, S. *Phys. Rev. A*, 1988, *38*, 434;
Puri, S.; Oono, Y. *Phys. Rev. A*, 1988, *38*, 1542.

[22] Russel. W.B.; Saville, D.A.; Schowalter, W.R.; *Colloidal Dispersions*; Cambridge University Press: Cambridge, 1989.

[23] Kraynik, A.M.; *Ann.Rev.Fluid Mech.*, 20, *1988*, 325;
Reinelt, D.A.; Kraynik, A.M.; *J.Fluid Mech.*, 215, *1990*, 431.

Rheology, Dielectric, Relaxation, and Adhesion of Thermodynamically Confined Diblock Copolymer Chains

H. Watanabe and T. Kotaka

Department of Macromolecular Science, Faculty of Science,
Osaka University, Toyonaka, Osaka 560, JAPAN

abstract: Various thermodynamic requirements exist in strongly segregated A-B diblock copolymer systems. Some of them, e.g., an *osmotic* requirement of uniform segment density distribution in microdomains, impose a constraint on the block conformation. On the other hand, there always exists an *elastic* requirement of randomizing the conformation. These contradicting requirements *thermodynamically confine* the copolymer chains, thereby providing them some unique properties. Those properties are briefly reviewed and their thermodynamic origins are discussed.

1. INTRODUCTION

Strongly segregated A-B diblock copolymer chains exhibit unique microdomain structures, e.g., spheres, cylinders, bicontinuous double-diamonds, and lamellae [1-3]. These structures are determined by a balance of various thermodynamic requirements [2-7]: Some of them, e.g., an *osmotic* requirement of maintaining uniform segment density (or concentration) distribution in each microdomain and an *interfacial* requirement of reducing A-B contacts, impose a constraint on the block conformation and tend to decrease the conformational entropy. On the other hand, there always exists an *elastic* requirement of randomizing the conformation and increasing this entropy. These contradicting requirements *thermodynamically confine* the copolymer chains in solutions and melts, thereby providing them not only the unique structures but also some properties like plasticity [8-12] and highly retarded dielectric relaxation [13,14] that are never observed for homogeneous homopolymer systems.

Interesting properties are found also for block copolymers adsorbed on some substrates, in particular when one of the blocks is selectively adsorbed and the non-adsorbed blocks have tethered (end-grafted) conformations. For such cases, the copolymer chains are subjected to an energetic interaction with the substrate in addition to the thermodynamic requirements explained above [15-23]. As in bulk systems, a balance of all requirements determines unique structures and properties of adsorbed copolymer chains: a stretched conformation and long-range repulsions in solutions [15-23] and a flattened conformation and enhanced adhesion at dry state [24,25].

As explained above, some unique properties of strongly segregated and/or adsorbed block copolymers commonly have thermodynamic origins. From this point of view, we here make a brief review of those properties and discuss their origins.

A. Teramoto, M. Kobayashi, T. Norisuje (Eds.)
Ordering in Macromolecular Systems
© Springer-Verlag Berlin Heidelberg 1994

2. RHEOLOGY AND STRUCTURE OF MICELLAR SOLUTIONS

2-1. Plasticity of Micellar Solutions

Diblock copolymer chains form micelles in a selective solvent that dissolves only one block but precipitates the other [8-12,26]. For example, styrene-butadiene (PS-PB) diblock copolymers having rather short PS blocks form spherical micelles with PS-cores and PB-corona in a PB-selective solvent, n-tetradecane. Such micellar solutions have unique rheological properties.

As an example, Figure 1 shows steady flow behavior at various temperatures for a 20 wt% C14 solution of a PS-PB 16-36 copolymer with $M_{PS} = 16 \times 10^3$ and $M_{PB} = 36 \times 10^3$ [9,11,12].

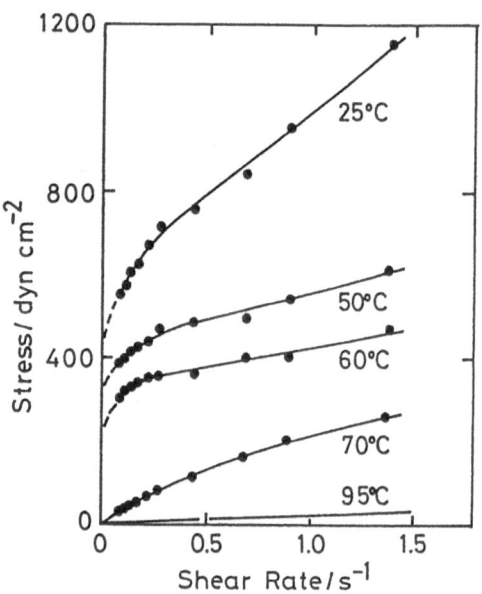

As seen in Figure 1, the PS-PB/C14 solution at low temperatures $T \le 60^\circ C$ exhibits plastic flow that is never observed for homogeneous homopolymer solutions. Although not shown here, the solution also exhibited nonlinear responses against slowly oscillating strain [9,11,12]: When a small amplitude strain was applied and the resulting stress σ was smaller than the yield value σ_y, the Lissajou's pattern (stress-strain profile) was rectilinear and independent of frequency, i.e., elastic behavior prevailed. On the other hand, for large amplitude strain that induced $\sigma > \sigma_y$, we observed lozenge shaped patterns that corresponded to an elastic deformation followed by a

Figure 1
Steady flow behavior of a 20 wt% PS-PB 16-36/C14 solution obtained in a Couette geometry. The plastic flow behavior at $T \le 60^\circ C$ was found also in oscillatory measurements.

plastic flow being repeated twice in each cycle of oscillatory strain.

We also note in Figure 1 that the PS-PB/C14 solution loses its plasticity and behaves as a non-Newtonian viscous liquid as T is increased up to 70°C. At higher $T \ge 95^\circ C$, the solution becomes a Newtonian liquid with low viscosity. These rheological changes correspond to structural changes, as elucidated from small-angle x-ray scattering (SAXS) measurements. The SAXS profiles are shown in Figures 2, where no desmearing procedure was made and the origins of the profiles were diagonally shifted for easy comparison.

As seen in Figure 2, the PS-PB/C14 solution at 25°C exhibits the first, second, and third order scattering peaks at angles of 11.9, 16.5, and 20.8 min, respectively. This result indicates that the micelles are arranged on a so-called *macrolattice* having a long-range order [9,11,12].

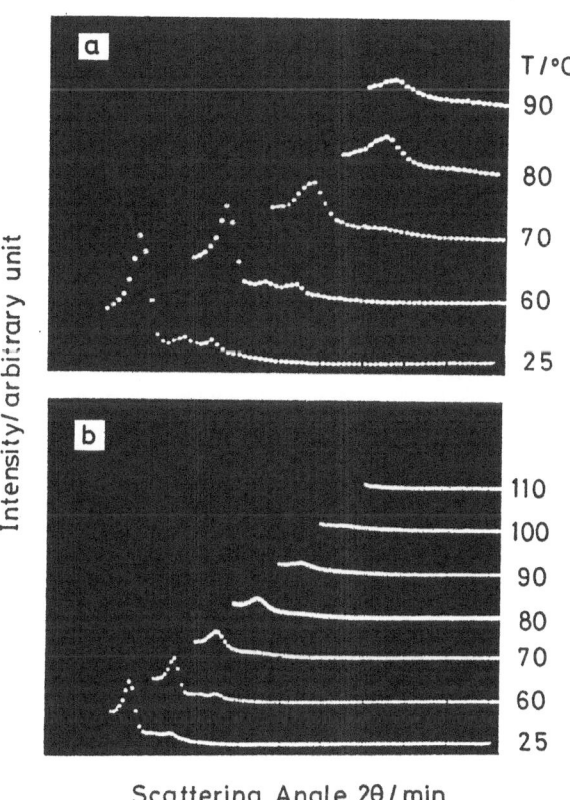

Figure 2
SAXS profiles for a 20 wt% PS-PB 16-36/C14 solution at (a) T = 25-90 ºC and (b) T = 25-110 ºC. One channel corresponds to 0.7 min of arc, and the origins of the profiles are shifted right and upward for easy comparison.

(Detailed analysis of a desmeared SAXS profile suggested that the macrolattice was most likely a simple-cubic lattice, rather than a body- or face-centered cubic lattice [27].) A lattice spacing = 450 Å evaluated from the peak angles is smaller than a size of micelles = 620 Å [9,27], indicating a rather deep overlapping of the corona PB blocks of neighboring micelles. Up to 60°C the SAXS profile remains essentially the same, namely, the long-range order of the macrolattice is preserved. At 70°C, the higher order peaks disappear and the first order peak is broadened, indicating that the macrolattice is disordered but the micelles themselves are still preserved. Finally, at higher temperatures ($\geq$ 100°C) the micelles disappear due to mixing of PS and PB blocks, as noted from the disappearance of the first order peak. These results indicate that the plasticity of the PS-PB/C14 solution at low T is attributed to the macrolattice of micelles: The macrolattice elastically deforms for sufficiently small σ while it flows without disrupting micelles when σ exceeds the lattice strength. (A constitutive equation was recently proposed for the macrolattice systems [28].)

The plasticity of the PS-PB/C14 spherical micellar system may look somewhat similar to an almost infinitely slow terminal relaxation of lamellar systems of block copolymers [29,30]. However, the latter is most likely due to defects of lamellar alignments and is different in nature from the former. In fact, an application of oscillatory shear with adequate amplitudes and frequencies considerably removes the defects and reduces the low-frequency moduli of lamellar systems [29,30]. On the other hand, the spherical micellar system should behave as a solid even if no defect exists, because any small deformation imposed on the system distorts the most stable cubic-symmetry of the micelle arrangement and thus induces a thermodynamic restoring force.

2-2. Driving Force for Macrolattice Formation

The macrolattice of PS-PB micelles is formed most likely due to a balance of the thermodynamic requirements explained earlier. In the micellar solution examined in Figures 1 and 2, the PS-cores are glassy at low T. For such cases, the number of PS-PB contacts should be essentially the same for the two cases of randomly dispersed and regularly arranged micelles, and the interfacial requirement should have no effects for the macrolattice formation. However, a balance of the osmotic and elastic requirements should still take place for the PB blocks that are tethered on the rigid PS cores. The macrolattice would have been formed so that the PB concentration (c_{PB}) distribution in the matrix phase became as uniform as possible while the PB block conformation became as random as possible.

This hypothesis was examined through rheological measurements on ternary systems composed of the PS-PB 16-36 copolymer, C14, and homo-PB with $M = 2 \times 10^3$ [10-12]. Figure 3 shows the steady flow behavior at 25°C. In the systems examined, the PS-PB content was always 20 wt% and the homo-PB content w_{hPB} in the remaining 80 wt% was varied. The PS-PB copolymers always formed spherical micelles with the PS cores and PB-corona, and the homo-PB molecules were involved in the PB/C14 matrix phase.

As seen in Figure 3, the plastic viscosity (slope of the flow curve) increases on addition of homo-PB because of the increase in the content of polymeric components (PS-PB + homo-PB). More importantly, the yield stress decreases with increasing w_{hPB} and finally vanishes for w_{hPB} = 50 wt%. This loss of plasticity was observed also in dynamic tests at low frequencies: Lissajou's patterns became elliptic and the stress level decreased on addition of homo-PB [10-12]. These results indicate that the homo-PB chains destroy the macrolattice, as was also confirmed from SAXS measurements [10-12]. This behavior of homo-PB supports the above hypothesis: A spatial variation of c_{PB} due to randomization of PB block conformation can be compensated by the homo-PB chains that have no tethered ends and can move in the PB/C14 matrix phase without a large burden on their conformations. Thus, in the ternary systems with large w_{hPB}, the osmotic and elastic requirements are no longer contradicting for the PB blocks and no driving force for macrolattice formation emerges.

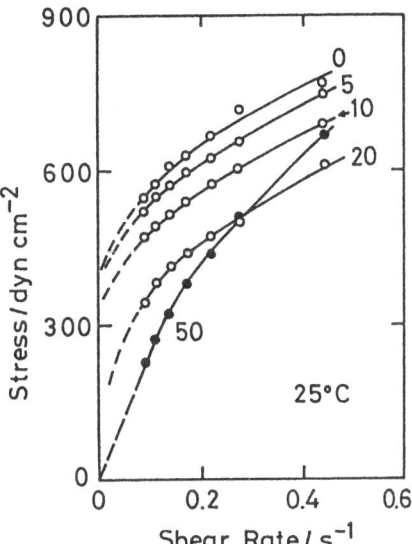

Figure 3
Steady flow behavior of PS-PB 16-36/homo-PB/C14 ternary systems at 25°C obtained in a Couette geometry. The PS-PB content is 20 wt% for all systems, and the numbers indicate the content of homo-PB w_{hPB} (wt%) in the PB/C14 matrix phase.

As demonstrated in Figures 1-3, slow rheological responses of the PS-PB/C14 micellar solution are dominated by the plasticity, that is in turn attributed to the macrolattice formed by a balance of the osmotic and elastic requirements for the tethered PB blocks. (A viscoelastic stress of PB blocks is negligibly small for sufficiently slow deformation.) Magnitudes of those requirements fundamentally determine the equilibrium modulus G of the macrolattice [31,32], although effects of lattice defects need to be also considered. In particular, for moderately concentrated PS-PB/C14 solutions, G was found to be proportional to the PB volume fraction in the matrix phase ϕ_{PB} (not to ϕ_{PB}^2 as expected for an entanglement relaxation [33]). For those solutions (with $\phi_{PB} \geq 20$ vol%), an osmotic compressibility is considerably small [34] and the osmotic requirement overwhelms the elastic requirement. Then, a small strain applied on the solutions should hardly change the concentration distribution in the PB/C14 matrix phase but force each PB block to statically deform as an entropic strand [31], so that G is proportional to the number of the PB blocks per unit volume, i.e., to ϕ_{PB} [31]. This result and those seen in Figures 1 and 3 clearly demonstrate that the slow rheological responses (elasticity before yielding and plasticity) of the micellar solutions have a thermodynamic origin.

3. DIELECTRIC RELAXATION OF TETHERED BLOCK COPOLYMERS

As an extension of the study explained above, it is of interest to examine thermodynamic effects on dynamic features of strongly segregated block copolymers. In principle, the effects can be elucidated through viscoelastic measurements at adequate frequencies. However, all blocks (and homopolymers if they exist in the system) contribute to the mechanical stress and it is not easy, for some cases, to evaluate a stress component due only to a particular block.

Another convenient method to examine dynamics of a particular block is a dielectric spectroscopy. For a polymer chain having type-A dipoles that are aligned parallel along the chain contour, the net polarization is proportional to the end-to-end vector so that the global chain motion induces prominent dielectric relaxation [35]. On the other hand, for polymer chains having only type-B dipoles that are perpendicular to the contour, the global motion practically induces no dielectric relaxation [35]. *Cis*-polyisoprene (PI) chains have both type-A and type-B dipoles [36,37], while PS chains have only the latter. Thus, for PS-PI diblock copolymers at T well above the glass transition temperature T_g of the PI blocks but well below T_g of the PS blocks, we can observe dielectric relaxation due to global motion of the PI blocks that should be affected by the thermodynamic requirements explained earlier. (Except this dielectric feature, the properties of PS-PI copolymers are very similar to those of PS-PB copolymers examined in the previous section.)

Figure 4 shows frequency (ω) dependence of dielectric loss factor (ε'') at 0°C for a bulk system of a PS-PI 13-10 copolymer with $M_{PS} = 13 \times 10^3$ and $M_{PI} = 10 \times 10^3$ [13]. Alternating

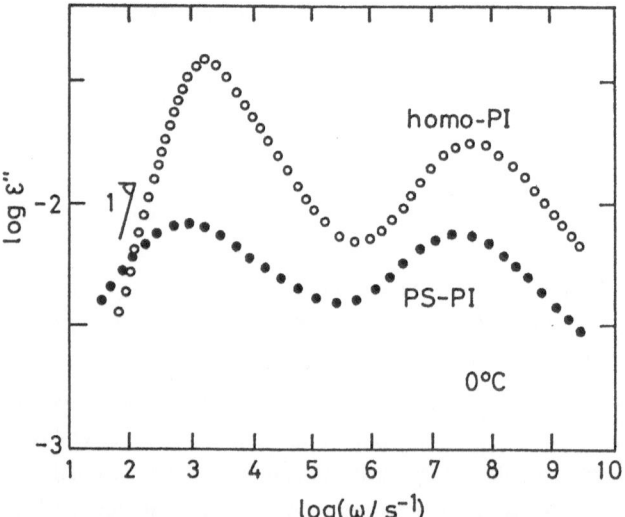

Figure 4
Comparison of dielectric loss curves for bulk systems of PS-PI 13-10 and homo-PI at 0°C. M_{PI} is exatcly the same for the PI block and homo-PI.

PS/PI lamellae parallel to the electrodes of a dielectric cell were formed in the system, and the PI blocks were tethered on the glassy PS domains. For comparison, Figure 4 also shows the behavior of a homo-PI that was obtained as a precursor for the PS-PI copolymer and had a molecular weight exactly the same as M_{PI} of the PI block [13].

In Figure 4, two relaxation processes are clearly seen for both PS-PI and homo-PI systems. The high-ω relaxation is due to the type-B dipoles of (block and homo) PI chains and is attributed to their segmental motion [13,35-37]. On the other hand, the low-ω relaxation due to type-A dipoles is related to the end-to-end vector fluctuation of the PI chains through equations 1 and 2 [13,35-38].

$$\varepsilon''(\omega) = -\Delta\varepsilon \int_0^\infty [d\Phi(t)/dt] \sin \omega t \, dt \quad (\Delta\varepsilon = \text{relaxation intensity}) \tag{1}$$

$$\Phi(t) = <R_n(t)R_n(0)>/<R_n^2> \qquad \text{(auto-correlation function)} \tag{2}$$

Here, R_n indicates the component of the end-to-end vector perpendicular to the PI lamellae. As can be seen from eqs 1 and 2, the shape (ω dependence) of the ε'' curve at low ω reflects a mode distribution for the global relaxation of the PI chains (represented by Φ).

As seen in Figure 4, the shape and location of the ε'' curve at high ω are nearly the same for the PS-PI copolymer and homo-PI systems. This result suggests that the segmental motion is nearly the same for the free homo-PI chains and the PI blocks confined in the lamellae. (The difference in the ε'' peak height is due to the difference of the PI content in the system [13].) On the other hand, a comparison of the curves at low ω suggests that the mode distribution for the global relaxation is much broader for the PI blocks than for the homo-PI chains.

A global relaxation of PI chains should complete at low $\omega < 1/\tau_1$, with τ_1 being their longest relaxation time. This terminal relaxation behavior characterized by a relation, $\varepsilon'' \propto \omega$, is clearly seen in Figure 4 for the homo-PI chains at $\omega \leq 10^3$ s^{-1} but not for the PI blocks: ε'' for the latter is almost proportional to $\omega^{1/2-1/3}$ even at the lowest ω examined (= 30 s^{-1}). These results indicate that τ_1 is at least 30 times longer for the PI blocks than for the homo-PI chains having the same M_{PI}, although the τ_1 value for the former was too large and not determined in our experiments.

For linear homo-PI chains, entanglements do not affect the slow dielectric mode distribution [39]. Interestingly, this was the case also for the PI blocks [13]: The shape of the ε'' curve was nearly the same for PI blocks with M_{PI} ranging from $0.8M_e$ to $8M_e$ ($M_e = 5 \times 10^3$; entanglement spacing for PI [33]). Consequently, the slow and broad global relaxation of the PI blocks seen in Figure 4 is not due to entanglements but essentially has a thermodynamic origin: The osmotic requirement is extremely strong in the *bulk* PI domains, so that the motion of the PI blocks should be highly cooperative to maintain the uniform PI segment distribution in the domains. During this cooperative motion, the PI blocks having tethered ends should violate the elastic requirement and pay some entropic penalty to take distorted conformations. These cooperativity and penalty most likely lead to the very slow and broad relaxation behavior of the PI blocks. The relaxation of the PI blocks looks similar to that of entangled star chains [40] in a sense that a retardation takes place due to an entropic penalty, although the penalty emerges from the entanglement effect for the latter but not for the former.

From the above argument, we expect that the dielectric behavior of PS-PI copolymers becomes closer to that of homo-PI chains when a PI-selective solvent is added, because the solvent increases the osmotic compressibility and weaken the above requirement of cooperativity. This change of the behavior was confirmed experimentally [14]. Another consequence of the above argument is for the M_{PI} dependence of τ_1 of the PI blocks. In our experiments, τ_1 were too long to be determined. However, from a similarity of the entropic penalties considered for the PI blocks and for star chains [40], one may expect τ_1 for the PI blocks to increase exponentially with M_{PI} [41]. A test for this expectation is an interesting future work.

In Figure 5, the shape of ε'' curves at low ω are compared for (a) a homo-PI ($M_{PI} = 10 \times 10^3$), (b) a 5/90/5 (wt/wt/wt) blend of this homo-PI, PS-PB 9-9 copolymer ($M_{PS} = M_{PB} = 9 \times 10^3$), and homo-PS ($M_{PS} = 11 \times 10^3$), (c) a PS-PI 13-10/PS-PB 9-9 blend (20/80 wt/wt),

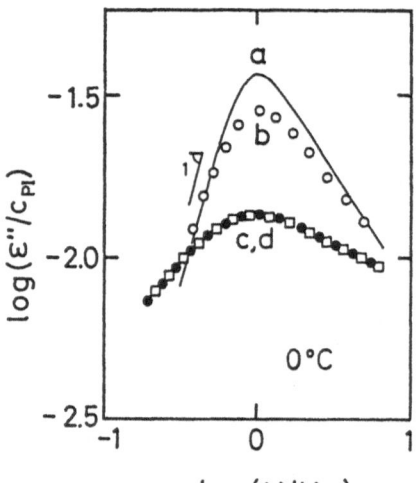

Figure 5
Comparison of the shape of the ε'' curves at 0°C for (a) homo-PI, (b) a ternary blend of homo-PI, PS-PB 9-9, and homo-PS, (c) a binary blend of PS-PI 13-10 and PS-PB 9-9 (squares), and (d) bulk PS-PI 13-10 (filled circles). ε'' are reduced by the PI content in the systems and plotted against a reduced frequency ω/ω_m. ($\omega_m = \varepsilon''$ peak frequency.)

and (d) bulk PS-PI 13-10 copolymer [13]. The diene (PB+PI) content is nearly the same ($\cong 50\%$) for all systems b-d. The blends b and c are similar to each other, except that the PI chains involved in the former have no tethered ends.

The PS and PB chains having no type-A dipoles are dielectrically inert at low ω. Thus, the dielectric relaxation processes of the blends b and c observed in Figure 5 are attributed to the global motion of the homo-PI and/or block PI chains involved. For ε'' of those PI chains in the blends, the temperature coefficients (shift factors) a_T were identical to a_T for PB blocks [13]. This result indicates that (PB+PI)/PS lamellae are formed in the blends examined and the motion of the PI chains (the *dilute* component in the diene lamellae) is induced by the motion of the surrounding PB blocks. Such PI chains experience a thermodynamic field created by the PB blocks. This field is similar to that in the PS-PI copolymer system, as suggested from the fact that the shape of the ε'' curve is the same for the blend c and bulk PS-PI 13-10 (cf., Figure 5).

Comparing the curves a and b in Figure 5, we note that the dielectric mode distribution of the homo-PI chains is a little broader in the blend than in their bulk state. Thus, a small thermodynamic effect exists even for motion of the PI chains without tethered ends. However, as seen for the curves b and c, the effect is much larger for the PI blocks with tethered ends. This result suggests that the osmotic and elastic requirements contradict much more strongly for tethered chains than for free chains, indicating an important role of the tethered ends for motion of thermodynamically confined chains.

4. ADHESION OF THIN AND DRY BLOCK COPOLYMER LAYERS

When solutions of block copolymers are exposed to a substrate that has a high affinity for one block, a selective adsorption takes place and the copolymer chains form a layer structure on the substrate. For example, 2-vinylpyridine (PVP)-PI and PVP-PS diblock copolymers form such layers on mica from toluene solutions [15,16,21,24,25]: The selectively adsorbed PVP blocks (not dissolved in toluene) are at the bottom of the layer, and the PI and/or PS blocks at the top are tethered on mica by the PVP blocks.

For such adsorbed layers *in solution phases*, experimental and theoretical work has been extensively made and several interesting features were elucidated [15-24]: The adsorbed amount is essentially determined by a balance of an energetic (van der Waals) interaction between the adsorbed blocks (e.g., PVP) and the substrate (e.g., mica), an osmotic interaction between the tethered blocks (e.g., PI), and an elastic free energy for each tethered block [19-21]. In most of the experimental situations, that amount is large enough to induce lateral overlapping of the tethered blocks. For such cases, a balance of the osmotic and elastic free energies forces the tethered blocks to take stretched conformations and form a polymer brush on the substrate [15-24]: The brush height is proportional to the molecular weight of the tethered blocks, and a long-range repulsion emerges on compression of two brushes.

A situation is entirely different for adsorbed block copolymer layers *at dry state*. In general, those layers have smooth surfaces (because of an interaction with atmosphere) and their thicknesses are smaller than the unperturbed sizes of the copolymer chains. For such cases, the copolymer chains are forced to take highly flattened conformations. We naturally expect thermodynamic effects for properties of such flattened chains having a large elastic free energy. In this section, we examine the effect for adhesion of dry PVP-PI copolymer layers adsorbed on mica.

For five PVP-PI layers (cf., Table 1) in nitrogen atmosphere, we used an Israelachvili-type surface force apparatus [42] and measured at room temperature the layer thicknesses L and pull-off forces F_{ad}, the latter being necessary to separate two identical layers brought into molecular contact [24,25]. Interactions between the PI blocks at the top of the two layers determine the F_{ad} values, and the glassy PVP blocks at the bottom behave just as a rigid glue that bound the PI block ends on mica. When the layers were kept in contact for > 10 s, F_{ad} was insensitive to the contact time. This result suggests that F_{ad} has no kinetic contribution (due to entanglements, for example) and is related to thermodynamic changes for the PI blocks on contact. Thus, applying the Johnson-Kendall-Roberts relation to F_{ad} [43,44], we evaluated a decrease γ of the surface free energy density on contact. Structural quantities, a thickness L_{PI} of a PI layer and a number v of adsorbed chains per unit area, were calculated from L and molecular characteristics of the copolymer used.

Table 1 summarizes the γ_{PVP-PI}, L_{PI}, and v values for the adsorbed layers of the PVP-PI copolymers with various M_{PI} and M_{PVP} [24,25]. We note that γ_{PVP-PI} for the PI blocks are larger than $\gamma_{homo-PI}$ ($\cong 31$ erg cm^{-2} [45]) for bulk homo-PI. This result suggests that two kinds of interactions contribute to γ_{PVP-PI}, one being the van der Waals interaction (common to the PI block and homo-PI) and the other

Table 1 Characteristics of thin adsorbed layers of PVP-PI copolymers.

Code*	γ/erg cm^{-2}	L_{PI}/Å	$10^{-12}v$/chains cm^{-2}
15-73	67	25	1.8
34-73	42	17	1.3
72-73	41	15	1.2
72-338	63	19	0.31
30-217	55	14	0.34

*: The sample code numbers indicate $10^{-3}M_{PVP}$-$10^{-3}M_{PI}$.

being an interaction specific to the PI blocks. As can be seen from Table 1, L_{PI} is significantly smaller than an unperturbed end-to-end distance $R_{\theta,PI}$ of the PI blocks: For example, $L_{PI} = 25$ Å and $R_{\theta,PI} = 220$ Å [46] for the PVP-PI 15-73 copolymer. Such flattened PI blocks have a small conformational entropy S_{PI}. Thus, an elastic requirement of increasing S_{PI} should induce interpenetration of the PI blocks that belong to the two layers brought into contact. From this argument, we may attribute a difference $\Delta\gamma = [\gamma_{PVP-PI} - \gamma_{homo-PI}]$ to an entropy gain due to interpenetration. ($\gamma_{homo-PI}$ is considered to be the van der Waals contribution.)

Following the above argument, we here examine a relationship between the PI block length and the free energy gain per chain, $v^{-1}\Delta\gamma$. For this purpose, we consider each PI block to be composed of *blobs* [47] of a size L_{PI}. Then, the thin PI layer is regarded as a 2-dimensional lattice for the blobs. Before contact of the layers, the number of possible conformations of a PI block composed of N blobs is estimated to be $\Omega_0 \cong \Omega_{blob} z_0^N$, with z_0 and Ω_{blob} being a lattice coordination number and the number of conformations in each blob, respectively. On contact, each PI chain obtains a freedom to interpenetrate into the other layer so that Ω_0 would increase to $\Omega \cong \Omega_{blob} z^N$ with $z > z_0$. Thus, we expect a proportionality between $v^{-1}\Delta\gamma = kT \ln(\Omega/\Omega_0)$ and N.

We assumed an unperturbed conformation of the PI block in each blob, and estimated N from $R_{\theta,PI}$ [46] and L_{PI} (Table 1) as $N = (R_{\theta,PI}/L_{PI})^2$. In Figure 6, we have used these N values and made the $v^{-1}\Delta\gamma$ *vs* N plots. We note that $v^{-1}\Delta\gamma$ is roughly proportional to N, in harmony with the above expectation. (As seen in Table 1, L_{PI} was not very different for the copolymers layers examined. Thus, a proportionality was found also for $v^{-1}\Delta\gamma$ and M_{PI}.)

The above calculation of Ω and Ω_0 is too simple to be quantitatively applied to the data. However, the result found in Figure 6 strongly suggests that the gain of conformational freedom on contact is essential for the enhanced adhesion of thin PI block layers. In this regard, it is important and interesting to examine a temperature dependence of $\Delta\gamma$. $\Delta\gamma$ should be proportional to T if it has an entropic origin. A test for this hypothesis is now being attempted.

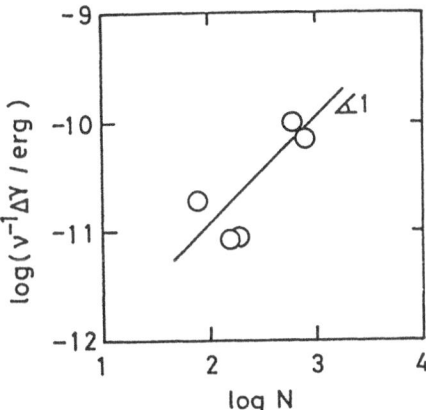

Figure 6
Plots of $v^{-1}\Delta\gamma$ against the number N of blobs per PI block for dry adsorbed layers of PVP-PI copolymers.

5. REFERENCES

1   Molau GE (1970) In: Aggarwal SL (ed) *Block Copolymers*. Plenum, New York
2   Hashimoto T (1981) In: Kotaka T, Ide F, Ogino K (ed) *Polymer Alloy*. Kagaku Dojin, Tokyo.
3   Bates FS, Fredrickson GH (1990) *Ann Rev Phys Chem* 41:525
4   Meier DJ (1969) *J Polym Sci* C26:81
5   Helfand E, Wasserman ZR (1976) *Macromolecules* 9:829; (1978) *ibid* 11:960
6   Helfand E, Wasserman ZR (1982) In: Goodman I (ed) *Developments in Block Copolymers-1*. Applied Sci, New York
7   Semenov AN (1985) *Soviet Phys TETP* 61:733
8   Kotaka T, White JL (1973) *Trans Soc Rheol* 17:587
9   Watanabe H, Kotaka T, Hashimoto T, Shibayama M, Kawai H (1982) *J Rheol* 26:153
10 Watanabe H, Kotaka T (1983) *J Rheol* 27:223
11 Watanabe H, Kotaka T (1984) *Polym Eng Rev* 4:73
12 Kotaka T, Watanabe H (1987) In: Ottenbrite RM, Utracki LA, Inoue S (ed) *Current Topics in Polymer Science-II*. Hanser, New York
13 Yao ML, Watanabe H, Adachi K, Kotaka T (1991) *Macromolecules* 24:2955
14 Yao ML, Watanabe H, Adachi K, Kotaka T (1991) *Macromolecules* 24:6175
15 Patel S, Tirrell M, Hadziioannou G (1988) *Colloid Surf* 31:157
16 Patel S, Tirrell M (1989) *Ann Rev Phys Chem* 40:597
17 Milner ST, Witten TA, and Cates ME (1988) *Macromolecules* 21:2610
18 Milner ST (1991) *Science* 251:905
19 Marques C, Joanny JF, Leibler L (1988) *Macromolecules* 21:1051
20 Halperin A, Tirrell M, Lodge TP (1992) *Adv Polym Sci* 100:31
21 Parsonage E, Tirrell M, Watanabe H, Nuzzo RG (1987) *Macromolecules* 24:1987
22 Taunton HJ, Toprakcioglu C, Fetters LJ, Klein J (1988) *Nature* 332:712
23 Taunton HJ, Toprakcioglu C, Fetters LJ, Klein J (1990) *Macromolecules* 23:571
24 Watanabe H, Tirrell M (1989) *Polym Prepr* 30:387; submitted to *Macromolecules*, 1993
25 Watanabe H, Matsuyama S, Mizutani Y, Kotaka T; in preparation.

26 Sadron C, Gallot B (1973) *Macromol Chem* 164:301
27 Shibayama M, Hashimoto T, Kawai H (1983) *Macromolecules* 16:16
28 Doi M *OUMS proceedings* presented in this issue
29 Larson RG, Winey KI, Patel SS, Watanabe H, Bruinsma R (1993) *Rheol Acta* in press
30 Koppi KA, Tirrell M, Bates FS, Almadal K, Colby R (1993) *J Phys (Paris)* in press
31 Hashimoto T, Shibayama M, Kawai H, Watanabe H, Kotaka T (1983) *Macromolecules* 16:361
32 Watanabe H, Kotaka T (1983) *Macromolecules* 16:1783
33 Ferry JD (1980) *Viscoelastic Properties of Polymers. (3rd ed)* Wiley, New York
34 Noda I, Kato N, Kitano T, Nagasawa M (1981) *Macromolecules* 14:668
35 Stockmayer WH (1969) *Pure Appl Chem* 15:539
36 Adachi K, Kotaka T (1984) *Macromolecules* 17:120
37 Imanishi Y, Adachi K, Kotaka T (1988) *J Chem Phys* 89:7585
38 Cole RH (1965) *J Chem Phys* 42:637
39 Watanabe H, Yamazaki M, Yoshida H, Adachi K, Kotaka T (1991) *Macromolecules* 24:5365
40 Doi M, Edwards SF (1986) *The Theory of Polymer Dynamics*. Clarendon, Oxford
41 Witten TA, Leibler L, Pincus PA (1990) *Macromolecules* 23:824
42 Israelachvili JN, Adam GE (1976) *Nature (London)* 262:774; (1978) *J Chem Soc Faraday Trans-1* 74:975
43 Johnson KL, Kendall K, Roberts AD (1971) *Proc Royal Soc London A* 324:301
44 Horn RG, Israelachvili JN, Probac F (1987) *J Colloid Interface Sci* 115:480
45 Shafrin EG (1975) In: Brandrup J, Immergut EH (ed) *Polymer Handbook. (2nd ed)* Wiley, New York
46 Tsunashima Y, Hirata M, Nemoto N, Kurata M (1988) *Macromolecules* 21:1107
47 de Gennes PG (1979) *Scaling Concepts in Polymer Physics*. Cornell University Press, Ithaca

Crystallization Kinetics and Microdomain Structures for Blends of Amorphous-Crystalline Block Copolymers with Amorphous Homopolymers

K. Sakurai[*1], D. J. Lohse[2], D. N. Schulz[2], J. A. Sissano[2], M. Y. Lin[2], M. Agamalyan[3], and W. J. MacKnight[1]

[1]Department of Polymer Science and Engineering, University of Massachusetts, Amherst, MA 01003, [2]Corporate Research Laboratories, Exxon Research & Engineering Company, Annandale, NJ 08801, and [3]Brookhaven National Laboratories, Upton NY 11973

ABSTRACT

Blends of a symmetrical diblock poly(ethylene-b-propylene) (DEP) and an atactic polypropylene (APP) were studied by calorimetry, small angle neutron scattering, and electron microscopy. Addition of APP to DEP drastically changed the crystallization kinetics of the polyethylene block. The results were interpreted on the basis of the morphological studies of the microdomain structure in both the crystalline state and the melt state of the blends.

INTRODUCTION

The particular chemical structure of symmetrical diblock copolymers results in a lamellar microdomain morphology and blending them with the corresponding homopolymer causes a structural transition from the lamellar to a cylindrical and finally to a spherical structure.[(1,2)] It would be interesting to extend these results to the blends of a diblock copolymer containing a crystallizable block. Those blends should display some unique physics in the relation between the spatial ordering of the microdomain structure and the molecular ordering from crystallization of the crystalline block. This study may lead to the development of methods to control the supermolecular structures as well as the crystallization kinetics and, it is hoped, will result in the ability to design a new type of polymer blend.

We have examined binary polymer blends consisting of an atactic polypropylene (APP) and a diblock poly(ethylene-b-propylene) (DEP). This paper will review our recent results of the crystallization kinetics of the blends[(3)] and the morphological studies on the blends by small angle neutron scattering[(4)] and electron microscopy.[(5)]

SAMPLE PREPARATION AND CHARACTERIZATION

We synthesized poly(butadiene-b-2-methyl-1,3-pentadiene) by sequential anionic polymerization and then hydrogenated the resulting diblock copolymer using $Pd/CaCO_3$ as a catalyst to form DEP.[(3)] We also prepared three samples of APP with different molecular weights by hydrogenation of poly(2-methyl-1,3-pentadiene) according to the Fetters method.[(6)] Furthermore a polyethylene (PE) sample was made by hydrogenation of polybutadiene.

---

* Present address: Polymer Research Lab., Research & Development Center, Kanebo Ltd., 1-5-90, Tomobuchi-cho, Miyakojima-ku, Osaka, Japan

A. Teramoto, M. Kobayashi, T. Norisuje (Eds.)
Ordering in Macromolecular Systems
© Springer-Verlag Berlin Heidelberg 1994

IR and NMR measurements confirmed that the hydrogenation was complete. Molecular weight, ethylene content in PE block, and other molecular characteristics for these samples are listed in Table I, as well as the nomenclatures used in this paper.

Table I Nomenclatures and Molecular Characteristics

type of sample	nomenclature	M_w[a]	M_w/M_n[a]	f[b]	ethyl branch[c] mol %
DEP	DEP113	113,000	1.12	0.48	3.0
	DEP99	99,100	1.07	0.50	3.0
	DEP61	61,300	1.05	0.50	3.0
	d-DEP65[d]	64,800	1.07	0.52	3.0
APP	APP15	15,100	1.05	0	-
	APP39	39.300	1.04	0	-
	APP190	190,000	1.10	0	-
polyethylene	PE43	43.000	1.10	1	3.0

(a) calculated from unhydrogenated precursors' results which were measured by GPC
(b) symmetric factor defined as ethylene composition (vol %), determined by [13] C NMR.
(c) from [13] C NMR and FT-IR.
(d) used d-butadiene instead of butadiene

ESTIMATION OF THE EXTENT OF SEGREGATION

It is essential to determine whether DEP forms a microdomain structure in the melt state, and, if it does, to evaluate how strongly the phases are segregated. Small angle x-ray scattering and electron microscopy are not feasible in this case because of the small electron density difference between polyethylene and polypropylene in the melt state. Although not a direct method for determining morphology, rheological behavior has been shown to be related to the presence or absence of a microphase in block copolymers.[7]

Figure 1 compares master-curves of the frequency dependence of G' and G'' for DEP99, APP39, and PE43. Superposition appeared to be valid within experimental error, indicating that no order-disorder transition takes place in this temperature range. The dynamic elastic and loss responses at low reduced frequencies exhibited a limiting frequency dependence of $G' \sim G'' \sim \omega^{0.5}$. This is in contrast to the typical terminal zone behavior of $G' \sim \omega^2$ and $G'' \sim \omega^1$ observed for homopolymers of APP39 and PE43. According to the experimental result of Rosedale and Bates,[7] and also to the theoretical calculation by Rubinstein and Obukhov,[8] the exponent of 0.5 for both elastic and loss moduli at the low frequency limit is characteristic of the presence of a lamellar structure. Although the data are not shown here, the presence of long time relaxation was also observed for a DEP99 solution containing 85 wt% of 1,2,4-trichlorobenzene. These features in the rheology suggest that DEP99 is strongly segregated in the melt state.

CHANGES OF DSC THERMOGRAMS UPON BLENDING

Figure 2 shows cooling thermograms of DSC for the blends, illustrating the influence of the molecular weight and composition of APP on the crystallization of PE block. DEP113 showed a relatively broad main crystallization peak with an onset temperature of 94°C. Upon blending with APP190, the main peak (peak I) shifted to lower temperatures and above 50 wt% another peak (peak II) appeared around 70°C. However, the overall changes were relatively smaller compared to APP15 blends. Adding APP15 up to 45 wt% caused peak I to shift to lower

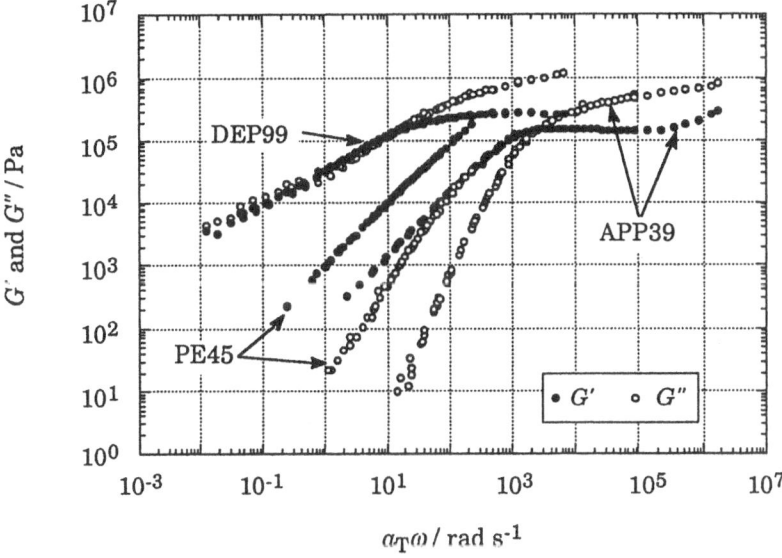

Figure 1. Master-curves obtained from dynamic mechanical shear measurements for DEP99, APP39 and PE45. The refernce temperature is 140°C and the temperature range is 100-200°C. Here a_T is the shift factor.

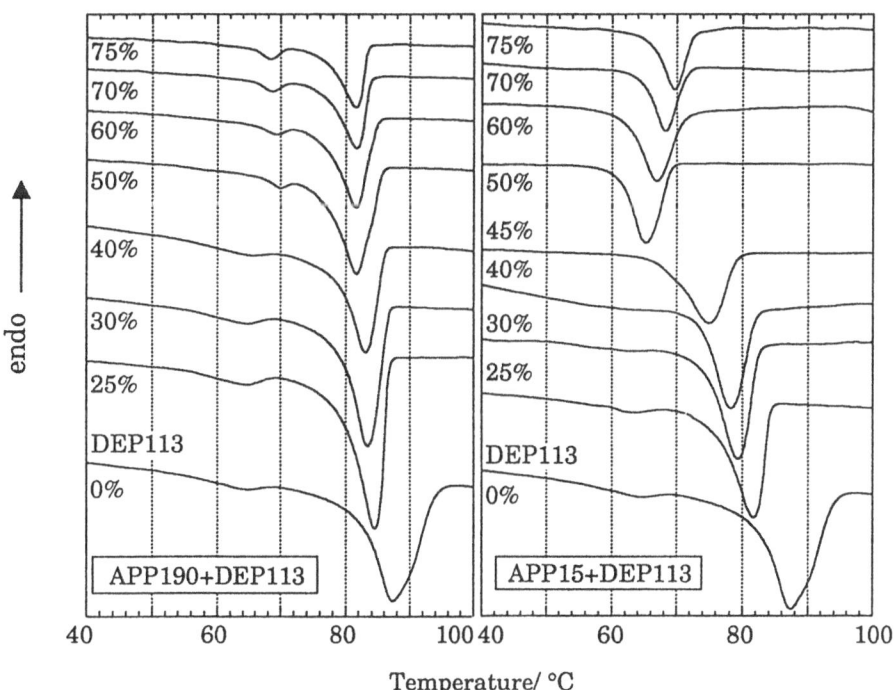

Figure 2. DSC cooling thermograms measured for two series of blends of DEP113 andAPP190, and DEP113 and APP15. The cooling rate is 10°C/min.

temperatures, which is the same trend as APP190 blends. Once the composition increased to 50 from 45 wt%, however, drastic changes occurred. Peak I disappeared completely and peak II became the main one. Strangely, with increasing APP15 composition, peak II shifted to the higher temperature side instead of the lower side. Although those data are not shown in this paper, DSC melting thermograms showed featureless additive behavior for all series and all the blends melted at about the same temperature of about 100°C. In summary, DSC measurements revealed that blending affects only crystallization and the effect is more pronounced for the blends with lower molecular weight of APP.

ISOTHERMAL CRYSTALLIZATION KINETICS

Isothermal crystallization measurements were carried out and the data were analyzed using the Avrami theory. The Avrami formulation at small degrees of crystallinity is given by [9]

$$m_c(t)/m_0 \propto t^n \tag{1}$$

Here, m_c/m_0 is the degree of crystallinity at a crystallization time t, and n is the Avrami exponent and usually ranges between 4-1, depending on the type of the nucleation mechanism and the dimensionality of crystallization growth geometry. In Figure 3 several of isothermal crystallization data are analyzed according to equation (1), and from the slopes the Avrami exponents can be evaluated without ambiguity. The values of n are plotted against APP composition in the blends in Figure 4. For the APP190 blend, n does not change upon blending. On the other hand, with increasing APP15 content, n decreases in two steps corresponding to the two different types of peaks (peak I and II) in cooling thermograms. The value of n is constant between 10 and 40 % and decreases abruptly by 0.5 around 45 %, remaining constant again between 50 and 70%.

The changes in n upon blending with APP15 and DEP113 can be interpreted in the following way. The PE microdomains provide spatial constraints to the crystallization growth of the PE, therefore, the microdomain structure should influence the crystallization growth. Blending DEP113 with APP15 is expected to cause a structural transition in the microdomain of PE in the melt state. Blending with the more APP15 results in the lower dimensionality of the PE microdomain. Although there may not be a strict correspondence, a lower geometrical dimensionality in the microdomain should give a higher spatial constraint, which means a lower dimensionality in the crystallization growth, and therefore a lower value of the Avrami exponent. To prove this hypothesis morphological studies are necessary.

SMALL ANGLE NEUTRON SCATTERING

We applied planar shear extension to the melt state of neat d-DEP65 and its blend with 10 wt% of APP15 and quenched the samples by immersion into liquid nitrogen. We assumed that this procedure could "align and freeze" the microdomain in the melt state. Needless to say, the PE block can crystallize even in quenching; however, it may not destroy the microdomain structure if

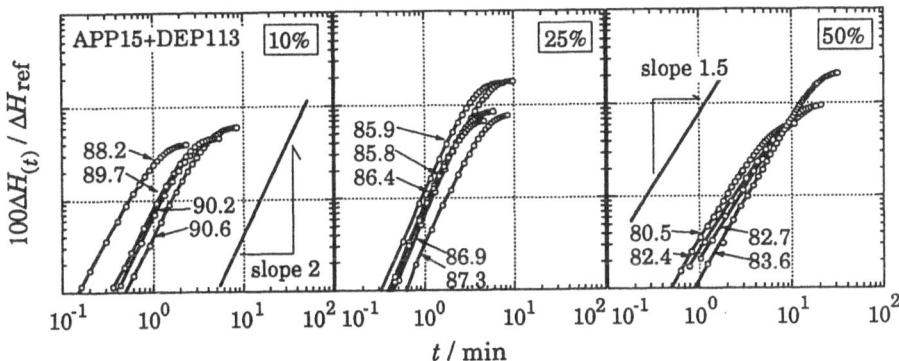

Figure 3 Plots of degree of crystallinity ($\Delta H(t)/\Delta H$ref) against crystallization time (t). Here, $\Delta H(t)$ is enthalpy of fusion measured by DSC, and ΔHref is the enthalpy of the perfect polyethylene crystal and assumed to be 290 J/g for our system.

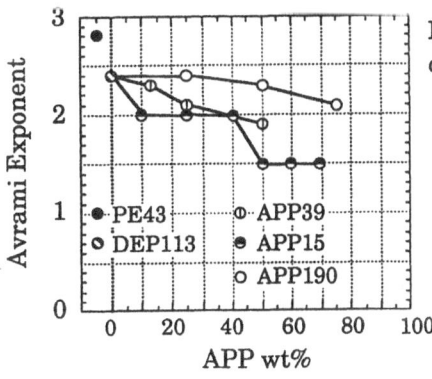

Figure 4 APP composition dependence of the Avrami exponent.

Figure 5 Parallel view of SANS from d-DEP65.

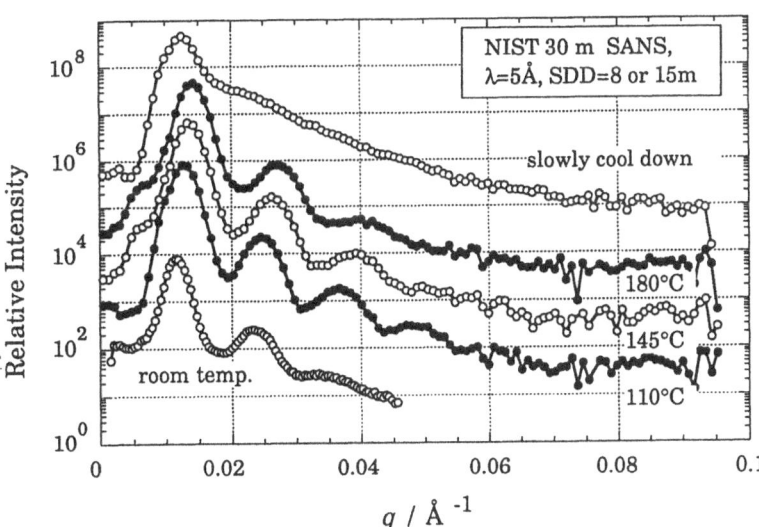

quenched fast enough.   Samples prepared in this manner were used in small angle neutron scattering (SANS) experiment at the Cold Neutron Facility of the National Institute of Standards Technology in Gaithersburg, Maryland, USA.   In the neutron scattering cell at NIST, the samples were examined from 24°C to 180C, *i.e.*, above the melting temperature of polyethylene.   Two kinds of samples, one with the shear direction perpendicular to the neutron beam and one parallel, were used.   The wavelength of the neutron was 5Å and sample detector distances of 8 and 15 m were used.

Figure 5 shows the temperature dependence of the scattering pattern for a parallel view of d-DEP66.   In the "aligned and frozen" state, three distinct orders of peaks are observed.   The scattering pattern can be assigned as the typical one from lamellae structures[10] and the domain spacing is evaluated to be 520Å. When heated up above the melting temperature the peak moves to higher q corresponding to 470Å at 110°C, and more higher order peaks are observed. These features indicate a shrinkage of the domain spacing and a more uniform distribution of the spacing.   When the sample was cooled down to room temperature during about 1 hour, the higher order peaks disappeared. A similar behavior was observed for the blend with 10 wt% of APP15 except that addition of the homopolymer increased the spacing in line with the added amount.

The neutron scattering confirms that DEP and 10 wt% blends with APP15 form lamellar structures in the melt state and that quenching by liquid nitrogen can preserve the domain structures.   Although there is a slight deformation due to crystallization, the quenching technique is good enough to examine the microdomain morphology in the melt state for the blends.

MORPHOLOGICAL STUDIES BY ELECTRON MICROSCOPY

Transmission electron microscopy (TEM) was carried in the bright field mode for ultrathin sections and thin films for DEP113 and the blends.[5] Staining was not necessary for our system because polyethylene crystals diffract the electron beam hence providing the needed diffraction contrast. Three types of microdomain structures were observed by TEM.   Lamellar structures were observed for neat DEP and the blends with up to 20 wt % of APP15.   A bicontinuous phase was observed for 25 - 45 wt% of APP15 and for all compositions with APP190 and APP39.   Furthermore, a discrete cylindrical or spherical morphology was observed for blends with more than 50 wt% of APP15.   TEM micrographs in (a) - (c) of Figure 6 show examples of those morphologies.   Scanning electron microscopy (SEM) was done for the blends which were annealed for 5 days, quenched, and immersed in toluene at room temperature for 1-12 hours.   We assumed that if a blend were macrophase separated, immersing the blend in toluene could remove only polypropylene from the blends and could form depressions on the surface.   Although the micrographs are not presented here, SEM observation could be used to judge whether macrophase separation was present or absent in the blends.   The observations by TEM and SEM are summarized in a morphological diagram shown in figure 8.   Since the microscopy was done on the frozen samples, the morphological diagram is considered to represent the melt state.

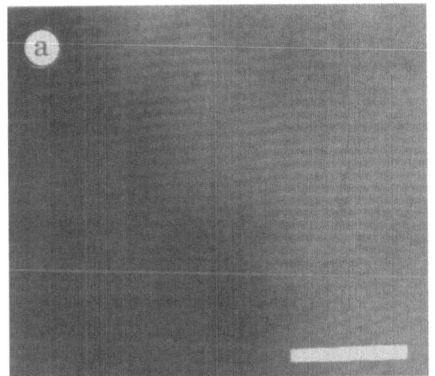
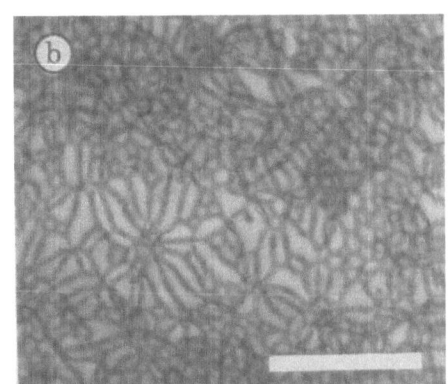
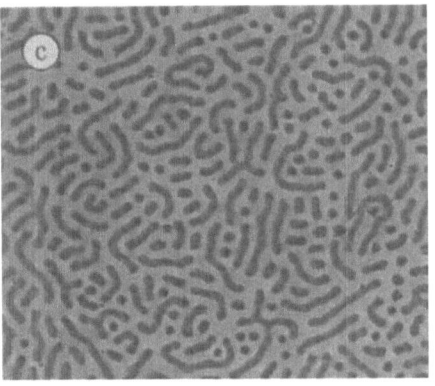

Figure 6 TEM bright field images (a) DEP113, a lamellar structure obtained for an ultrathin section perpendicular to the shear, (b) a bicontinuous structure obtained for a thin film of a blend with 35 wt% of APP39, (c) a discrete cylindrical structure in a thin film for a blend of 65 wt% of APP15. The white bars correspond to $1\mu m$.

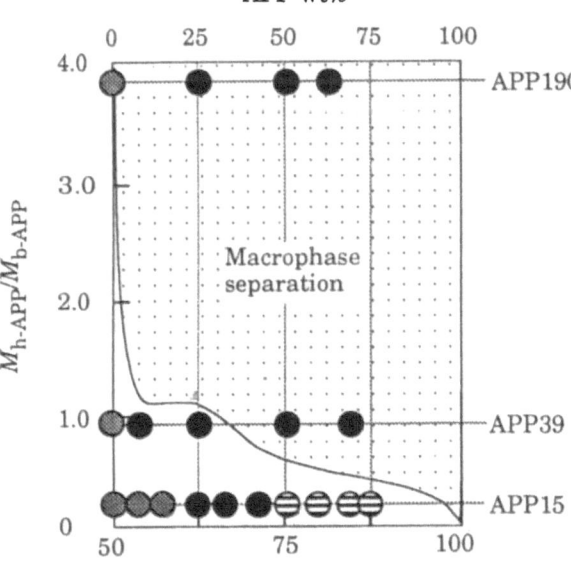

Figure 7 Morphological diagram constructed based on SANS, TEM, and SEM for blends with DEP113 and APP. Lamella; ◐, bicontinuous cylinders; ●, micelles; ⊖.

MICRODOMAIN STRUCTURES AND CRYSTALLIZATION KINETICS

Winey and coworkers have studied blends of amorphous block copolymers and the corresponding homopolymers extensively, and constructed a morphological diagram.[2] Our morphological diagram presented in figure 7 is consistent with their results in the major features. According to them and also to some theoretical work, the phase behavior can be rationalized in terms of the ratio of molecular weight between the block and the homopolymer, for our case, APP and polypropylene block in DEP ($M_{h\text{-APP}}/M_{b\text{-APP}}$). In the case of $M_{h\text{-APP}}/M_{b\text{-APP}} > 1$, macrophase separation takes place in almost the entire composition range. This is because the size of the APP chain is too large to merge into the microdomain of APP block. On the other hand, in the case of $M_{h\text{-APP}}/M_{b\text{-APP}} < 1$, macrophase separation is suppressed and blending causes a morphological transition of the microdomain. The reason for this is that the lower molecular weight homopolymer can enter into the A rich phase in the microdomain and thus change the volume fraction of the domain and the interfacial free energy, and so induce the morphological transition.

The morphology in the melt state of the blends clearly corresponds to the crystallization behavior described above. The crystallization kinetics of APP15 blends changes at 50 wt%, which is the same composition for the transition from a bicontinuous to a micelle structure by TEM. This agreement provides an evidence for the hypothesis that different microdomain structures are responsible for the different crystallization behavior; peak I and II and the different Avrami exponents.

SUMMARY AND CONCLUSION

Three series of blends of DEP113 and APP with different molecular weights were studied by calorimetry. Blending causes enormous changes in the crystallization behavior, and the results can be interpreted in the framework of the currently accepted microphase separation model for homopolymer/diblock copolymer blends.

ACKNOWLEDGMENT

We thank Mr. E. Habeeb and Mr. R. Krishnamoorti for their assistance with hydrogenation of the polydienes. We also acknowledge Prof. M. Muthukumar, Prof. S. Kumar, Prof. Winter and Dr. L. J. Fetters for helpful discussions. K. S. extends appreciation to Kanebo Ltd. for providing him the opportunity to work on this project. Acknowledgment is made to the UMASS Material Research Laboratory, funded by the National Foundation for support through the use of the central facilities.

REFERENCES

1 Hashimoto T, Tanaka H, Hasegawa H (1988) In: Nagasawa M (ed.) Molecular Conformation and Dynamics of Macromolecules in Condensed System. Elsevier Science Publishers B.V.
2 Winey K I, Thomas E L, Fetters L J (1992) Macromolecules 25:2645
3 Sakurai K, MacKnight W J, Lohse D J, Schulz D N, Sissano J A (1993) submitted to Macromolecules
4 (1) Sakurai K, MacKnight W J, Lohse D J, Lin M Y, Schulz D N, Sissano J A, Agamalyan M, (1992) abstracts for ACS fall meeting, (2) We are planing to submit to Macromolecules.

5   (1) Sakurai K, MacKnight W J, Lohse D J, Schulz D N, Sissano J A (1993) Macromolecules
    26:3236 (2) We are planing to submit a full paper to Macromolecules.

6   Xu Z, Mays J W, Chen X, Hadjichristidis N, Schilling F, Bair H E, Pearson D S, Fetters L
    J (1985) Macromolecules 18:2560

7   Rosedale J H, Bates F S (1990) Macromolecules 23:2329

8   Rubinstein R, Obukhov S P, (1993) Macromolecules in press

9   Alamo R G, Mandelkern L (1991) Macromolecules 24:6480

10 Hashimoto T, Tanaka H, Hasegawa H (1990) Macromolecules 23:4378

Reversible Heteropolymer Gelation
- Phase Diagrams of Mixed Networks -

F. Tanaka

Department of Physics, Faculty of General Education
Tokyo University of Agriculture and Technology
Fuchu-shi, Tokyo 183, Japan

Abstract: This paper presents possible phase diagrams of binary polymer mixtures A/B in which mixed networks are formed. To ensure the thermal equilibrium of the system, we consider reversible cross-links which are created and destroyed by thermal motion of the chains. On the basis of the lattice-theoretical picture, we develop a statistical-mechanical theory of network-forming polymer solutions. We derive possible phase diagrams of *alternating networks, interpenetrating networks* and *randomly mixed networks* with special attention to the interference effects among different phase transitions.

INTRODUCTION

Polymers exhibit a variety of condensed phases when some of their segments are capable of forming weak bonds which can be created and destroyed by thermal motion [1]~[3]. Transition from one phase to another caused by such *"segment association"* is reversible by the change of the temperature and the concentration, so that it is called *"reversible phase transition"*. What types of reversible phase formation are possible for a given associative interaction? What is the most fundamental laws which govern the competition between molecular association and phase separation? This paper surveys, as typical examples of reversible phases, macroscopic phase separation and gelation, with special emphasis on the new features which arise when mixed networks are formed.

We consider a mixture of A-chains and B-chains, each carrying f-functional and g-functional groups. They are assumed to form pairwise bonds whose binding energies are comparable to the thermal energy. In the case where either of the functionalities exceeds three, a cluster grows to macroscopic dimensions as soon as a threshold in temperature or in composition is reached[4]. Above the threshold a network comprized of the two components is formed. Most general structure of a cluster is depicted in Fig.1, in which three types of pairings, *i.e.* AA, BB, and AB, are allowed. The strength of the bonds can be expressed as

$$\lambda_{AA} = \exp(-\beta\Delta f_{AA}), \quad \lambda_{BB} = \exp(-\beta\Delta f_{BB}), \quad \lambda_{AB} = \exp(-\beta\Delta f_{AB}) \qquad (1)$$

in terms of the free energies Δf of bonding, where $\beta \equiv 1/k_B T$ is the inverse temperature.

A. Teramoto, M. Kobayashi, T. Norisuje (Eds.)
Ordering in Macromolecular Systems
© Springer-Verlag Berlin Heidelberg 1994

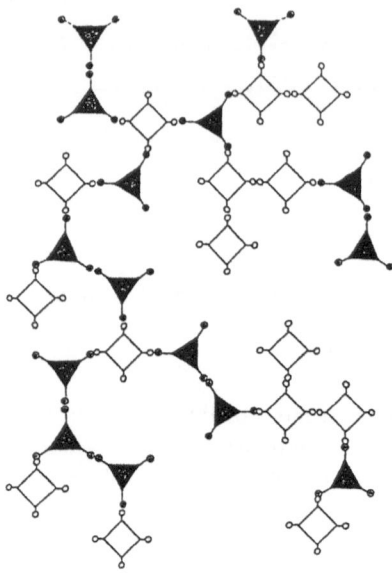

Fig.1 A giant cluster made up of functional A- and B-molecules

Typical examples of the mixed networks are:

(1) *Alternating networks* (AN) — Cross-links between different species are allowed, i.e.,$\lambda_{AA} = \lambda_{BB} = 0$. We refer to this case A·B. Because the formed clusters (of finite or infinite size) are multi-block copolymers, the system may undergo microphase separation. Hence, macrophase separation, microphase separation and gelation interfere each other. Special example of this case is coterminous cross-linking; B-chains carrying associative groups at their both ends coterminously cross-link A-chains. If we have solvent molecules in place of B-chains, coterminous cross-links are regarded as network junctions formed by solvent complexation [4].

(2) *Interpenetrating networks* (IPN) — Polymer chains A and B are cross-linked each other within the same species, but don't form bonds between different species, i.e.,$\lambda_{AB} = 0$. We refer to this case as A·A/B·B. The formed A-network and B-network are nonbonded but unseparable, because they are topologically connected. IPN's usually phase separate due to their small entropy of mixing, but, for sufficiently strong association, there is a possibility for the existence of thermally stable IPN [5].

(3) *Randomly mixed networks* (RMN) — If strength of the associative force in three combinations AA, BB, and AB are on the same order, the cluster formation progresses randomly. The resultant networks can be regarded as the macroscopic random copolymers.

THERMODYNAMICS OF ASSOCIATING POLYMERS

In what follows, we assume that the number of statistical units on a chain is n_A for an A chain and n_B for a B chain. In a thermal equilibrium, associative forces among the groups form intermolecular clusters with a wide spectrum of the aggregation number.

To derive the free energy of our system, we develope a lattice theoretical description of the binary polymer mixtures. We devide the total volume V of the system into small cells of the size a [4]. We then have total number $\Omega \equiv V/a^3$ of the microscopic cells. We first specify the part of the system containing only the finite-size clusters, which will be refered to *sol*. The total volume fraction of A-chains in the sol is given by $\phi_A^S = n_A \sum_{l,m} l\nu_{l,m}$, where $\nu_{l,m}$ is the number of clusters consisting of l A-chains and m B-chains (specified by the symbol (l,m)), and similarly $\phi_B^S = n_B \sum m\nu_{l,m}$ for B-chains. Contribution from the macronetwork is not included in the sum. The total volume fraction of the sol in the system is given by $\phi^S = \phi_A^S + \phi_B^S$. This should be equal to unity for nongelling systems or pregel regime, but becomes smaller than unity as soon as an infinite network (refered to as *gel*) appears (postgel regime). In the postgel regime we have the relations $\phi_i^S + \phi_i^G = \phi_i$ for $i = A, B$, where ϕ_i is the volume fraction of the species i, which is experimentally controlled. Since we have an identity $\phi_A + \phi_B = 1$, we can take ϕ_A as an independent variable and write it simply as ϕ. The volume fraction of B is then given by $\phi_B = 1 - \phi$.

In order to study the thermodynamic properties, we start from the *reference state* in which unconnected A-chains and B-chains are prepared separately. We first consider the free energy change ΔF_{rea} to bring the system to a fictitious intermediate state in which the chains are connected in such away that the cluster distribution is exactly the same as the real one. It is given by

$$\beta\Delta F_{rea}/\Omega = \sum_{l,m} \Delta_{l,m}\nu_{l,m} + \delta_A \nu_A^G + \delta_B \nu_B^G, \tag{2}$$

where $\beta \equiv 1/k_B T$ is the inverse temperature, and $\Delta_{l,m}$ the change in the internal free energy produced in the process in which a single (l,m) cluster is formed from l A-chains and m B-chains. Similarly δ_i (i =A,B) are the free energy change produced when an isolated chain of the species i becomes a part of the gel network. Let $\mu_{l,m}^\circ$ be the internal free energy of a (l,m) cluster. The free energy difference $\Delta_{l,m}$ is then given by $\Delta_{l,m} = \mu_{l,m}^\circ - l\mu_{1,0}^\circ - m\mu_{0,1}^\circ$. Similarly we have $\delta_A = \mu_A^{\circ G} - \mu_{1,0}^\circ$, and $\delta_B = \mu_B^{\circ G} - \mu_{0,1}^\circ$. Under a constant pressure, $\mu_{l,m}^\circ$ is equivalent to the internal free energy due to the combination, configuration and the bond formation of the constitutional chains.

In the second step, we mix these clusters to form a real mixture. According to the lattice theory [6] of polydisperse polymer mixtures, the mixing free energy ΔF_{mix} in this process is given by

$$\beta\Delta F_{mix}/\Omega = \sum_{l,m} \nu_{l,m} \ln \phi_{l,m} + \chi\phi_A\phi_B, \tag{3}$$

where $\phi \equiv (n_A l + n_B m)\nu_{l,m}$ is the volume fraction of (l,m)-clusters, χ the Flory χ-parameter which specifies the strength of the van der Waals type contact interaction between the monomers of different species. χ-parameter varies with the temperature. The total free energy from which our theory starts is given by the sum of the above two parts:

$$\Delta F = \Delta F_{rea} + \Delta F_{mix}. \tag{4}$$

By differentiation with respect to the number of molecules or clusters, we find the chemical potentials for the clusters.

Pregel Regime

Having obtained the chemical potentials, we now impose the *multiple equilibrium* conditions to ensure the equilibrium distribution of the cluster. These are

$$\Delta\mu_{l,m} = l\Delta\mu_{1,0} + m\Delta\mu_{0,1} \tag{5}$$

for all possible combinations of the integers (l,m). Upon substitution of the chemical potentials, we find the volume fractions of the clusters to be given by $\phi_{l,m} = K_{l,m}x^l y^m$, where we have written as x and y for simplicity for the unimer concentrations $\phi_{1,0}$ and $\phi_{0,1}$. The new constant $K_{l,m}$ (called *association constant*) is defined by $K_{l,m} = \exp(l+m-1-\Delta_{l,m})$, which depends only on the temperature through $\Delta_{l,m}$. The total volume fraction ϕ^S is found by taking the infinite sum:

$$\phi^S(x,y) = \sum_{l,m} K_{l,m}x^l y^m. \tag{6}$$

The total number of clusters $\nu^S(x,y)$ is similarly found.

In order to study the thermodynamic properties, we must express the unimer concentration as a function of the total composition ϕ, which is controllable in the experiments. In the pregel regime this can be done by solving the coupled relations $n_A x(\partial\nu^S(x,y)/\partial x) = \phi$, and $n_B x(\partial\nu^S(x,y)/\partial y) = 1 - \phi$. Once we find x and y as functions of ϕ (and the temperature T), we can express all physical quatities in terms of these two independent thermodynamic variables. The followings are impotant properties on which we will discuss in detail for specific systems:

(1) The osmotic pressure π of the A component is given by

$$\beta\pi/n_B a^3 = -\beta\Delta\mu_B/n_B = (1 + \log y)/n_B - \nu^S(x,y) + \chi\phi^2. \tag{7}$$

In a polymer solution in which B component is a low molecular weight solvent ($n_B = 1$), this definition reduces to the osmotic pressure in the conventional meaning.

(2) Two-phase equilibrium can be found by the balance of the chemical potential of each component:

$$\Delta\mu_A(\phi', T) = \Delta\mu_A(\phi'', T), \tag{8a}$$

$$\Delta\mu_B(\phi', T) = \Delta\mu_B(\phi'', T), \tag{8b}$$

where ϕ' and ϕ'' are the composition of the dilute A phase and concentrated A phase respectively.

(3) The thermodynamic stability limit or *spinodal* can be found for a binary system by a single condition $(\partial \Delta \mu_A / \partial \phi)_T = 0$. Our result for $\Delta \mu_A$ leads to the equation

$$\frac{\kappa_A(\phi)}{n_A \phi} + \frac{\kappa_B(\phi)}{n_B(1-\phi)} - 2\chi = 0, \tag{9}$$

where the new functions are defined by $\kappa_A(\phi) \equiv \phi \partial \log x / \partial \phi$ and $\kappa_B(\phi) \equiv -(1-\phi) \partial \log y / \partial \phi$.

Sol-to-Gel Transition

So far we have tacitly assumed that the infinite double summation in ϕ^S (and hence for ν^S) converges. For a solution capable of gelling, a borderline exists which separates the unit square on the (x, y) plane into a convergent and divergent region of the infinite sum. Exactly on the boundary line, the sol composition ϕ^S takes a finite value smaller than unity, but they diverge outside this line. Since the radius of convergence generally depends on the total composition, let us express the boundary by a parametric representation $(x^*(\phi), y^*(\phi))$ for $0 < \phi < 1$.

In the postgel regime, a chain participating in the gel network is in a chemical equilibrium with an isolated chain of the same species. This imposes additional conditions $x^* = \exp(\delta_A - 1)$ and $y^* = \exp(\delta_B - 1)$, if they are gelling. In the postgel regime x and y in the physical quantities must be replaced by these values. As a result we have, for example, a function κ^* which is different from κ. The function κ^* corresponding to a gelling component vanishes because it is proportional to the weight-average molecular weight.

INTERNAL FREE ENERGY

In order to obtain the intra-cluster free energy $\Delta_{l,m}$, we have to count the number of ways to form a (l, m) cluster from the constituent l A-chains and m B-chains. This is a familiar combinatorial problem we encounter in the classical theory of chemical gelation[7]. If clusters of tree type only — that is, those which have no internal loops — are allowed, the exact counting is possible. The most general solution of such tree statistics was found by Fukui and Yamabe [8]. We here employ their result. The key idea is that subclusters of one species, for example B, in a giant cluster are identified as multiple "junctions" which connect A subclusters. Our problem is thus mapped onto the thermoreversible gelation with junctions of variable multiplicity, the details of which is given in ref.[9].

SPECIFIC SYSTEMS

For the numerical calculation of the phase diagrams we introduce the reduced temperature $\tau \equiv 1 - \Theta_0/T$, where the unperturved theta temperature Θ_0 is the temperature which satisfies the equation $\chi(\Theta_0) = 1/2$, and hence the unrenormalized second virial coefficient of the osmotic pressure vanishes at this temperature. We then have $\chi(T) = 1/2 - \psi_1\tau$. For the association constant, we rewrite λ_{AB} as $\lambda_{AB}(T) = \lambda_0 \exp\{\gamma(1-\tau)\}$ in terms of the dimensionless binding energy $\gamma \equiv \Delta\epsilon_{AB}/k_B\Theta_0$ and the constant $\lambda_0 \equiv \exp(\Delta s/k_B)$, which is related to the entropy change Δs_{AB} in bonding. Other two constants are similarly rewritten.

Alternating Networks

When $\lambda_{AA} = \lambda_{BB} = 0$, chains are alternatingly connected. The resultant clusters are basically multiblock copolymers, so that the system may undergo microphase separation as well. Fig.2(a)∼(c) show examples of the phase diagrams for gels of low molecular weight ($n_A = n_B = 3$) trifunctional molecules ($f = g = 3$). When the association constant λ_0 is small, for instance $\lambda_0 = 0.1$ as in Fig.2(a), gelling region is completely included in the miscibility gap. The solid line shows the spinodal, and the broken line sol-to-gel transition line. Spinodal has a critical point CP on the top. Gel in this figure has no physical significance as far as equilibrium state is concerned, since it lies completely inside the unstable region. In non-equilibrium state, however, it largely affects the morphological evolution of the mixture. The spinonal decomposition is arrested by gelling in some stage, thus leading to the pinning of the intermediate state. As the association constant is increased, the gel region expands and eventualy touch the spinodal as is shown in Fig.2(b). The miscibility is improved in the central concentration region because of the existence of the AB clusters. At the same time, the critical point split into two parts. As λ_0 is still increased, thermally stable gel region appears on the top of the spinodal line as is seen in Fig.2(c).

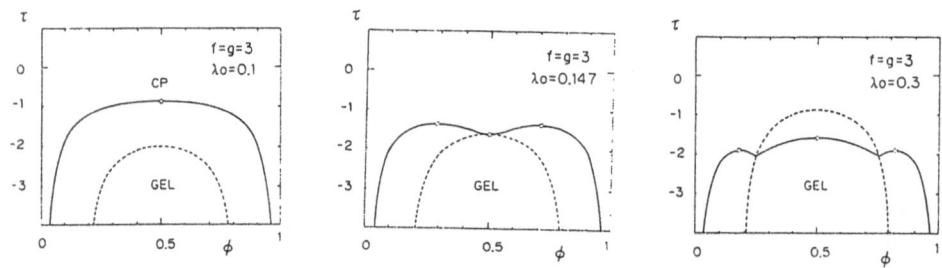

Fig.2 Phase diagrams of alternating networks (a)$\lambda_0 = 0.1$ (b)$\lambda_0 = 0.147$ (c)$\lambda_0 = 0.3$

As the second example of the alternating case, we consider cross-linking by the solvent comlexation. Assume that A is a long polymer chain ($n_A = n \gg 1$) carrying a large number $f \gg 1$ of functional groups, while B is a solvent molecule ($n_B = 1$) with functionality $g = 2$. Thus solvent complexation is represented by coterminous cross-linking mediated by a small solvent molecule. Fig.3 shows the comparison of the predicted sol-gel transition line with the experimental data. Experimental data were found for atactic polystyrene in carbon disulfide (a-PS/CS$_2$) by differential scanning calorymetry (DSC), i.e., from the anomaly in heat capacity[10]. They exhibit a maximum at concentration $\phi \sim 0.2$ at which gelation is most enhanced. Solid lines show the theoretical calculation done by assuming that CS_2 molecule is stuck in between the two phenyle groups on the separate PS chains. The association strength λ_0 is changed from curve to curve. To fit the data a very large entropy loss (small value of λ_0) is neccessary. Another conclusion we can draw from the calculation is that a large value of γ (~ 8) is required to fit the theoretical curve in dilute region. The experimental data even indicate a finite transition temperature when extrapolated into the vanishing polymer concentration. Since this is clearly unphysical, theoretical consideration poses doubt on DSC data in dilute region.

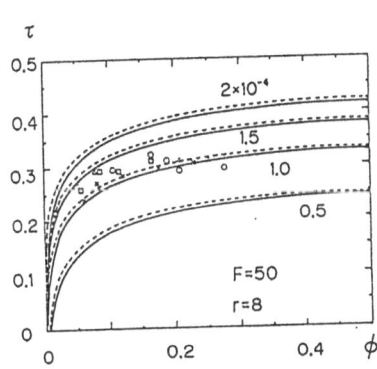

Fig.3 Sol-gel transition line of at-PS/CS$_2$

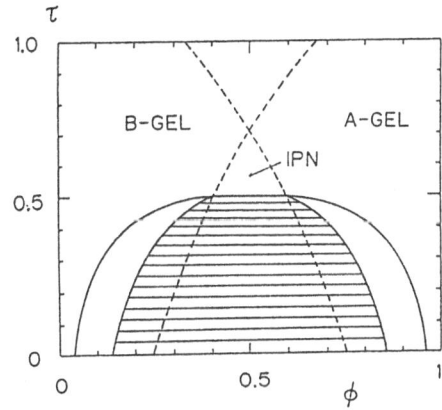

Fig.4 Predicted phase diagram of stable IPN

Interpenetrating Networks

Interpenetrating polymer networks (IPN's) are defined as a nonbonded but unseparable combination of two polymers, each in network form. Like most other multicomponent polymer materials, IPN's usually phase separate due to their small entropy of mixing, but the presence of the cross-links leads to the reduction of the domain size. As a consequence of the complex dynamic balance between the two opposite tendencies, the synthesis of IPN's can produce materials ranging from molecularly homogeneous ones to microscopically phase separated ones with phase domains of various sizes, hence yeilding a unique method of controlling the morphology and mechanical properties. Chemistry and physics of IPN are extensively reviewed in Ref.[11]. There are two main methods of synthesizing IPN's: simultaneous IPN (sim-IPN) and sequential IPN (seq-IPN). For sim-IPN, functional monomers (or primary chains) of both species are mixed together and polymerized. For seq-IPN, a polymer network of A species is synthesized, and functional monomers B are swollen into the network and polymerized.

Fig.4 shows a typical result for the sim-IPN of low molecular-weight symmetric sim-IPN ($n_A = n_B = 1$) with functionality $f = g = 3$[5]. Thick lines show the binodals and the spinodals, while the broken lines the sol-to-gel transition lines. In the temperature region below the broken lines, we have a gel of each species; B gel in low concentration region and A gel in high concentration region. In the overapping region of the two, we have IPN. For a small λ_{AA} (as $\lambda_{AA} = 1$ shown in this figure) the overlap region lies inside the spinodal, showing that the IPN cannot stay as a stable phase. But as the entropy parameter λ is increased, the overlap region is enlarged and comes closer to the spinodal, until eventually the CP disappears; the top part of the overlap region goes out of the miscibility gap. Fig.4 shows the case $\lambda_{AA} = 3$ as an example. In this small triangular region, the IPN stays as a stable homogeneous phase. The intersections between the gelation lines and the spinodal (and also the binodal) are the *tricritical points* (TCP)[12]. Because a larger entropy loss Δs_{AA} (< 0) gives a larger value of λ_{AA}, existence of a stable IPN is more probable for the associative force which produce a large entropy loss when a bond is formed. Orientational restriction in the course of the bond formation, such as seen in the hydrogen bonding for instance, can lead to a large entropy loss. The asymmetric high molecular-weight sim-IPN whose n_A is much larger than n_B for instance, can also be studied in a similar way. The miscibility gap shifts to lower concentration region as in the uncrosslinked polymer solutions, but the triangular IPN region remains essentially unchanged.

Randomly Mixed Networks

Finally we discuss the most complex case where all types of associations are allowed. Randomly mixed networks appear in the postgel regime. Their statistical structure depends on the relative strength of the three λ's, but the fundamental features in the phase diagrams are basically the combination of Fig.2 and Fig.4. Fig.5 shows an example of low molecular-weight primary molecules carrying associative groups whose bonds have the same strength $\lambda_{AA} = \lambda_{BB} = \lambda_{AB}$. Mixed gel, A-gel, B-gel and IPN simultaneously coexist on the temperature-concentration phase plane. Relative position of each region to the miscibility gap changes according to the competition between λ and χ.

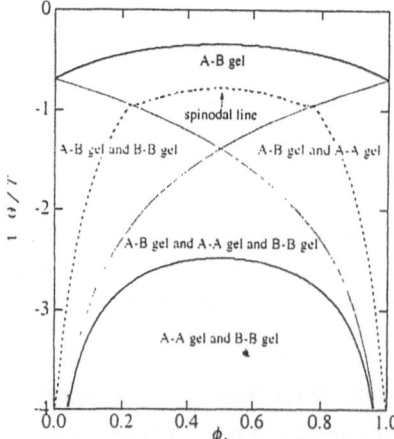

Fig.5 Phase diagrams of randomly mixed networks

CONCLUSION

It has been shown that competing two terms — reaction and mixing — must be considered in the free energy to describe associating polymer blends and solutions. Several new concepts have been introduced that can be reached only if both synthetic aspect and material-properties are simultaniously taken into consideration. But dynamical phenomena of the phase formation still remain unsolved. Polymer association is a problem which is quite new and fundamental, providing many questions yet to be theoretically and experimentally studied.

[1] Russo, R. S. *"Reversible Polymeric Gels and Related Systems"*; American Chemical Society: Washington,DC, 1987; Vol.350.

[2] Kramer, O. *"Biological and Synthetic Polymer Networks"*; Elsevier Applied Science: London and New York, 1988.

[3] Utracki, L. A.; Weiss, R. A. *"Multiphase Polymers: Blends and Ionomers"*; American Chemical Society: Washington, DC, 1989; Vol. 395.

[4] Tanaka, F. Macromolecules 1990, *23*, 3784; 3790.

[5] Tanaka, F. Phys. Rev. Lett. 1992, *68*, 3188.

[6] Flory, P. J. J. Chem. Phys. 1942, *10*, 51; Huggins, M. L. J. Chem. Phys. 1942, *46*, 151.

[7] Flory, P. J. J. Am. Chem. Soc. 1941,*63*, 3091;3096; Stockmayer, W. H. J. Chem. Phys. 1943, *11*, 45; 1944, *12*, 125.

[8] Fukui, K; Yamabe. T, Bulletine Chem. Soc. Japan 1967, *40*, 2052.

[9] Stockmayer, W. H. Macromolecules 1991, *24*, 6367; Tanaka, F.; Stockmayer, W.H.; to appear in Macromolecules.

[10] Francois, J.; Gan, J. Y. S.; Guenet, J. M., Macromolecules 1986, *19*, 2755; Klein, M.; Guenet, J. M., Macromolecules 1989, *22*, 3716.

[11] Sperling, L. H., *Interpenetrating Polymer Networks and Related Materials*; Plenum Press, New York 1981.

[12] Tanaka, F. Macromolecules 1989, *22*, 1988; Tanaka, F.; Matsuyama, A. Phys. Rev. Lett. 1989, *62*, 2759.

[13] Tan, H. M.; Moet, A.; Hiltnet, A.; Baer, E. Macromolecules 1983, *16*, 28.

SANS Studies of Early Stage Spinodal Decomposition

T. Hashimoto, H. Jinnai, H. Hasegawa, C. C. Han*

Division of Polymer Chemistry, Graduate School of Engineering, Kyoto University, Kyoto 606-01, Japan

*Polymers Division, National Institute of Standards and Technology, Gaithersburg, Maryland 20899, USA

Abstract: Time-resolved small-angle neutron scattering (SANS) experiments were performed on the self-assembling process of a binary mixture of deuterated polybutadiene and protonated polybutadiene at the critical composition. Specimens held in the single-phase state at an initial temperature (T_i) were quenched to a point inside the spinodal phase boundary at a final temperature (T_f) to induce phase separation via spinodal decomposition (SD). The effect of the thermal concentration fluctuations on the SD was examined by changing T_i for a fixed T_f. The wave number and temperature dependence of Onsager kinetic coefficient were also investigated as a function of T_i and T_f.

I. INTRODUCTION

Over the last decades a considerable number of studies have been made on self-assembling processes via spinodal decomposition (SD) in binary polymer blends [1-13]. Generally, the coarsening process via SD can be classified, at least, into three stages: (i) early stage; (ii) intermediate stage; (iii) late stage [4]. Most studies with polymers have concluded that the linearized Cahn-Hilliard theory [14] (CH) can describe very well the early stage SD at relatively deep quench depths. In particular, the growth of the Fourier modes of the concentration fluctuations with small wave numbers, q, satisfying $qR_g \ll 1$ (R_g is the average radius of gyration of the constituent polymers), is well described by the theory. The CH theory predicts that the wave number $q_m(t;T)$ is independent of time t, where $q_m(t;T)$ corresponds to the peak position of the structure factor at t and at a phase separating temperature T [15]. Here, q is given by

$$q = (4\pi / \lambda)\sin(\theta / 2), \tag{1}$$

with θ and λ being the scattering angle and the wavelength of the incident beam (laser light or neutron) in the medium, respectively.
Most studies have been carried out using time-resolved light scattering (LS). This technique is very useful for investigating the time-evolution of structures with small q or large length scales (0.1 to 10 μm). However, because

A. Teramoto, M. Kobayashi, T. Norisuje (Eds.)
Ordering in Macromolecular Systems
© Springer-Verlag Berlin Heidelberg 1994

of the limited q-range accessible by LS, there are still several problems to be solved in the early stage of SD, as listed below:

(1) The CH theory predicts $q_m(0;T) \sim R_g^{-1}$ in deep quench conditions, where $q_m(0;T)$ and R_g are, respectively, the wave number of the Fourier mode of the concentration fluctuations which grows at a maximum rate in the early stage SD at T and an average radius of gyration of constituent polymers. Since R_g typically is of the order of hundreds Å, $q_m(0;T)$ is of the order of 0.01Å^{-1}, which is much larger than the wave numbers covered by LS ($q \lesssim 9 \times 10^{-4} \text{ Å}^{-1}$).

(2) Because of this small q-range covered by LS, there is a fundamental problem that remains unclear in regard to how the wave number $q_m(t;T)$ at the maximum of scattering intensity, which characterize the size of the phase-separated structure at T, changes with t in the very beginning of SD. Few time-resolved LS works have observed the variation of $q_m(t;T)$ with time in the early stage SD.

(3) Izumitani and Hashimoto [6] found that their mixtures of SBR (styrene-butadiene random copolymer) and HPB (polybutadiene) at deep quenches exhibit the time-independent scattering maximum in the early stage SD at the large q-limit ($q \cong 9 \times 10^{-4} \text{ Å}^{-1}$) covered by their LS experiment. However, in their "Cahn plot", the critical q value, $q_c(T)$, at which the growth rate of concentration fluctuations changes sign from positive to negative with increasing q was not directly observed, though it gives important information.

(4) From a study of the SD kinetics for a binary blend of polystyrene (HPS)/poly(vinylmethylether) (PVME), Okada and Han [5] concluded that the effect of thermal concentration fluctuations is important even at small q covered by their LS experiment at shallow quenches. Namely, the linearized theory of Cahn-Hilliard-Cook (CHC) which *includes the effect of thermal fluctuations* [16] can better describe the early stage of SD in the shallow quench region than the CH theory which omits the thermal fluctuation effect. Needless to say that the effect becomes increasingly important with increasing q. The question of whether the effect is important or not for the SD at deep quenches has been left unanswered. The effect was regarded to be insignificant at the small q covered by LS [3, 6-9, 12, 13]. However, it has not been investigated so far over a wide q-range, including the q-range covered by small-angle neutron scattering (SANS). It is also interesting to investigate how the initial thermal concentration fluctuations affect the growth of concentration fluctuations in SD, and how the thermal noise in the single phase is transformed into the concentration fluctuations as predicted by CH and CHC. Since the dominant Fourier modes of the fluctuations in the very early stage of SD seemingly have large q, the process leading to the early stage SD are best studied by the time-resolved SANS.

(5) Binder [17] and Pincus [18] discussed q-dependence of Onsager kinetic coefficient, $\Lambda(q;T)$. So far it has been assumed that $\Lambda(q;T)$ is independent of q and equal to $\Lambda(q=0)$ for any q satisfying $qR_g \ll 1$. Only a few works have addressed this issue [9, 19]. Obviously more works are

deserved along this line, because this is a problem unique and inherent to polymer systems.

The present paper reports our studies of these problems by the time-resolved SANS method for a critical blend of normal (protonated) and deuterated polybutadienes. The blend chosen is convenient for time-resolved SANS measurement because it undergoes very slow SD. The typical characteristic time t_c is of the order of 1000 sec. The deep quenching was employed so that the scattering peak may become observable in the q-range accessible to our SANS experiments.

We observed the time-evolution of S(q,t;T) after the quenching from three different initial temperatures (T_i = 102.3, 123.9, and 171.6°C) to a single temperature (T_f = -7.5°C) to investigate how much and how long the effect of initial concentration fluctuations in the system affects the subsequent SD (**the memory effect or the initial temperature dependence**). We performed another set of experiments in which the specimens were held at T_i = 123.9°C and quenched to three different final temperatures (T_f = -7.5, 1.1, and 10.5°C), for the q-dependence and T-dependence of $\Lambda(q;T)$ (**final temperature dependence**).

II. EXPERIMENTAL SECTION

Both perdeuterated polybutadiene (DPB) and protonated polybutadiene (HPB) used in the present study were synthesized by living anionic polymerization. Table I summarizes their molecular characteristics. The two polymers are almost identical with respect to density and statistical segment length. As shown elsewhere [20], this blend has a nearly symmetric phase diagram. The glass transition temperature (T_g) of an HPB having similar microstructure is about -100°C, so that in the temperature range of SANS experiments reported here both DPB and HPB and also the blend are far above their corresponding T_g and hence behave like liquid at a long time scale.

Specimens of the DPB/HPB blend having 46.6% of DPB and 53.4% of HPB by volume were prepared according to the procedure detailed elsewhere [20]. This composition of the blend corresponds to a critical volume fraction calculated according to the Flory-Huggins lattice theory.[21] Analysis of the SANS data on the blend in the single-phase state showed that this blend has an upper critical solution temperature (UCST) type phase diagram with the spinodal temperature (T_s) at the critical composition being 99.2°C. The cell containing the blend specimen was left standing for a sufficient period of time in a heating block controlled at T_i which was well above T_s so that the specimen was equilibrated in the single phase state. Temperature-drops (quenching) were carried out by manually transferring the sample at T_i into a copper block controlled to the measuring temperature T_f. It took about 4 to 5 min. for the sample to settle down to T_f after the quenching.

A time-resolved SANS profile at a given t was obtained over the counting time between (t-1.5) and (t+1.5) in min. SANS data were taken by using the 30-m SANS instrument at the NIST Cold Neutron Research Reactor. The detailed

Table I. Polymer Characteristics

Sample Code	$M_n \times 10^{-4}$ [a]	Z_w/Z_n [b]	Z_n [c]	microstructure, [d] %		
				1,2	cis-1,4	trans-1,4
DPB	37.4	1.28	6223	21.5	36.3	42.2
HPB	27.3	1.06	5039	29.3	70.7 [e]	

a)   Determined by membrane osmometry.
b)   Determined by size exclusion chromatography equipped with light scattering.
c)   Number-average degree of polymerization.
d)   Determined by ^{13}C-NMR.
e)   Sum of cis-1,4 and trans-1,4.

experimental condition and data acquisition procedure were described in detail in a separate publication [20].

III. RESULTS

Figure 1 shows the time-evolution of SANS profiles for our critical mixture after the quench from 123.9 to -7.5°C. Here, $S(q,t;T_f)$ is plotted against wave number q. The solid line in the lower part (a) of the figure shows the measured profile in the single-phase at 123.9°C. Similar data were obtained for all of the experiments with various combination of T_i and T_f, although not shown here. As seen in Fig. 1 (a), the profiles show a maximum shortly after the quenching, and the maximum height $S_m(t;T_f)$ increases with t. In what follows, the q value at the maximum is denoted by $q_m(t;T_f)$. Surprisingly, $q_m(t;T_f)$ shifts to *larger values* at first (see the vertical segments of lines marked on the profiles from 27.3 to 111.2 min.), stays constant for a while (see the profiles from 111.2 to 248.7 min. and the constant value is defined as $q_m^M(T_f)$ hereafter) and eventually decreases with t (see the profiles after 248.7 min.). The initial increase in $q_m(t;T_f)$ seems to contradict the results from computer simulations of the CHC linearized theory by Strobl [22], who showed that $q_m(t;T_f)$ either stayed unchanged or became smaller depending upon the parameters. Experimentally, Motowoka et al.[23] and Schwahn et al.[19] found this "initial peak-shift" to smaller q with t for their deuterated polycarbonate/poly(methylmethacrylate) and deuterated polystyrene/poly(methylmethacrylate) blends, respectively. The mechanism of the unusual shift of $q_m(t;T_f)$ in the *very early stage of SD* is discussed later in section IV-3.

The wave number satisfying $q = 1/\overline{R}_g$ is shown by an arrow in Fig. 1, where $\overline{R}_g$ is the mean radius of gyration of the mixture. Our deep quench experiments gave $q_m^M(T_f)$ that comes relatively close to $1/\overline{R}_g$, but $q_m^M(T_f)$ is *not* always equal to $q_m(0;T_f)$ because of the contribution of the thermal noise to

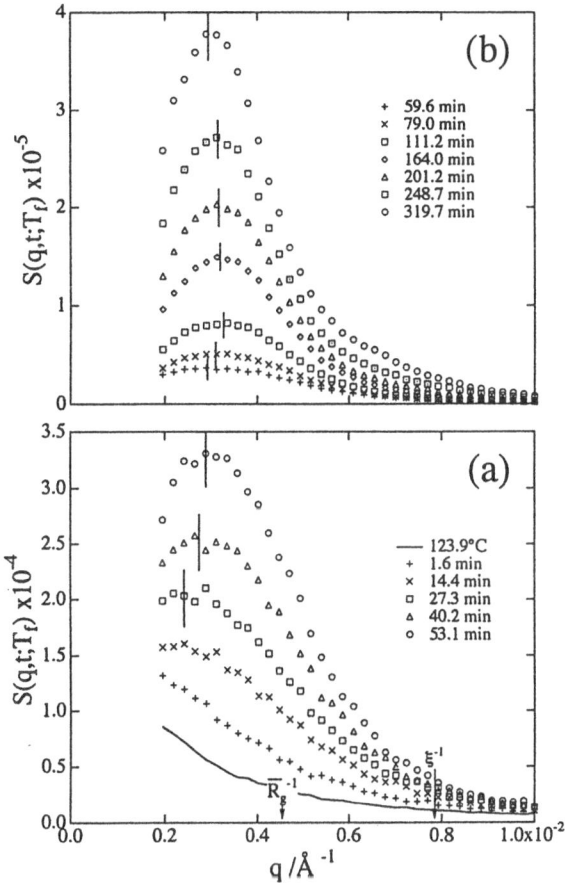

Figure 1. Evolution of SANS profiles for the critical DPB/HPB mixture after the quench from 123.9°C to -7.5°C. (a) the early stage SD; (b) the intermediate stage SD. Vertical bars display the peak positions at different t. The wave numbers corresponding to $q = 1/R_g$ and $q = 1/\xi$ are shown by arrows.

$S(q,t;T_f)$, as will be discussed in section IV-3 (Fig. 5). The true $q_m(0;T_f)$ is the wave number at which the growth rate of concentration fluctuations, $R(q;T)$, reaches a maximum. The correlation length ξ for thermal concentration fluctuations is another important parameter. ξ at -7.5°C was calculated to be 127.4 Å [20]. The wave number satisfying $q = \xi^{-1}$ is shown in Fig. 1 by another arrow.

Figure 2 shows plots of $q_m(t;T_f)$ and $S_m(t;T_f)$ vs. t. As time goes on, $q_m(t;T_f)$ increases, stays unchanged, and then decreases, while $S_m(t;T_f)$ keeps increasing. The other important observation is that, depending upon T_i at which the specimen is held prior to the quenching, $q_m(t;T_f)$ and $S_m(t;T_f)$ follow different time-dependencies. Remarkably, this memory effect persists for a considerably long period of time in the subsequent SD process; in fact, it eventually disappears (at about 300 min. after the onset of SD). It is also observed that the higher the T_i, hence the smaller the thermal concentration fluctuations before the quenching, the sooner $q_m(t;T_f)$ reaches $q_m^M(T_f)$ and the smaller the shift of $q_m(t;T_f)$ in the very early stage becomes.

$q_m(t;T_f)$ and $S_m(t;T_f)$ for the experiments of the final temperature dependence show the trends similar to those as described above, although they are not shown here. We note that as the quench depth increases (or as T_f is lowered) the time at which $q_m(t;T_f)$ becomes maximum shifts to a longer time (i.e., the kinetics becomes slower). Our blend has an UCST type phase diagram so that the lower the temperature T_f, the deeper the quench depth. If the kinetics is governed by thermodynamic driving force, it should become faster as

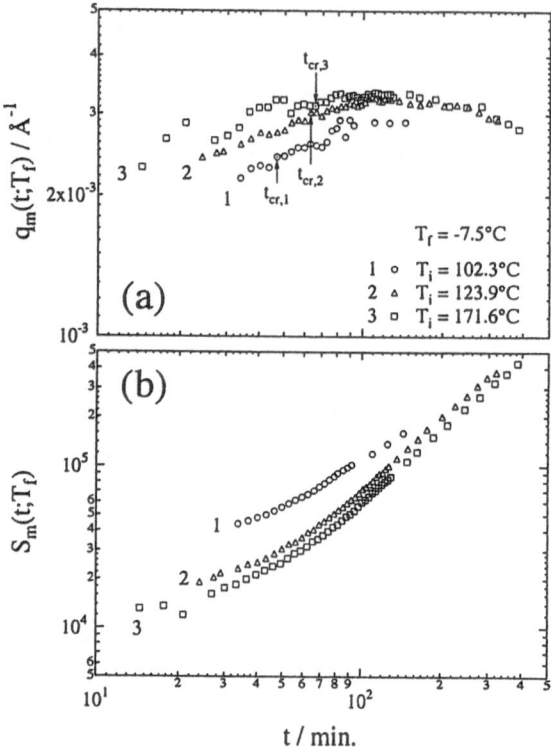

Figure 2. Evolution of peak position $q_m(t;T_f)$ (part a) and peak intensity $S_m(t;T_f)$ (part b) for three different T_i as given in the figure. The crossover time $t_{cr,i}$ from the early to intermediate stage for each experiment is shown by an arrow.

T_f decreases, whereas if the kinetics is controlled by diffusion, the reverse should be true. Thus, the rate of phase separation of a blend having an UCST type phase diagram is determined by the two competing factors. Therefore, the SD process of our blend must be diffusion controlled. It was also observed that $q_m{}^M(t;T_f)$ is larger for a deeper quench which is in accordance with the CH theory. The evolution of $S_m(t;T_f)$ is much simpler than $q_m(t;T_f)$. The higher the T_f is, the larger the $S_m(t;T_f)$ becomes over the time range of an experiment. This is another piece of evidence for SD being diffusion-controlled at the quenches treated in the present work.

IV. ANALYSIS WITH CAHN-HILLIARD-COOK THEORY

1. THEORETICAL BACKGROUND

The time-evolution of the structure factor for a binary mixture undergoing SD is generally described by the nonlinear differential equation well known as Time-Dependent Ginzburg-Landau equation.

In the early stage SD, the concentration fluctuations are so small that higher order terms in the differential equation may be neglected. Then, it is reduced to an inhomogeneous linear differential equation [14, 16], and its solution reads

$$S(q,t;T_f) = S_T(q;T_f) + \left[S(q,0;T_f) - S_T(q;T_f) \right] \exp\left[2R(q;T_f)t \right], \qquad (2)$$

where

$$R(q;T_f) = -\Lambda(q;T_f)q^2 / S_T(q;T_f). \qquad (3)$$

Here $S_T(q;T_f)$ is the structure factor at $T = T_f$ and is called the virtual structure factor. $S(q,0;T_f)$ is the structure factor at the onset of SD at $T = T_f$. It is worth noting that all terms in Eq. 2 refer to T_f only but, as will be seen, they depend on the structure factor at the "initial" temperature T_i. We will return to this point later.

Since the system at $T = T_f$ cannot stay in one-phase, $S_T(q;T_f)$ is not directly measurable. Moreover it becomes negative at $q < q_c$ and hence it is called "virtual" structure factor. It has to be inferred. One method is experimental. It is known that $S_T(q;T)^{-1}$ in the single-phase region become parallel straight lines with respect to T when they are plotted against q^2. $S_T(q;T)$ at $T = T_f$ can be obtained by an extrapolation of data at the high q-range in this plot to the small q [19]. The other is theoretical. It invokes a theory for $S_T(q;T)$. Here we used the latter, taking the extended version of de Gennes' theory for a binary polymer mixture consisting of the constituent polymers having a Schultz-Zimm molecular weight distribution and different monomer molecular volume [24]. It reads

$$S_T(q;T) = \left[\frac{v_0}{\phi_A \langle Z_A \rangle_n v_A S_A(q)} + \frac{v_0}{\phi_B \langle Z_B \rangle_n v_B S_B(q)} - 2\chi(T) \right]^{-1} \qquad (4)$$

where

$$S_i(q) = \frac{2}{X_i^2} \left[\left(\frac{h_i}{h_i + X_i} \right)^{h_i} - 1 + X_i \right] \qquad (i = A \text{ or } B) \qquad (5)$$

with

$$X_i = q^2 \langle R_{gi}^2 \rangle_n = q^2 \langle z_i \rangle_n b_i^2 / 6 \qquad (6a)$$

and

$$h_i = (\langle Z_i \rangle_w / \langle Z_i \rangle_n) - 1)^{-1} \qquad (6b)$$

$\langle Z_i \rangle_w$ and $\langle Z_i \rangle_n$ denote, respectively, the number- and weight-average degree of polymerization for the ith component (i = A and B), and ϕ_i is the volume fraction of the ith component with molar volume v_i and statistical segment length b_j. v_0 is the molar volume of the reference cell defined as $v_0 = (\phi_A/ v_A + \phi_B/ v_B)^{-1}$, and χ is Flory's segmental interaction parameter. The temperature dependence of χ was determined by the analysis of SANS data in the single-phase region and is given by [20]

$$\chi(T) = -5.34 \times 10^{-4} + 0.314 /T. \qquad (7)$$

The value of $\chi(T_f)$ needed to calculate $S_T(q;T_f)$ from Eq. 4 was estimated by assuming that Eq. 7 is valid down to T_f. With $\chi(T_f)$ thus obtained, the correlation length ξ for the thermal concentration fluctuations at $T = T_f$ can be calculated [20].

In Fig. 3, $S(q,t;T_f)$ at the respective q are plotted against t in semilogarithmic scale for the experiment with $T_i = 123.9°C$ and $T_f = -7.5°C$, where data from different q are shifted vertically except for data at $q = 1.953 \times 10^{-3}$ Å^{-1} for viewing clarity. The CH theory predicts that (i) $q_c(T_f) \sim \xi(T_f)^{-1}$ and (ii) $S(q,t;T_f)$ at $q < q_c$ exponentially increases with t, while that at $q \geq q_c$ exponentially decreases. Looking at data at $q = 7.929 \times 10^{-3}$ Å^{-1} which satisfies $q \geq q_c(T_f) \sim \xi(T_f)^{-1}$, the intensity increases with t contrary to the prediction from the CH theory. This disagreement demonstrates that the CH theory is invalid at the high q-range, indicating the necessity of incorporating the thermal concentration fluctuations term in the theory. All the plots in Fig. 3 indicate some curvatures even in the time domain shorter than $t_{cr} \cong 3600$ sec where the linearized theory of CHC is expected to be applicable as will be discussed later in conjunction with Fig. 8.

Equation 2 representing the CHC theory contains three unknown functions $S_T(q;T_f)$, $S(q,0;T_f)$, and $R(q;T_f)$. One of the main aims of our work was to determine their actual forms and look at their features. Conveniently we call this operation the CHC analysis. We have already determined the first function $S_T(q;T_f)$ using the extended version of de Gennes' theory (Eq. 4). Once $S_T(q;T_f)$ was calculated, the remaining quantities in the CHC analysis can be determined as illustrated in Fig. 4. Here, $\ln[S(q,t;T_f)-S_T(q;T_f)]$ at various q are plotted against t (in sec) for the quench experiment from 123.9 to -7.5°C. For clarity, the plots except that for $q=1.953 \times 10^{-3}$ Å^{-1} have been shifted vertically. The data points at any given q are seen to initially follow a straight line, indicating that there is a regime where Eq. 2 holds. Deviations from the linearity at a later time are believed to be due to the onset of the later stage SD (intermediate and late stage SD [4, 25]). The slope and intercept (at $t = 0$) of each straight line allow $R(q,T_f)$ and $S(q,0;T_f)$ to be independently evaluated.

2. DATA ANALYSIS WITH THE CHC THEORY

Figure 5 (a) shows the "initial temperature dependence" of $R(q;T_f)$ while Fig. 5 (b) shows the "final temperature dependence". The lines in the Fig. 5 (a) and (b) represent best-fits of the data to the Pincus theory to be discussed in

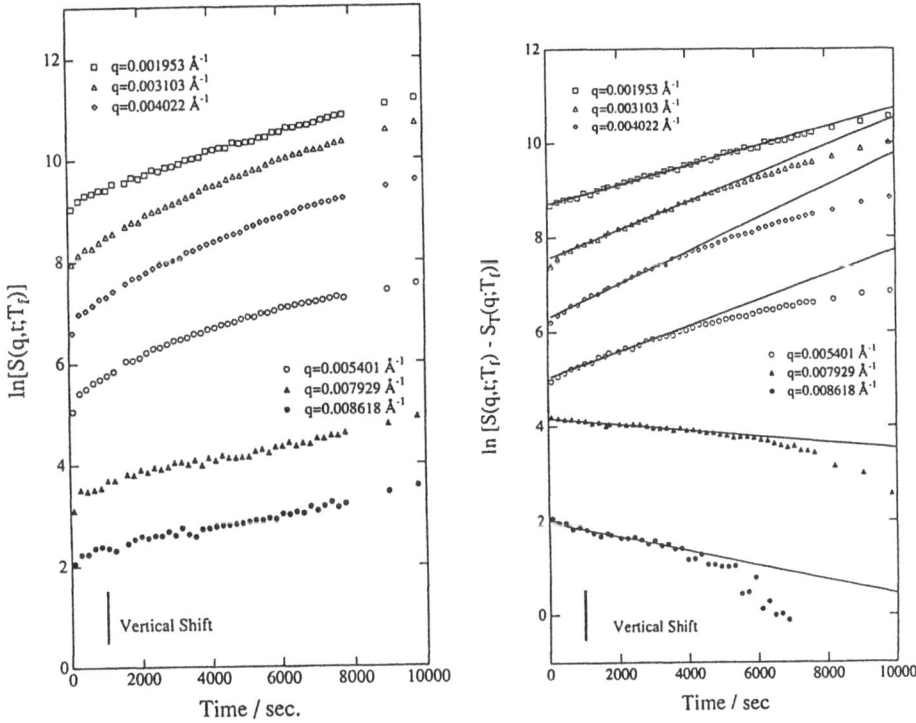

Figure 3. (left) Plot of $S(q,t;T_f)$ vs. t at six different q for the mixture quenched from 123.9 to -7.5°C. **Figure 4. (right)** Plot of $\ln[S(q,t;T_f) - S_T(q;T_f)]$ vs. t at six different q for the critical DPB/HPB mixture quenched from 123.9 to -7.5°C.

$$\varepsilon(T_f) = \left(\chi(T_f) - \chi_s\right)/\chi_s . \tag{8}$$

section V. Two important features can be seen from Fig. 5. First, $R(q;T_f)$ changes sign from positive to negative at the same wave number, $q_c(T_f)$, regardless of T_i. Secondly, $R(q;T_f)$ does not essentially depend on T_i but only on T_f within experimental error. In Fig. 5 (b), we see that $R(q;T_f)$ changes sign at $q = q_c(T_f)$ depending on T_f. Actually, $q_c(T_f)$ decreases as T_f increases (i.e., as the quench depth $\varepsilon(T_f)$ decreases). This behavior of $q_c(T_f)$ is in agreement with the prediction from the CHC theory. $\varepsilon(T_f)$ is defined in terms of χ by. Here χ_s is the χ value at T_s. The values of $q_m(0;T_f)$ and $q_c(T_f)$ determined from Fig. 5 are listed in Table II, together with the corresponding $\varepsilon(T_f)$ which have been calculated by using Eqs. 7 and 8.

The *"effective initial structure factor"* $S(q,0;T_f)$ determined from Fig. 4 for the quenching from T_i (=123.9°C) to T_f (=-7.5°C) is shown by unfilled circles in Fig. 6. The solid line (curve 1) representing $S_T(q;T=123.9°C)$ appears

Table II. Parameters characterizing the early stage of SD

$T_i/°C$	$T_f/°C$	$q_c(T_f) \times 10^3 / Å^{-1a)}$	$q_m(0;T_f) \times 10^3 / Å^{-1b)}$	$\varepsilon(T_f)$
102.3	-7.5	7.44	3.90	1.09
123.9	-7.5	7.44	3.79	1.09
171.6	-7.5	7.44	3.71	1.09
123.9	1.1	6.96	3.65	0.973
123.9	10.5	6.59	3.43	0.851

a)   The wave number at which $R(q;T_f)$ changes from a positive value to a negative value with increasing q.
b)   The wave number at which $R(q;T_f)$ at a given T_f becomes a maximum value.

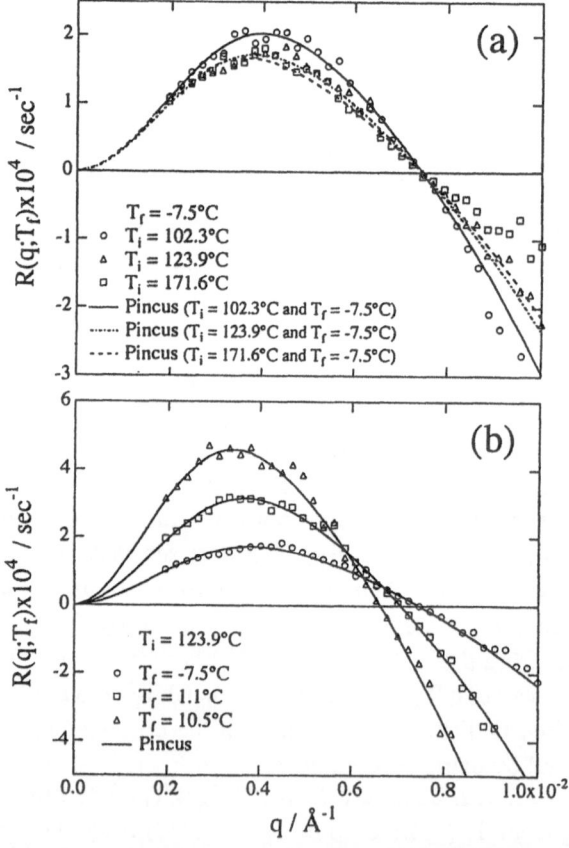

Figure 5. Q-dependence of $R(q;T_f)$ for the indicated condition of T_i and T_f.

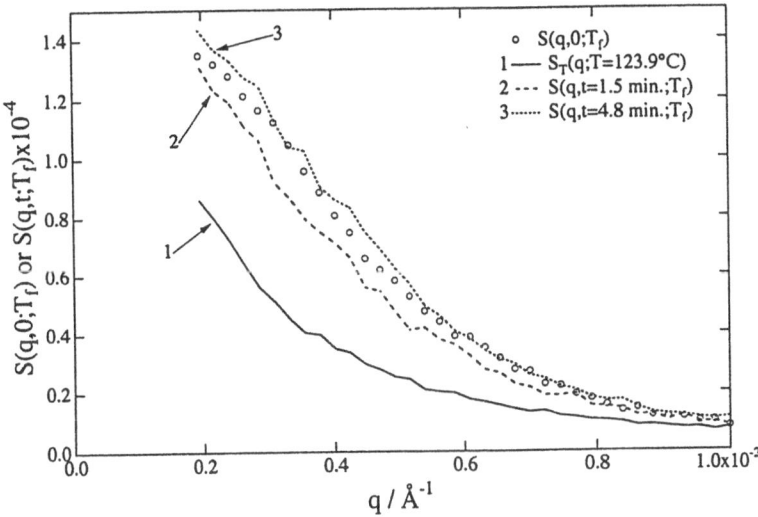

Figure 6. Plot of the "effective" initial structure factor, $S(q,0;T_f)$ vs. q for the critical mixture encountered by quench from 123.9 to -7.5°C together with $S_T(q;T=123.9°C)$ (curve 1) and the time-dependent SANS structure factor at t=1.5 (curve 2) and 4.8 min. (curve 3).

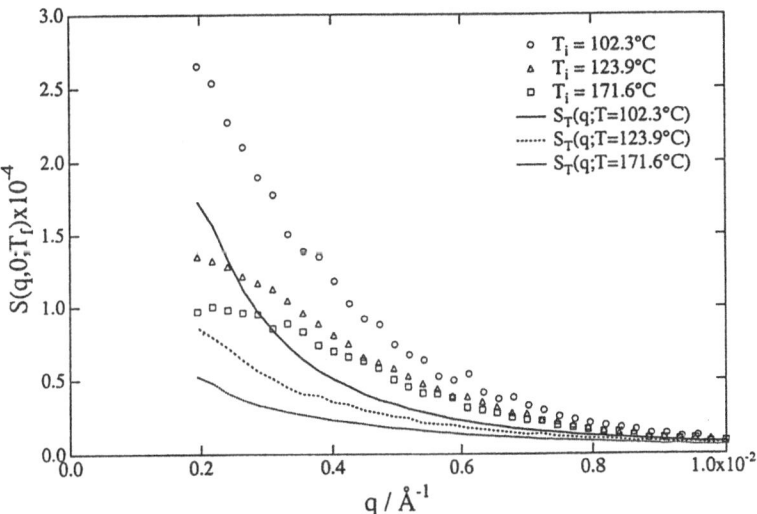

Figure 7. "Effective" initial structure factor $S(q,0;T_f)$ for three different $T_i = 102.3$, 123.9, and 171.6°C are plotted against q together with the corresponding $S_T(q;T=T_i)$.

below $S(q,0;T_f)$. The same trend has been observed for all other quench experiments. If, in the quenching from 123.9 to -7.5°C, the temperature of the specimen had reached T_f instantaneously, $S(q,0;T_f)$ would have been very close to $S_T(q;T=123.9°C)$. The mechanism which caused $S(q,t;T_f)$ to deviate significantly from the measured structure factor, $S_T(q;T=123.9°C)$, is to be discussed below. In Fig. 6, the structure factors at t = 1.5 and 4.8 min. are also plotted against q by a dashed (curve 2) and a dotted line (curve 3), respectively. Interestingly, $S(q,0;T_f)$ appears in between these two lines, indicating that the specimen settled down to T_f at a time somewhere between 1.5 and 4.8 min. after the quench. Actually, this time was about 4 to 5 min.

Data for $S(q,0;T_f)$ from the experiments with three different T_i are displayed together with the corresponding $S_T(q;T_i)$ in Fig. 7. Again we see that $S(q,0;T_f)$ always appears above $S_T(q;T=T_i)$. Importantly, there is a parallelism between $S(q,0;T_f)$ and $S_T(q;T=T_i)$: both are largest for $T_i=102.3°C$, intermediate for $T_i=123.9°C$, and smallest for $T_i=171.6°C$. Thus the initial fluctuations at T_i crucially affect $S(q,0;T_f)$ in such a way that the larger the initial thermal fluctuations, the higher the $S(q,0;T_f)$ becomes. Thus, mathematically, $S(q,0;T_f)$ should be considered as a function of T_i as well as q and T_f. It is in order to note that $S(q,0;T_f)$ from the experiments with three different T_f are roughly the same as that in Fig. 6 (unfilled circle) because T_i for these experiments were same.

3. ORIGIN OF THE PEAK SHIFT IN THE EARLY STAGE SD

We now proceed to consider why the peak of $S(q,t;T_f)$ shifts to higher q in the very beginning of SD. Though it may look unreasonable at a glance, the behavior is not unexpected if we consider that $S(q,t;T_f)$ depends on three q-dependent functions $S_T(q;T_f)$, $S(q,0;T_f)$, and $R(q;T_f)$ through Eq. 2. To confirm this, we calculated $S(q,t;T_f)$ for a series of t by substituting the numerical data as obtained above for these three functions into Eq. 2. The solid lines in Fig. 8 (a), (b), and (c) illustrate the results thus obtained for the quench from $T_i = 102.3°C$, 123.9, and 171.6°C to $T_f = -7.5°C$. We note that the zig-zag nature of the calculated profiles in Fig. 8 reflects those of estimated $S(q,0;T_f)$ and $R(q;T_f)$. It is seen that they reproduce the corresponding observed behavior up to, at least, 46.6, 62.8, and 65.4 min. for (a), (b), and (c), respectively. These times may be identified as the times t_{cr} at which SD crosses over from the early stage to the intermediate stage. Remarkably the peaks of the calculated profiles in Fig. 8 shift toward higher q as t increases in the early stage of SD, just as the experimental data show. Although not shown here, same result was obtained for other T_f [26]. Thus, we can conclude that the CHC theory expressed by Eq. 2 is well consistent with our experimental findings on the early stage SD and that the time dependence of $q_m(t;T_f)$ is irrelevant to validity of the linearized theory of SD.

The above-mentioned t_{cr} values are indicated by arrows in Fig. 2 where $t_{cr,i}$ (i = 1 to 3) denote the t_{cr} for T_i = 102.3, 123.9, and 171.9°C, respectively. Looking at the arrows one notices that the time domain where $q_m(t;T_f)$ is independent of t belongs to the intermediate stage SD rather than the early stage SD. The increase in $q_m(t;T_f)$ in the early stage SD for a given T_i is due to the thermal noise effect put in the sample as a memory. The appearance of the

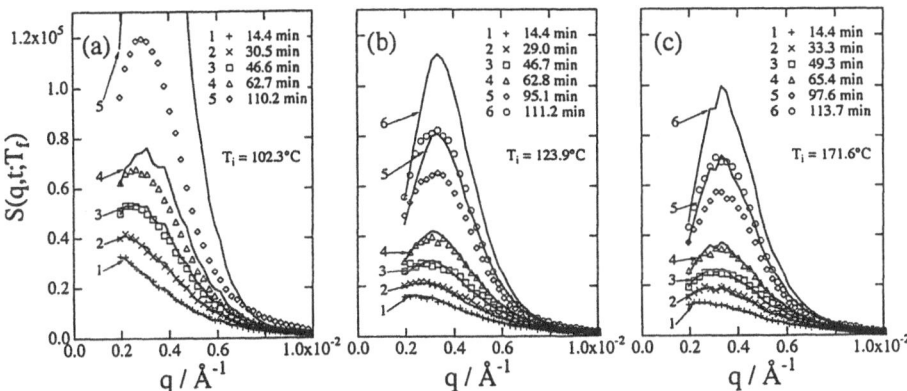

Figure 8. $S(q,t;T_f)$ vs. q for T_i = 102.3, 123.9, and 171.6°C (shown by various symbols) and the predicted $S(q,t;T_f)$ according to the CHC theory (solid lines).

time-independent $q_m(t;T_f)$ is a result of the cancellation of the above effect by the nonlinear effect [25] which tends to decrease $q_m(t;T_f)$.

We have demonstrated that the CHC theory can reproduce the observed peak-shift toward larger q in the very early time of SD. This initial peak-shift, however, can be understood from the following qualitative considerations. The quench from T_i to T_f takes a finite time, so that concentration fluctuations in the specimen begin to grow before reaching T_f after crossing the phase boundary. Owing to this effect, the excess scattering adds up to $S_T(q;T=T_i)$ and makes $S(q,0;T_f)$ different from and greater than $S_T(q;T=T_i)$. Thus $S(q,0;T_f)$ should have a memory which consists of the concentration fluctuation at $T=T_i$ and the path of the temperature quench. If the change in temperature of the specimen during the quench were precisely measured, it would be possible to quantify the memory effect on the basis of the CHC theory by solving a coupled time evolution equations for the concentration fluctuations and for temperature (or the quench depth).

The alternative way to analyze data is to choose $S(q,t;T_f)$ at the third scan as an initial structure factor, i.e. $S(q,t=t_3;T_f)=S(q,t_a=0;T_f)$, where t_a is a new time defined by $t_a = t - t_3$ with t_3 being the time at the third scan. We confirmed that at the beginning of the third scan the temperature of the specimen had already settled down to $T = T_f$. In this way the initial state can be well defined. We have shown that this alternative method were able to describe the evolution of the structure factor including the initial q-shift very well [20]. Therefore, we conclude that once $S(q,0;T_f)$ in the CHC theory is accurately determined, the theory is capable of describing the time-change in the scattering profile even when the quench from T_i to T_f cannot be carried out rapidly.

V. Q-DEPENDENCE OF $\Lambda(q;T_f)$

One of the main purpose of the present study is to investigate the q-dependence of $\Lambda(q;T_f)$ [26]. Our analysis is based on the linear response theory given by Eq. 3. As shown in the previous section, $S_T(q;T_f)$ can be calculated (Eqs. 4-7) and $R(q;T_f)$ is known from Fig. 5. Therefore $\Lambda(q;T_f)$ may be obtained from Eq. 3. The results so obtained will be compared below with the theories of Pincus and Binder for $\Lambda(q;T_f)$.

The Pincus theory [18] deals with a binary mixture of polymers having the same degree of polymerization ($Z_A = Z_B = Z$) (symmetrical blend) and incorporates reptation dynamics. Since the blend used in this work was nearly symmetrical and monodisperse, we expect the Pincus theory to be applicable to a good approximation. Note that $R_{g,DPB}/R_{g,HPB} \cong 1.1$ where $R_{g,k}$ is the radius of gyration of k (DPB or HPB). Pincus gives $\Lambda(q;T_f)$ as follows:

$$\Lambda(q;T_f) = \phi(1-\phi)\left(Wd^2/Z\right)\left[1-\exp(-q^2R_g^2)\right]/q^2R_g^2, \tag{9}$$

where W and d are an elementary monomer jump rate and the tube diameter in the reptation model, respectively. The front factor in Eq. 9 is related the mutual diffusion coefficient, $D_{app}(T_f)$. Using Eqs. 3 and 9 with the definition of $D_{app}(T_f)$, we get [26]

$$D_{app}(T_f) = 4\phi(1-\phi)\left(W\,d^2/Z^2\right)\varepsilon(T_f). \tag{10}$$

Binder [17] has derived $\Lambda(q;T_f)$ that is somewhat different from Eq. 9. It is also based on the reptation model and reads

$$\Lambda(q;T_f) = 2c\phi(1-\phi)\left(W\,d^2/Z\right)\left[1-\frac{1}{q^2R_g^2}\left\{1-\exp\left(-q^2R_g^2\right)\right\}\right]/q^2R_g^2, \tag{11}$$

where c is a constant of order unity. In this theory, Eq. 11 can be used to derive

$$D_{app}(T_f) = 4\,c\phi\,(1-\phi)\left(W\,d^2/Z^2\right)\varepsilon(T_f), \tag{12}$$

which agrees with Eq. 10 except for a constant factor c. Eqs. 10 and 12 relate the parameter Wd^2 to the experimentally measurable $D_{app}(T_f)$ and $\varepsilon(T_f)$.

Figure 9 shows double-logarithmic plot of $\Lambda(q;T_f)$ against q for the same combinations of T_i and T_f as in Fig. 5. The plots clearly demonstrate that $\Lambda(q;T_f)$ has considerable q-dependence in the q-range accessible by SANS

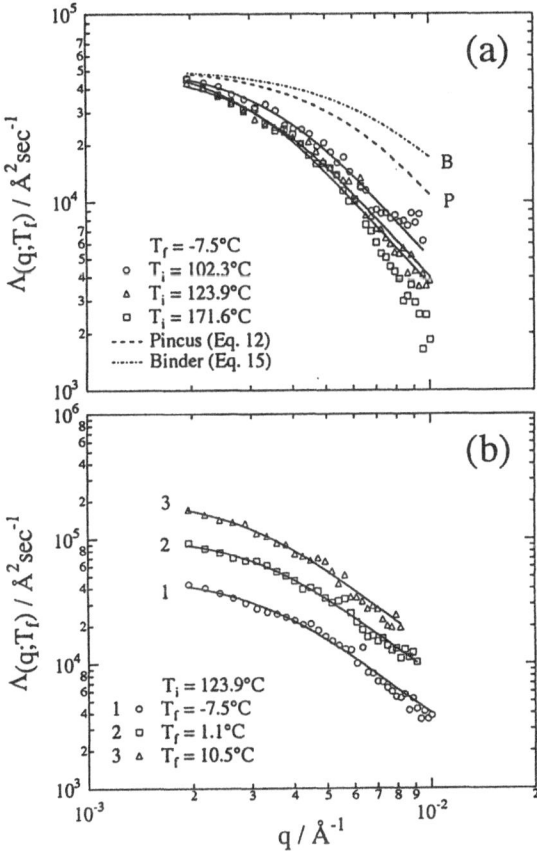

Figure 9. Fitting Pincus' and Binder equations to experimental data for $\Lambda(q;T_f)$. (a), quench from one T_i to three T_f, (b), quench from three T_i to one T_f.

$(q\overline{R}_g \geq 1)$. The $\Lambda(q;T_f)$ for different T_i but an identical T_f are roughly equal as seen in Fig. 9 (a), indicating that $\Lambda(q;T_f)$ is *not* a strong function of T_i. On the other hand, from Fig. 9 (b), $\Lambda(q;T_f)$ depends appreciably on T_f, and its magnitude increases as T_f increases, as naturally expected. Equation 9 for $\Lambda(q;T_f)$ consists of a q-independent prefactor which is related to $D_{app}(T_f)$ multiplied by a function of a reduced wave number qR_g alone. Hence in the double logarithmic plot the magnitude of $D_{app}(T_f)$ is expressed as a vertical shift, while the shape of $\Lambda(q;T_f)$ is determined by R_g. $\Lambda(q;T_f)$ determined from Pincus' theory and Binder's theory, using Eqs. 9 and 11 with $\overline{R}_g = 218\text{Å}$, are shown by the dashed line labeled P and B, respectively, in Fig. 9 (a). Since we are mainly interested in the q-dependence of $\Lambda(q;T_f)$, we arbitrarily shifted vertically the theoretically-predicted lines. It is seen that the experimental data show a stronger q-dependence than the theoretical lines. This disagreement may

Table III. Parameters evaluated by fittings of theoretical $\Lambda(q;T_f)$ to the data.

$T_i/°C$	$T_f/°C$	$R_g/Å^{a)}$	$R_g/Å^{b)}$	$D_{app}(T_f)/Å^2sec^{-1 a)}$	$D_{app}(T_f)/Å^2sec^{-1 b)}$	$t_c/sec^{c)}$
102.3	-7.5	330	447	36.1	38.2	1821
123.9	-7.5	362	499	34.4	37.2	2021
171.6	-7.5	402	562	38.5	42.4	1885
123.9	1.1	368	506	66.8	71.8	1124
123.9	10.5	405	561	116.7	126.9	728

a)   From fitting by Pincus' equation (Eq. 9).
b)   From fitting Binder's equation (Eq. 11).
c)   Characteristic time $t_c(T_f)^{-1} = q_m(0;T_f)^2 D_{app}(T_f)$, computed with $D_{app}(T_f)$ from the analysis by Pincus' equation.

indicate that the experimental $\Lambda(q;T_f)$ is *not* scaled by the bare $\bar{R}_g$ as the theories predict.

With $\phi(1-\phi)Wd^2$ and R_g as floating parameters, a non-linear regression fit routine to Eq. 9 and the data was performed. The solid lines in Fig. 9 show the results. From the fitting parameters and Eq. 3, the solid curves for $R(q;T_f)$ in Fig. 5 have been calculated. The estimated values $\phi(1-\phi)Wd^2$ have been used in Eq. 10 to compute $D_{app}(T_f)$ by use of Eq. 8 for $\varepsilon(T_f)$. Table III lists the values of $D_{app}(T_f)$ along with those of R_g.

Similar fits to Eq. 11 led to similar values of $D_{app}(T_f)$ and R_g which are also listed in Table III. The parameter values obtained from the two theories are somewhat different from one another. Table III shows $R_g = 370 \pm 40$ Å for Pincus' theory and $R_g = 510 \pm 50$ Å for Binder's theory. When these are compared with the bare $\bar{R}_g$ for our polymer blend, we get $R_g = (1.70 \pm 0.2)\bar{R}_g$ for the former and $R_g = (2.35 \pm 0.2)\bar{R}_g$ for the latter. Applying an analysis similar to that described above, Schwahn et al. founded that $R_g = (5-7)\bar{R}_g$, and $R_g = (2-3)\bar{R}_g$ for their DPS/PVME [19], and DPS/HPS blends [10], respectively. The apparent values of R_g, obtained by the best-fitting procedures, are much larger that the bare radius $\bar{R}_g$ in all cases. The explanation of this disagreement deserves future work.

VI. SCALING ANALYSIS

With the values of $D_{app}(T_f)$ and $q_m(0;T_f)$ given in Table III, it is possible to estimate the characteristic time $t_c(T_f)$ of our mixture by

$$t_c(T) \equiv [q_m(0;T)^2 D_{app}(T)]^{-1}, \qquad (13)$$

as well as the reduced wave number Q_m and reduced peak intensity $\tilde{S}_m$ [4, 25]. The estimated values $t_c(T_f)$ from Pincus' theory are also listed in Table III. The knowledge of $t_c(T_f)$ makes it possible to calculate a reduced time τ and hence to test the scaling postulate [25].

We found that the time-evolution of the reduced wave number and the reduced intensity at the peak of the scattering structure factor in the early and intermediate stages of SD were found to be scalable in terms of τ when T_i was fixed and T_f was varied, but not when T_f was fixed and T_i was varied [26]. The failure of the scaling law in the latter instance may be attributed to the fact that the concentration fluctuations at the onset of SD are different. Thus, the memory of the concentration fluctuation affects the subsequent SD over an extended period of time. After sufficiently long times when the memory effect decays out, the scaling law is found to be recovered [26].

VII. SUMMARY

The self-assembling process was investigated for a binary mixture of deuterated and protonated polybutadiene at the critical composition. The system examined had an UCST type phase diagram with a critical temperature of 99.2°C. The evolution of structure factor, $S(q,t;T_f)$, after quenching from an initial temperature T_i to a final temperature T_f was followed using the time-resolved small-angle neutron scattering (SANS) method.

It was found that $S(q,t;T_f)$ clearly showed scattering maxima expected in the early stage SD (Fig. 1). It was found in all our experiments that the peak wave number $q_m(t;T_f)$ increased at the beginning of the phase separation, leveled off, and then decreased with t. For the experiments with a fixed T_f but various T_i, $q_m(t;T_f)$ and the peak intensity $S_m(t;T_f)$ of $S(q,t;T_f)$ followed different time-changes in the beginning of SD depending upon T_i (Fig. 2). However, both $q_m(t;T_f)$ and $S_m(t;T_f)$ for different T_i tended to merge about 300 min. after the onset of SD. This experimental observation showed that thermal concentration fluctuations present in the system before quenching affect the early stage SD and the early phase of the intermediate SD and hence the coarsening process.

The time-evolution of $S(q,t;T_f)$ was successfully described by the Cahn-Hilliard-Cook (CHC) theory. We found that even the sift of $q_m(t;T_f)$ toward a large wave number with t is accounted for by the theory. The CHC analysis led to the following findings. (i) The growth rate $R(q;T_f)$ of the concentration fluctuations is essentially independent of T_i and there is a wave number q_c such that $R(q;T_f)$ is positive at $q < q_c$, zero at $q = q_c$, and negative at $q > q_c$ (Fig. 5). (ii) The effective initial structure factor at $T = T_f$, $S(q,0;T_f)$, was always larger than $S_T(q;T=T_i)$. The latter was interpreted in terms of the athermal SD occurring during the quench, because of a finite time needed for the quench.

We found a correlation between $S(q,0;T_f)$ and T_i: the higher the T_i, the smaller the $S(q,0;T_f)$ is. This experimental finding evidences that thermal concentration fluctuations in the single-phase state before the quench contribute to $S(q,0;T_f)$. (iii) The Onsager kinetic coefficient, $\Lambda(q;T_f)$, were estimated for various combination of T_i and T_f. The results were compared with the theories proposed by Pincus and by Binder. It was found that $\Lambda(q;T_f)$ at the high q-range covered by SANS had a considerable q-dependence. Although reasonably good fits were found, the value of R_g estimated by fitting the theoretical $\Lambda(q;T_f)$ to the experimental data were about two times larger than $\overline{R}_g$:

$$R_g = (1.70 \pm 0.2)\overline{R}_g \text{ from Pincus' theory and } R_g = (2.34 \pm 0.2)\overline{R}_g \text{ from Binder's}$$

theory. This discrepancy indicates that the experimental $\Lambda(q;T_f)$ *may not be scaled with the bare* $\overline{R}_g$ as the theories predict.

V. REFERENCES AND NOTES

1 Hashimoto T (1988) Phase Transitions 12: 47
2 Nose T (1989) In: Tanaka F, Doi M, Ohta T (ed) Space-Time Organization in Macromolecular Fluids. Springer, Berlin
3 Hahsimoto T, Kumaki J, Kawai H (1983) Macromolecules 16: 641; Snyder HL, Meakin P, Reich S (1983) Macromolecules 16: 757
4 Hashimoto T, Itakura M, Hasegawa H (1986) J. Chem. Phys. 85: 6118; Hashimoto T, Itakura M, Shimidzu N (1986) J. Chem. Phys. 85: 6773
5 Okada M, Han CC (1986) J. Chem. Phys. 85: 5317
6 Izumitani T, Hashimoto T (1985) J. Chem. Phys. 83: 3694
7 Kyu K, Lim DS (1991) J. Chem. Phys. 92: 3944
8 Takenaka M, Hashimoto T (1992) J. Chem. Phys. 96: 6177
9 Higgins JS, Fruitwala HA, Tomlins PE, (1989) Br. Polym. J. 21: 247
10 Schwahn D, Hahn K, Streib J, Springer T, (1990) J. Chem. Phys. 93: 8383
11 Meier H, Strobl GR (1987) Macromolecules 20: 649
12 Hashimoto T, Takenaka M, Jinnai H (1991) J. Apply. Cryst. 24: 457
13 Bates FS, Wiltzius P (1989) J. Chem. Phys. 91: 3258
14 Cahn JW (1961) Acta Metall. 9: 795; Cahn JW (1965) J. Chem. Phys. 42: 93
15 The time t should be the one at a sufficiently long time in the early stage so that the scattering arising from the concentration fluctuations developed by SD is significantly larger than that from the initial concentration fluctuatations existing before SD.
16 Cook HE (1972) Acta Metall. 15: 287
17 Binder K (1983) J. Chem. Phys. 79: 6387
18 Pincus P (1981) J. Chem. Phys. 75: 1996
19 Schwahn D, Janssen S, Springer T (1992) J. Chem. Phys. 97: 8775
20 Jinnai H, Hasegawa H, Hashimoto T, Han CC J. Chem. Phys., in press
21 Flory PJ, J. Chem. Phys. (1941) 9: 660; Haggins ML J. Chem. Phys. (1941) 9: 440
22 Strobl GR (1985) Macromolecules 18: 558
23 Motowaka M, Jinnai H, Hashimoto T, Qiu Y, Han CC J. Chem. Phys., in press
24 Warner M, Higgins JS, Carter AJ (1983) Macromolecules 16: 1931
25 Langer JS, Bar-on M, Miller HD (1974) Phys. Rev. A 11: 1417
26 Jinnai H, Hasegawa H, Hashimoto T, Han CC J. Chem. Phys., in press

Dynamic Interplay Between Wetting and Phase Separation in Geometrically Confined Polymer Mixtures

Hajime Tanaka

Institute of Industrial Science, University of Tokyo, Minato-ku, Tokyo 106, Japan.

Abstract

Here we study the interplay between wetting and phase separation for binary polymer mixtures confined in one-dimensional (1D) or two-dimensional (2D) capillaries. It is found that near the symmetric composition, the hydrodynamics unique to bicontinuous phase separation plays a crucial role on the wetting dynamics and the surface effect is strongly delocalized by the interconnectivity of the phases. The wetting dynamics is discussed on the basis of the hydrodynamic coarsening, focussing on the effect of the dimensionality of the geometrical confinement. We also discuss the possibility that the quick reduction of the interface area spontaneously causes double phase separation.

1 Introduction

Phase separation phenomena have been extensively studied in the past two decades from both the experimental and the theoretical viewpoints [1]. Since the finding of critical wetting phenomena by Cahn [2], wetting phenomena have been widely studied mainly in the stable region [3,4]. Both phenomena have long histories of the research, but they have been studied almost independently. Phase separation causes two semi-macroscopic phases having different wettability, and the wettability causes the interaction between the phases and the solid surface. Thus the phases are spatially rearranged to lower the total energy including the solid-liquid interactions. The late stage of pattern evolution is governed by the competition between the liquid-liquid and solid-liquid interfacial energies [5-13]. Polymer mixtures are suitable for studying this interplay since the interfacial energy between two different polymer phases is extremely small compared to low molecular weight systems [10] and thus the wetting effect is enhanced.

Here we discuss the wetting dynamics of a mixture undergoing phase separation in a one-dimensional (1D) or a two-dimensional (2D) capillary. We also discuss the general feature of the interplay between wetting and phase separation, focussing on the effects of the kind of the geometrical constraint on the pattern evolution and the effects of the composition symmetry. Unusual wetting-induced double phase separation will also be discussed.

2 Wetting in 1D Capillary

Figure 1 shows the pattern evolution of the poly(vinyl methyl ether) (PVME)/water (7/93) (weight ratio) mixture confined in a 1D capillary ($r_0 = 79 \mu m$) at 33.3°C. In this mixture water-rich phase is more wettable to glass. The phase separation temperature was 33.1°C. Since this composition was symmetric, bicontinuous phase separation was observed in the initial stage. Then the macroscopic wetting layer was rapidly formed. After the formation of the wetting layer, the Rayleigh instability led to the formation of bridges [12]. This instability is a result of the competition between the destabilizing effect of the transverse curvature at long wavelengths and the stabilizing effect of the longitudinal curvature at short wavelengths. This instability is characteristic of a 1D capillary. There is no analogous instability in a 2D capillary since the area of planar interface can only be increased by small perturbations.

The coarsening dynamics in off-symmetric mixtures is essentially different from that in symmetric mixtures. When the minority phase is less wettable to glass, nothing unusual happens

A. Teramoto, M. Kobayashi, T. Norisuje (Eds.)
Ordering in Macromolecular Systems
© Springer-Verlag Berlin Heidelberg 1994

and droplets grow mainly by Brownian coagulation mechanism. They eventually locates around the cyrindrical axis probably due to the van der Waals interaction between droplets and the wall.

The droplet size finally exceeds the pore size (the crossover from three dimensions (3D) to one dimension (1D)), and the droplets transform into capsules. However, before the dimensional crossover, the coarsening rate becomes extremely small since (1) there is a barrier for the translational motion of a large droplet due to both the 1D nature and the conservation law and further (2) the little difference in the curvature between droplets makes the evaporation-condensation mechanism [1] ineffective.

On the other hand, when the minority phase is more wettable to glass, the droplets gradually wet to glass by diffusion and form a wetting layer. The wetting layer is stable for strongly asymmetric case, but it becomes unstable for slightly asymmetric case and transforms into a bridge structure. Hydrodynamic stability analysis indicates the layer or unduloid configuration is stable for a thinner film.

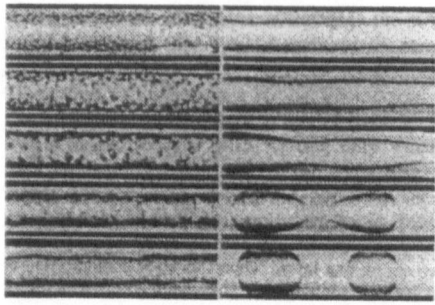

Figure 1: Phase separation of PVME/water (7/93) in a 1D capillary. Photographs correspond to 0.5s, 1.5s, 2.5s, 3.5s, 4.5s, 8.5s, (left column) 58.5s, 148.5s, 178.5s, 238.5s, 246.5s, and 268.5s (right column) from top to bottom, respectively, after the temperature jump from 32.5°C to 33.3°C. The bar is 200μm in length.

In this case, the formation of the wetting layer is extremely slower than that for symmetric mixtures since the former is governed by diffusion but the latter by hydrodynamics (see Table 1 in Ref.[12]).

3 Wetting in 2D Capillary

Here we describes the pattern evolution in a 2D capillary for symmetric mixtures. Figure 2 shows the pattern evolution of the PVME/water (7/93) mixture confined in a 2D capillary ($d \sim 19\mu m$) at 33.7°C. In the very initial stage, the bicontinuous pattern was observed (see (a) and (b)). Then, the wetting layer formation and the subsequent formation of disklike droplets from the tubes bridging the upper and lower plates were observed (see (c) and (d)). The latter is unique to a 2D system. In the late stage, the disklike droplets were attracted with each other by the wetting-induced interaction due to the capillary instability (see (e) and (f)). After this stage the droplet pattern finally transformed into the bicontinuous pattern because of an increase in the apparent in-plane symmetry reflecting the thinning of the wetting layer [13]. The completely similar behavior was commonly observed for symmetric binary mixtures confined in 2D capillaries [11,13].

Such behavior is never observed in asymmetric compositions, where droplet phase separation occurs. In the latter, droplets gradu-

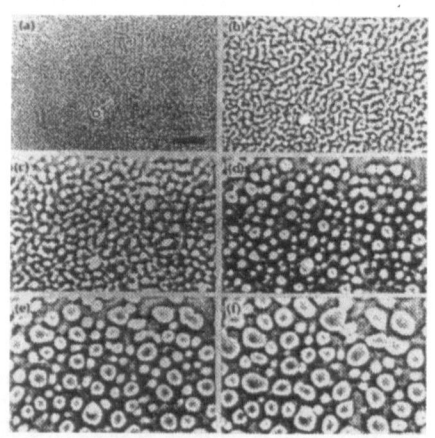

Figure 2: Phase separation of PVME/water (7/93) in a 2D capillary ($d = 19\mu m$). (a) 0.7s, (b) 1.7s, (c) 2.2s, (d) 2.7s, (e) 3.7s, and (f) 4.7s after the temperature jump from 32.8°C to 33.7°C. The thickness was about 19μm. The bar is 100μm in length.

ally wet to the walls and the wetting speed is much slower than in bicontinuous phase separation as in the case of 1D capillaries. For strongly asymmetric mixtures where the minority phase is more wettable to glass, droplets gradually disappear and finally form complete wetting layers which is supported by the fact that in optical microscopic observation all the droplets completely disappear. In 2D systems these wetting layers are stable against any fluctuation, and thus glass surfaces are completely covered by the wettable phase.

4   Composition Symmetry and Final Equilibrium Structure

Here we discuss the composition effect on the final equilibrium configuration under the influence of the wetting effect.

4.1   Confinement in 2D Capillary

In the late stage of phase separation, we need consider only the free energy originating from the interface. There are two possible final configurations for a 2D capillary: (1)layer structure (complete wetting configuration) and (2)disklike droplet structure (partial wetting configuration) (see Fig.3). The free energies per unit area for the configurations of complete and partial wetting are given by

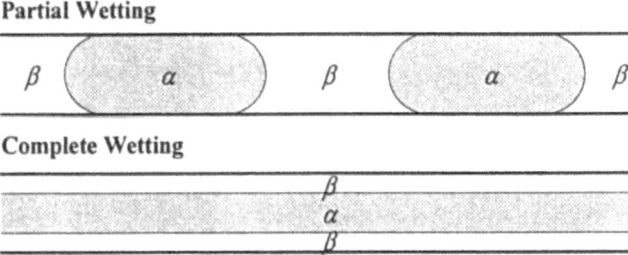

Figure 3: Final equilibrium structures for partial and complete wetting

$F_{cw} = 2(\gamma_\alpha + \sigma)$ and $F_{pw} = 2\phi_\alpha\gamma_\alpha + 2\phi_\beta\gamma_\beta + \sigma f(d)$, respectively. σ is the interfacial tension between the α and β phases and γ_i is the interfacial tension between the i phase and the wall. ϕ_α and ϕ_β are the volume fraction of the α and β phase, respectively. Here the α phase is more wettable to the wall. $f(d)$ is the total interface area between the α and β phases per unit area. Neglecting the last term in F_{pw}, $F_{cw} - F_{pw} = 2\sigma(1 - \phi_\beta\Delta\gamma/\sigma)$, where $\Delta\gamma = \gamma_\beta - \gamma_\alpha$. The morphological transition occurs at $\phi_\beta = \sigma/\Delta\gamma$.

4.2   Confinement in 1D capillary

As discussed by Liu et al.[6], the free energies per unit length for the configurations of complete and partial wetting (tube and plug) (see Fig.3) for a 1D capillary are given by $F_{cw}/2\pi = r_c\sigma + r_0\gamma_\alpha$ and $F_{pw}/2\pi = r_0[\phi_\alpha\gamma_\alpha + \phi_\beta\gamma_\beta] + g(r_0)\sigma$, respectively. Here r_c is the radius of the fluid tube for the complete wetting configuration and $r_c^2 = \phi_\beta r_0^2$. $g(r_0)$ is the total interface area between the α and β phases per unit length for the partial wetting configuration. Provided that the last term in F_{pw} is negligible, $(F_{cw} - F_{pw})/2\pi = r_0\sigma[\phi_\beta^{1/2} - (\Delta\gamma/\sigma)\phi_\beta]$. The transition between the complete and partial wetting configurations occurs at $\phi_\beta = (\sigma/\Delta\gamma)^2$.

Since $\sigma = \sigma_0\epsilon^\mu$ ($\mu = 1.26$) and $\Delta\gamma = \Delta\gamma_0\epsilon^\delta$ ($\delta = 0.34$) where $\epsilon = (T - T_c)/T_c$, the transition compositions are given by $\phi_t = (\sigma_0/\Delta\gamma_0)\epsilon^{(\mu-\delta)}$ for a 2D capillary and by $\phi_t = (\sigma_0/\Delta\gamma_0)^2\epsilon^{2(\mu-\delta)}$ for a 1D capillary. Thus for both 1D and 2D capillaries, the partial wetting configuration is energetically favored near the symmetric composition except for a very shallow quench. This is consistent with our observation [12].

5 Wetting Dynamics Characteristic of Bicontinuous Pattern

Here we consider how the phase-separation pattern transforms from the initial, bulk bicontinuous pattern to the above final, equilibrium configuration affected by walls. Before discussing the wetting effect, we should understand a characteristic feature of bicontinuous phase separation near the symmetric composition. In this particular type of phase separation, the hydrodynamics due to the capillary instability [7,13,14] plays a drastic role in the coarsening under the influence of wetting. Near the symmetric composition, the wettability is probably important mainly in the very initial stage to establish the initial condition that the surface is covered by the wetting layer, to which the bicontinuous tubes are connected. This initial configuration establishes the anisotropic pressure gradient from the bicontinuous tubes in bulk to the wetting layer and the more wettable phase can be continuously supplied into the wetting layer through the percolated tubes until the tube network disappears. Because of the percolated structure, the wetting effect is not localized near the wall and strongly affects the whole sample (see Fig.1). Thus, phase separation could be seriously affected by wetting phenomena especially near the symmetric composition even though a sample size is macroscopic.

5.1 Dynamics of Wetting Domain Formation in 2D Capillary

Here we discuss the initial stage of the wetting droplet formation [13]. This problem was experimentally studied in a 2D capillary by Wiltzius and Cumming [8]. They found that the size of the wetting droplet l grows as $t^{3/2}$ although the growth mechanism was not clear. The anomalous, fast growth mode found by them could be explained as follows. There is a pressure gradient between a bicontinuous tube and its wetting part, reflecting the difference in the transverse curvature of the tube between them. Thus there should be a hydrodynamic flow from the tube to the wetting droplet. Since the pressure gradient between the tube and the wetting layer is $\sim \sigma/a$ over the distance a, the flux of this flow is estimated as $Q \sim (\sigma/\eta)a^2$, where η is the viscosity and a is the characteristic size of the tube.

Provided that the limiting process of the droplet spreading is the supply of the more wettable phase from the tubes by the hydrodynamic flow, we get the relation $ldl/dt \sim Q$. The assumption is probably valid for the case of *strong wettability* and it is supported by the 2D nature of the droplet growth experimentally confirmed [8]. Using the Siggia's growth law, $a \propto (\sigma/\eta)t$, we obtain the relation $l \sim [(\sigma/\eta)t]^{3/2}$. This is consistent with their observation. The prefactor $(\sigma/\eta)^{3/2}$

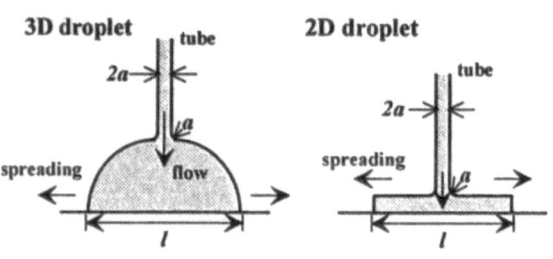

Figure 4: Schematic figure of spreading wetting droplet for D=3 and D=2

is roughly proportional to $(\Delta T)^{3\nu}$, where ΔT is the quench depth, ν is the critical exponent for the correlation length ξ, and $\nu \sim 0.63$. This dependence of the prefactor on ΔT seems to be consistent with the experimental results [8].

For a weak wetting case, the droplet spreading might not be purely two-dimensional. For three-dimensional droplet growth, for instance, we obtain $l^2 dl/dt \sim Q$. Thus we can generalize the relation $l \sim [(\sigma/\eta)t]^{3/D}$, where D is the spational dimensionality of wetting droplet. For half-spherical droplet growth ($D = 3$) $l \sim t$, while for disklike droplet growth ($D = 2$) $l \sim t^{3/2}$ (see Fig.4). The transitional behavior of the exponent from 1 to 3/2 observed by Cumming et al. could be explained by the above idea.

5.2 Thickning Dynamics of Wetting Layer in 1D Capillary

Next we discuss the thickening dynamics of the wetting layer in a 1D capillary [12,13]. The thickening rate of the wetting layer $(d/dt)d_w$ should be proportional to the flux from the bicontinuous tubes after the formation of the wetting layer. The flux from a single tube Q can be estimated as $Q \sim (\sigma/\eta)a^2$ as before. The number of tubes per the area S is proportional to S/a^2. Thus the total flux Q_t is proportional to $S(\sigma/\eta)$. Since the thickening of the wetting layer is caused by this flux, $S(d/dt)d_w \sim S(\sigma/\eta)$. Therefore, $d_w = k_w(\sigma/\eta)t$, where k_w is the proportional constant. This linear-growth behavior has been experimentally confirmed for 1D capillaries.

5.3 Effect of Spatial Dimensionality of Confinement on Wetting

Here we discuss the difference in the late stage of pattern evolution between 1D and 2D capillaries. For a 1D capillary, the tube radius of the bicontinuous pattern a can never exceed the capillary tube radius r_0 because of the geo-

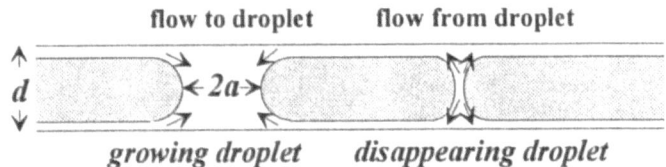

flow to droplet   **flow from droplet**

d $\leftarrow 2a \rightarrow$

growing droplet   *disappearing droplet*

Figure 5: Growing and disappearing droplets in 2D capillary

metrical constraint. Thus the complete layer configuration is first formed and then the Rayleigh instability causes a stable bridge structure [12]. In a 2D capillary, on the other hand, the tube diameter $2a$ can exceed the spacing d since there is no geometrical restriction parallel to the plates. Once $2a$ becomes larger than d, the longitudinal curvature of the tube ($\sim 2/d$) becomes larger than its transverse curvature ($\sim 1/a$). Thus the pressure in the tube becomes lower than the wetting layer. Therefore, the tube bridging the two walls, having a radius of $a(> d/2)$ starts to thicken with time by absorbing the wetting layers. On the other hand, the tube bridging the two walls whose radius a is smaller than $d/2$ disappears (see Fig.5). Actually, we have found that the 2D disklike droplet appearing in a 2D capillary always has the initial diameter $2a_0$ comparable to the thickness d [13]. The similar phenomena was first observed by Bodensohn and Goldburg [11]. The relation is likely universal for any binary liquid mixtures and also for any quench conditions.

With increasing d, the number of bridges (droplets) finally formed decreases. This can be explained by that the condition $2a > d$ becomes more difficult to be satisfied for a larger d since in a bicontinuous structure the number of percolated paths decreases with an increase in d and further bicontinuous tubes are spontaneously broken by any asymmetry during the coarsening. Thus the morphological evolution during the interplay is strongly dependent on the film thickness d.

6 Domain Growth in 2D Capillary

6.1 Initial Stage

Once the diameter $2a$ exceeds the thickness d, it grows linearly with time before the interaction between droplets starts to play a role. This linear growth behavior can be explained by that the hydrodynamic flow from the wetting layer into the 2D disklike droplet is largely dominated by the pressure gradient coming from the longitudinal curvature $2/d$. In this approximation, $da/dt \propto (\sigma/\eta)g(d, d_w)$ where $g(d, d_w)$ is a function of d and the wetting layer thickness d_w, and a grows linearly with time. The slowing down of the droplet growth is likely caused by the interdroplet interaction through the wetting layer. Since the wetting layer between neighboring droplets is finite, droplets cannot absorb the wetting layer independently in the late stage. In other words, $g(d, d_w)$ becomes a function of time through the change in d_w in this intermediate stage.

6.2 Interdroplet Interaction via Wetting Layers

In the intermediated stage the droplets are attracted with each other by absorbing the wetting layers and also by the capillary instability of the tube formed between neighboring droplets [15]. This behavior can be explained as follows. The wetting layers likely connect neighboring droplets which are composed of the more wettable phase. Thus the bow-shaped sides of the neighboring droplets and the upper and lower wetting layers between these droplets form a tube of fluid. The droplets are probably attracted to each other to reduce the interface energy of this tube. This phenomenon is very similar to the Rayleigh instability in a 1D capillary [12]. The behavior can be analysed on the basis of a Navier-Stokes equation,

$$\eta[\partial^2 v/\partial y^2 + \partial^2 v/\partial z^2] = dP/dx, \qquad (1)$$

where v is the flow velocity along x, η is the viscosity, and P is the pressure. The geometry is explained by Fig.6. Here x is along the tube axis, z is the thickness direction, and y is perpendicular to x and z. Here we define Δl_{min} as the minimum interparticle distance as shown in Fig.6. Thus the flow velocity v is related to the interface velocity $d\Delta l_{min}/dt$ as $v \sim (L/\Delta l_{min})d\Delta l_{min}/dt$ from the mass conservation. Reflecting the longitudinal change in the transverse curvature of the tube, the

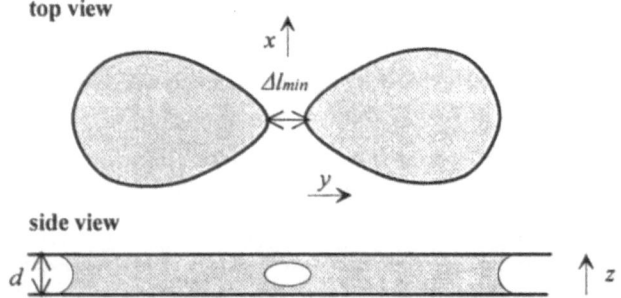

Figure 6: Droplets interacted via wetting layers

capillary pressure in the tube roughly varies by $\sigma/\Delta l_{min}$ (σ:interface tension between the two coexisting phases) over the characteristic length along the tube ($\sim L$). Here L can be estimated by the in-plane radius of the local curvature of the colliding droplet front. Thus, $dP/dx \sim \sigma/(\Delta l_{min}L)$. This pressure gradient causes the Poiseuille flow from the neck to the bulges of the tube. For $\Delta l_{min} < d$, v is estimated from the relation $\eta v/\Delta l_{min}^2 \sim dP/dx$ (see eq.(1)) as

$$v \sim \sigma \Delta l_{min}/(\eta L).$$

Provided that $L \sim \Delta l_{min}$,

$$\Delta l_{min} \sim (\sigma/\eta)\Delta t.$$

The linear dependence of L on Δl_{min} assumed above seems to be reasonable because (i) Δl_{min} is the only relevant length scale of the problem and further (ii) the curvature of the deforming droplet front tends to be balanced with the transverse curvature of the tube ($\sim 1/\Delta l_{min}$) (see Fig.3). L might also have a weak dependence on R since the curvature of the deforming droplet front should be affected by the original droplet curvature. For $\Delta l_{min} > d$, on the other hand, v is estimated from the relation $\eta v/d^2 \sim dP/dx$ (see eq.(1)) as

$$v \sim \sigma d^2/(\eta L \Delta l_{min}).$$

Provided again that $L \sim \Delta l_{min}$, the following equation is obtained:

$$\Delta l_{min} \sim (\sigma d^2/\eta)^{1/3}\Delta t^{1/3}.$$

Thus Δl_{min} is predicted to be proportional to Δt for $\Delta l_{min} < d$ and $\Delta t^{1/3}$ for $\Delta l_{min} > d$. This prediction is consistent with our experimental results [15].

6.3   Change in Apparent 2D Composition Symmetry : Temporal Decrease in Amount of Wetting Layer

We have found the unusual morphological transformation from an interconnected to droplet and back to interconnected pattern in a nearly symmetric binary mixture confined in a 2D capillary [15]. The overall behavior can be explained from the static aspect as follows: The composition symmetry initially leads to the bicontinuous pattern in bulk. Then the in-plane symmetry of the order parameter is broken by the existence of the wetting layer. This leads to the droplet morphology. Finally, the in-plane symmetry approaches to the bulk, composition symmetry since the wetting layers are absorbed by the domains bridging the glass plates. This is supported by the increase in the total droplet area with t_{sp}. Thus the morphology transforms from a droplet to interconnected pattern again. It should be noted that the temporal change in the total droplet area was first reported by Bodensohn and Goldburg [11].

Next we discuss the second morphological tansition on the basis of the kinetics of the elementary process. During phase separation, the deformation of a droplet shape is generally caused by interdroplet coalescences whose interval is characterized by the collsion interval (τ_c). The resulting shape relaxation process is characterized by the relaxation time τ_σ. The wetting effect drastically shortens τ_c compared to the usual case without wetting since (1) there is the wetting-induced, attractive interaction as already described and (2) the interdroplet distance becomes smaller with the phase-separation time, reflecting the increase in the in-plane symmetry. Further, the attractive interaction prevents the free shape relaxation and thus τ_σ is increased. The rate of the wetting-induced coarsening is mainly determined by the wetting layer and not strongly correlated with the particle size. On the other hand, the shape relaxation time is determined by the surface/volume ratio and proportional to R. Accordingly, the shape-relaxation time increases with the phase-separation time and becomes longer than the collision interval at a certain time. This crossover between the two characteristic times is also responsible for the morphological transformation from the isolated to interconnected structure.

7   Wetting-Induced Double Phase Separation

7.1   Experimental Evidences

First we show the experimental evidences of wetting-induced double phase separation. Figure 7 shows pattern evolution of the PVME/water (7/93) mixture confined in a 1D capillary ($r_0 = 79\mu m$) at 34.0°C. The quench depth ΔT was 0.8°C. Surprisingly the retarded, secondary phase separation was observed inside the macroscopically separated phases. This can be noticed around 1s after the quench from the fact that the two phases look cloudy. Then droplet pattern becomes visible around 20s after the quench. This unusual double phase separation (DPS) was not observed for a shallow quench ($\Delta T \leq 0.6K$) (see Fig.3 in Ref.[11]). DPS looks similar to the pattern caused by a temperature double quench [16]. The same behavior of double phase separation were also observed for a 2D capillary in a few binary mixtures.

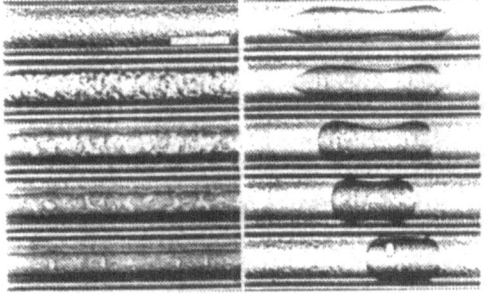

Figure 7:   Phase separation in a 1D capillary for PVME/water (7/93). Photographs correspond to 0.3s, 0.8s, 1.3s, 1.8s, 3.8s, (left column) 22.3s, 23.8s, 28.8s, 88.8s, and 479s (right column) from top to bottom, respectively, after the temperature jump from 32.5°C to 34.0°C. The bar corresponds to $200\mu m$.

We have also found that DPS is strongly suppressed by the strong spatial confinement. For a thick sample DPS was clearly observed, while for a thin sample ($d < 1\mu m$) it was never observed even though the other experimental conditions were the same. This fact indicates the importance of the bicontinuous phase separation in bulk. It should be noted that for any mixtures DPS was observed only near the symmetric composition, and even for a deep quench it was never observed in asymmetric compositions, where droplet phase separation occurs.

8 Physical Mechanism of Double Phase Separation

Here we show a possible scenario for DPS. In all the mixtures, the bicontinuous phase separation was observed in the initial stage (see the first photographs in Fig.7) because of the composition symmetry. In bicontinuous phase separation, the total interface area of the system is drastically reduced within a short time by the hydrodynamic coarsening originating from the coupling between the concentration and the velocity fields. According to the Siggia's mechanism [14], the interface area per unit volume s is estimated to decrease as $s \propto [(\sigma/\eta)t]^{-1}$, where σ is the interface tension and η is the viscosity. Since the hydrodynamic interface motion is much faster than the concentration diffusion, *the hydrodynamic flow due to the capillary instability causes only the geometrical coarsening and does not accompany the concentration change.* Thus the system cannot respond to the rapid decrease in the interface energy; namely the local equilibrium cannot be established. This likely causes a kind of double quench effect, which we call *interface quench*. In all the previous studies [1] the local equilibrium has been assumed in the hydrodynamic regime, but it is probably not true in the exact sense.

Here we mention the difference in the coarsening dynamics of symmetric mixture between phase separation in a confined geometry and bulk phase separation. For both cases the coarsening dynamics is dominated by capillary instability [13]. Only the difference is likely the prefactor k in the relation $R = k(\sigma/\eta)t$ [13]. Here $k = k_b$ or k_w for bulk and wetting phase separation, respectively. For phase separation under the influence of wetting, k_w is roughly estimated as ~ 0.1 from the Poiseuille's formula. For bulk phase separation, on the other hand, the tube flow is essentially caused by the fluctuation. San Miguel et al. [17] theoretically estimated k_b as 0.04 for the two phase fluids having similar viscosity, which was experimentally [18,19] supported. If we employ $k_b \sim 0.04$, the difference between k_b and k_w is likely within one order of the magnitude. However, the anisotropic ordering for wetting phase separation might accelerate the reduction of the interface area further.

9 Interface Quenh Effect

Next we estimate the *interface quench* effect. The free energy of the system can be described by the Ginzburg-Landau type free energy as

$$F = \int dr[-\frac{r}{2}\phi^2 + \frac{u}{4}\phi^4 + \frac{K}{2}(\nabla\phi)^2]. \quad (2)$$

Here ϕ is the concentration. Provided that the concentration profile can be approximated by the trapezoidal shape with a interface width of ξ [20],

$$F \sim [-\frac{r}{2}(\Delta\phi)^2 + \frac{u}{4}(\Delta\phi)^4] + \frac{K}{2}(\frac{\Delta\phi}{\xi})^2(\frac{\xi}{R(t)}), \quad (3)$$

where $\Delta\phi$ is the concentration of the phase measured from the average one. From the above local energy minimum condition,

$$\Delta\phi(t) = \Delta\phi_b(1 - \frac{K}{\xi r R(t)})^{1/2} = \Delta\phi_b(1 - \frac{2\xi}{R(t)})^{1/2}, \quad (4)$$

where $\Delta\phi_b = (r/u)^{1/2}$ is the equilibrium concentration for an infinite domain size. Here the relation $\xi^2 = K/2r$ is used. Reflecting the change in the domain size ($R(t) = k(\sigma/\eta)t$), $\Delta\phi(t)$ changes with time. However, the diffusion is not fast enough for the real concentration to follow this change in the local equilibrium concentration. Thus there could be a significant *interface quench* effect.

To check the above possibility, we have to study whether the concentration diffusion can follow this quick change in the local equilibrium concentration or not. The time required to hydrodynamically form the macroscopic phase having a domain size R is estimated as $\tau_h \sim R\eta/(k\sigma)$. On the other hand, the characteristic diffusion time for the domain size R is given by $\tau_D \sim R^2/D$, where D is the diffusion constant and $D = k_B T/(5\pi\eta\xi)$ (k_B: Boltzmann's constant). Thus the ratio between τ_h and τ_D is given by $\tau_h/\tau_D = D\eta/(k\sigma R)$. From the 2-scale-factor universality, $\sigma = A_\sigma k_B T/\xi^2$, where A_σ is the universal constant and $A_\sigma \sim 0.2$ in 3D [1]. Using this relation and the expression for D, we obtain the relation $\tau_h/\tau_D \sim \xi/(5\pi A_\sigma kR) \sim \xi/(3kR)$. For $\tau_h < \tau_D$, the concentration is different from its equilibrium value and the local equilibrium cannot be established. Thus the *interface quench* is likely initiated around $\tau_h/\tau_D \sim 1$. For wetting phase separation ($k_w \sim 0.1$), $R_t \sim 3\xi$ (R_t: the transient domain size when the *interface quench* is initiated.). For bulk phase separation ($k_b \sim 0.04$), on the other hand, we obtain $R_t \sim 10\xi$ or $\tau \sim 100$ ($\tau = t/\tau_\xi$, where $\tau_\xi = \xi^2/D$.) from the condition $\tau_h/\tau_D \sim 1$.

The beginning of the *interface quench* characterized by these values of R_t/ξ and τ is consistent with the crossover from the slow, diffusion growth to the fast, hydrodynamic growth in the scaled plots of $2\pi R/\xi$ against τ [1,18,19]. The *interface quench* probably brings the system into a new nonequilibrium (unstable or metastable) state and thus causes the retarded, secondary phase separation. The situation is schematically shown in Fig.8. Our interpretation of the phenomenon is consistent with the light scattering data by Wiltzius and Cumming [8], which indicates that (1) the secondary phase separation (in our notation) has the growth

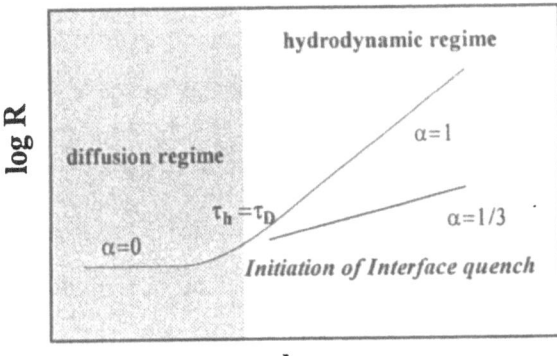

Figure 8: Schematic figure for the coarsening process of DPS.

law of $R \sim t^{1/3}$ unique to droplet pattern and (2) it starts to appear just after the hydrodynamic process (the fast mode in their notation [8,21]) starts to play a dominant role in coarsening (around $\tau_h = \tau_D$) (see Fig.13 in ref.[21]).

Finally we estimate the rate of the *interface quench*, which should likely be large enough to cause the secondary phase separation. Here we consider the equivalent temperature quench instead of the composition quench since the former is easier to understand. Since r is proportional to the quench depth ΔT, the change in $R(t)^{-1}$ has the same meaning as the change in $\Delta T(t)$, according to eq.(3). Using the relation $\Delta\phi_b = m\Delta T^\beta$ ($\beta \sim 0.33$, m:a constant), $\Delta T(t)$, which satisfies $\Delta\phi_b(\Delta T(t)) = \Delta\phi(\Delta T, R(t))$, is estimated as

$$\Delta T(t) = \Delta T(1 - 2\xi/R(t))^{1/2\beta}. \quad (5)$$

For example, the *interface quench* caused by the hydrodynamic coarsening from small domains ($R = 10\xi$) to large domains ($R = 100\xi$) is equivalent to the temperature double quench [15] composed of a first quench ($\Delta T_t = 0.88\Delta T$) and a second deeper quench ($\delta T = \Delta T - \Delta T_t = 0.12\Delta T$) around the time of $\tau_D = \tau_h$. Thus the average quench rate from $R = 10\xi$ to $R = 100\xi$ can be estimated as

$$\delta T/\delta t = (0.3\Delta T/90\xi)(k\sigma/\eta) = 4 \times 10^{-4}\Delta T/\tau_\xi. \quad (6)$$

The steep increase of the quench rate with an increase in ΔT is consistent with our experimental results that DPS was observed only for large ΔT.

10 Summary

In summary, we have found the universal wetting dynamics unique to a confined binary mixture. It has also been demonstrated that the wetting layer formation in bicontinuous phase separation is strongly dependent on the spatial dimensionality of the geometrical constraint. The hydrodynamic coarsening unique to bicontinuous phase separation delocalizes the wetting effect and drastically accelerates the wetting dynamics. Thus the macroscopic wetting structure is formed within a short time.

Further, *double phase separation* (DPS) has been found in confined symmetric binary mixtures. This phenomenon is likely universal for any confined, symmetric binary mixtures. We demonstrate that this unusual phenomenon is likely caused by the *interface quench* effect unique to bicontinuous phase separation. DPS might be observed even for bulk phase separation under a deep quench condition if the anisotropic ordering for wetting phase separation does not play a crucial role in the interface reduction.

11 Acknowledgment

This work was partly supported by a Grant-in-Aid from the Ministry of Education, Science, and Culture, Japan.

References

[1] J.D.Gunton, M.San Miguel, and P.Sahni, in *Phase Transition and Critical Phenomena*, edited by C.Domb and J.H.Lebowitz,(Academic, London, 1983), Vol.8.
[2] J.W.Cahn, J.Chem.Phys. **66**, 3667 (1977).
[3] D.Jasnow, Rep.Prog.Phys. **47**, 1059 (1984).
[4] P.G.de Gennes, Rev.Mod.Phys. **57**, 827 (1985).
[5] K.Williams and R.A.Dawe, J.Colloid Interface Sci. **117**, 81 (1987).
[6] A.J.Liu, D.J.Durian, E.Herbolzheimer, and S.A.Safran, Phys.Rev.Lett. **65**. 1897 (1990).
[7] P.Guenoun, D.Beysens, and M.Robert, Phys.Rev.Lett. **65**, 2406 (1990).
[8] P.Wiltzius and A.Cumming, Phys.Rev.Lett. **66**, 3000 (1991).
[9] F.Bruder and R.Brenn, Phys.Rev.Lett. **69**, 624 (1992).
[10] U.Steiner, J.Klein, E.Eiser, A.Budkowski, L.J.Fetter, Science **258**, 1126 (1992).
[11] J.Bodensohn and W.I.Goldburg, Phys.Rev. **A46**, 5084 (1992).
[12] H.Tanaka, Phys.Rev.Lett. **70**, 53 (1993).
[13] H.Tanaka, Phys.Rev.Lett. **70**, 2770 (1993).
[14] E.D.Siggia, Phys.Rev.**A20**, 1979 (1979).
[15] H.Tanaka, unpublished.
[16] H.Tanaka, Phys.Rev.E **47**, 2946 (1993).
[17] M.San Miguel, M.Grant, and J.D.Gunton, Phys.Rev. **A31**, 1001 (1985).
[18] P.Guenoun, R.Gastaud, F.Perrot, and D.Beysens, Phys.Rev. **A36**, 4876 (1987).
[19] F.S.Bates and P.Wiltzius, J.Chem.Phys. **91**, 3258 (1989).
[20] The trapezoidal profile is probably a good approximation in the late stage ($\tau = t/(\xi^2/D) \geq 200$) [18], and this assumption will not affect the main conclusion.
[21] A.Cumming, P.Wiltzius, F.S.Bates, J.H.Rosedale, Phys.Rev. **A45**, 885 (1992).

Wetting of Grafted Polymer Surfaces by Compatible Chains

Ludwik LEIBLER and Armand AJDARI,

Laboratoire de Physico-Chimie Théorique,
U.r.a. CNRS 1382,
E.S.P.C.I., 10 rue Vauquelin, 75231 Paris Cedex 05, France.

Ahmed MOURRAN, Ghislaine COULON and Didier CHATENAY,

P.S.I., Section de Physique et Chimie,
Institut Curie, 10 rue P. et M. Curie, 75231 Paris Cedex 05, France.

Abstract: We discuss the way a droplet of a polymer melt spreads on a surface densely grafted by chains of same chemical nature. If the melt chains are short they invade the grafted layer and complete wetting is expected. On the other hand interpenetration should be very limited for long melt chains and partial wetting the rule. These considerations should apply for an A melt on the surface of a segregated AB block copolymer, and are in qualitative agreement with recent developping experiments.

INTRODUCTION

It is an everyday experience that a droplet of an usual liquid completely wets a free surface of the same liquid. However this a priori trivial statement turns out to be questionable in the case of polymeric liquids the state of which is not solely described by the chemical nature of the monomers but also by the configurations of the macromolecules. We focus here on a specific problem: does a droplet of a given polymer spread on a surface protected by densely grafted chains of the same chemical structure (Fig.1)?

A. Teramoto, M. Kobayashi, T. Norisuje (Eds.)
Ordering in Macromolecular Systems
© Springer-Verlag Berlin Heidelberg 1994

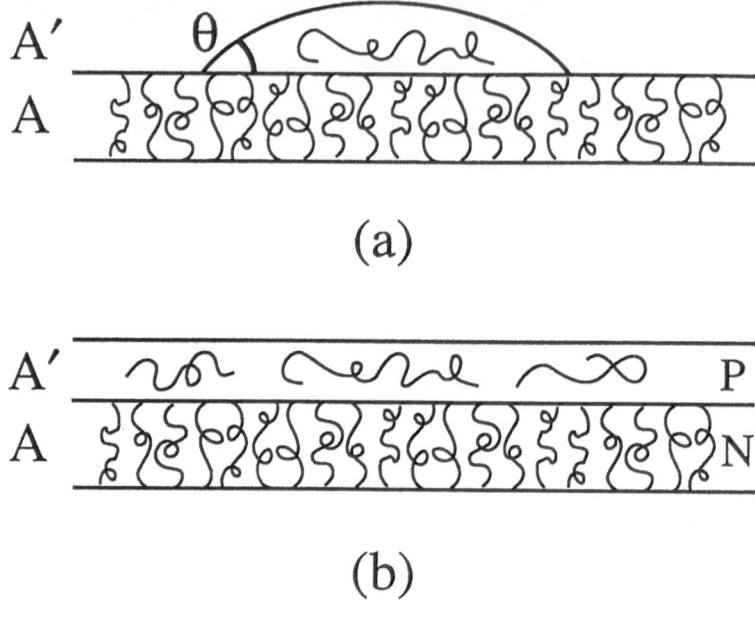

(a)

(b)

Figure 1: Schematic large-scale picture of the partial (a) and complete (b) wetting of an A-brush by an A'-homopolymer droplet. Possible interpenetration and swelling of the brush are not displayed.

These "brushy" surfaces are actually commonly encountered. An interesting case is the surface of a microphase-separated block copolymer. For example, a symmetric polystyrene/polyisoprene block copolymer exhibits a lamellar structure. When annealed, its surface is covered by a polyisoprene layer as polyisoprene has a lower surface tension. At room temperature, we therefore have a molten polyisoprene brush with chains anchored to a glassy polystyrene layer. A brushy layer can also be obtained by adsorbtion of copolymers at the surface of a solid or that of a polymer melt, or more directly when end-functionalized chains are chemically anchored on a given solid surface. Hence, as academic it may seem, the problem we adress is actually of some relevance in technological applications such as surface control, adhesion between a grafted surface and a melt, preparation and rheology of blends or composites (polymer + fillers,...).

To set notations we consider a surface on which A-chains of N monomers are grafted at a density Σ^{-1} such that even in the absence of a wetting solvent the chains are stretched (brush regime). If we set the monomer size a equal to the Kuhn length the brush regime corresponds to $\Sigma < N^{1/2}a^2$. On this surface we lay a certain amount of a polymer melt A' of same chemical nature (athermal solvent) and of polymerization index P.

In section I we recall that for short A' chains (P small), the melt is expected to swell considerably the brush and to wet the surface. In section II we show that, on the contrary, very long A' chains penetrate the brush only on a limited thickness. This slight interpenetration leads to a surface tension at the polymer/brush interface and the melt wets the surface only partially. The cross-over between these two limiting behaviour occurs for intermediate values $P \cong P^*$. In section III we discuss consequences of this analysis on the shape of the resulting droplet and on the kinetics of spreading.

I- WETTING BY A COMPATIBLE MELT OF SHORT CHAINS (GOOD SOLVENT).

In the Alexander-de Gennes picture [1,2], where all the brush chains are stretched by a same amount, an initially dry brush of thickness $h \cong Na^3/\Sigma$ brought into contact with a solvent of small compatible molecules (P≅1) is swollen so as to behave as a semi-dilute solution of blobs of size $\Sigma^{1/2}$. This leads to a brush thickness $L \cong Na^{5/3}\Sigma^{-1/3}$.

To estimate the free energy gain per chain $\Delta f_{ch} = f_{wet} - f_{dry}$, let us use a Flory-like approach and call $\varphi \cong h/L$ the average volume fraction of N-polymer in the wet brush. Then taking into account the solvent translational entropy which favors swelling and the chains configurational entropy which opposes it leads to:

$$\Delta f_{ch}/kT \cong (L^2/Na^2 - h^2/Na^2) + L\Sigma/a^3 \, (1-\varphi)Ln(1-\varphi) \tag{1}$$

Minimization with regards to φ leads to $\varphi \cong (h/Na)^{2/3} \ll 1$, which is equivalent to the result obtained by blob arguments (recall $\varphi \cong h/L$). Swelling results in a strong decrease of the free energy per chain due to the important gain in translationnal entropy of the small molecules:

$$\Delta f_{ch} \cong - NkT \, (1 - (h/Na)^{2/3}) \cong - N \, kT \tag{2}$$

This expression allows us to rederive the result of Halperin and de Gennes [3] for the spreading coefficient [4] of the small molecule solvent over the brush:

$$S = F_{dry} - F_{wet} \cong \gamma_{A'} + \gamma_{wAA'} - \gamma_{dA} + \Delta f_{ch}/\Sigma \qquad (3)$$

where $\gamma_{A'}$, γ_{dA}, and $\gamma_{wAA'}$ are the surface tensions respectively at the "air/melt A'", "air/dry brush A" and "wet brush A/melt A'" interfaces. At first order $\gamma_{A'}$ and γ_{dA} are equal (their difference scales as N^0) and $\gamma_{wAA'}$ is similarly negligible compared to the dominant term $\Delta f_{ch}/\Sigma$. Thus:

$$S \cong N \, kT/\Sigma > 0 \qquad (4)$$

A' wets the A-grafted surface due the high entropy of dissolution of the small A' molecules.

At this point we must mention that the preceeding description is limited by the assumption that all chain ends reach the surface in the brush. Since Semenov's seminal work [5] it is known that such is not the case. In a more sophisticated approach one should consider the chain statistics. As the chains in a brush are stretched, the usual Schrödinger-like equation describing the chains configuaration reduces to its classical counterpart that gives the distance $z(n)$ of the nth monomer of a chain from the surface as:

$$d^2z/dn^2 = - \, dV(z)/dz \qquad (5)$$

This equation describes the trajectory of a particle in the potential $V(z)$, actually representing the chemical potential of a monomer at altitude z over the surface. As this particle ends at "time" N on the surface whatever its starting point the potential has a parabolic shape (as in the classical case of the harmonic oscillator)[6]. In the dry brush case:

$$V(z) \cong V_0 - \pi^2/24N^2a^2 \, z^2 \qquad (6)$$

The chemical potential in this case ensures incompressibility. The chain end distribution in the brush $\rho(z)$ can be extracted from V and reads $\rho(z)$ = $(1/h\Sigma)[(z/h)^2/(1-(z/h)^2)^{1/2}$. The constant V_0 has been fixed by the incompressibility $V(h=Na^3/\Sigma) = 0$.

However these considerations do not alter the fact that the average stretching enthalpic penalty still scales as $f \cong L^2/Na^2$, and that the spreading coefficient is still dominated by the dissolution entropy and scales as in (5).

II- PARTIAL WETTING BY A COMPATIBLE MELT OF VERY LONG CHAINS.

We now focus on the opposite limit of an A' melt of very long chains $P \cong \infty$. In this limit the entropic gain associated with the penetration of an A'-chain in the brush is limited compared to the consequent stretching imposed to the brush chains. Therefore only short sections of A'-chains penetrate the brush in a kind of "quantum tunnelling" in the $V(z)$ potential of the dry brush [7,8,9] (Fig. 2). In this case one has to use the Schrödinger-like equation for the wave function $\Psi(z)$ (Ψ^2 is the probability density):

$$a^2/6 \; \Delta\Psi - V\Psi = 0 \qquad\qquad (7)$$

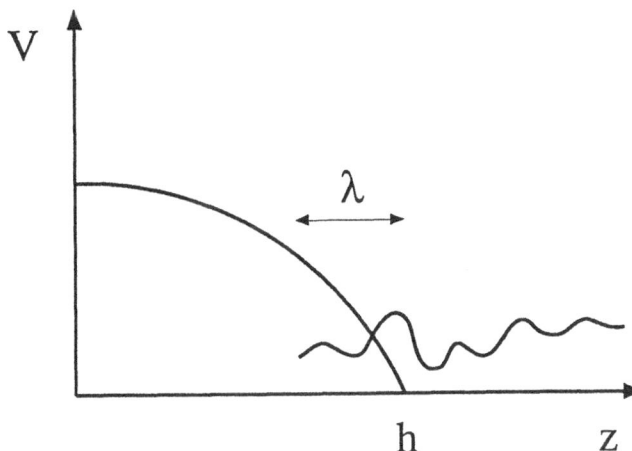

Figure 2: Schematic picture of an A'-chain tunneling into the potential V of the dry A-brush.

Using the expression (6) one can then derive the penetration length λ of A' sections in the A brush:

$$\lambda \cong (N^2 a^4/h)^{1/3} \cong (Na\Sigma)^{1/3} \qquad (8)$$

This scaling result can be obtained by writing that $a^2/\lambda^2 \cong V(h-\lambda)$ according to (7). However, a more complex derivation is needed to obtain more quantitative estimates as one should self-consistently take into account the deformation of the V potential due to the penetration of a few A' sections. The most practical parameter to use is $\varphi(z)$ the A monomer concentration at distance z from the surface ($\varphi_{dry}(z)=1$ for z<h in the dry case). The free energy density variation ΔF from a dry brush to a slightly swollen one can then be written at first order as in ref. [10]:

$$\Delta Fa^3/kT \cong a^2/8 \int (\nabla\varphi)^2/\varphi(1-\varphi) \ dz \ - \ \int V(z)(\varphi(z)-\varphi_{dry}(z)) \ dz \qquad (9)$$

Plugging into (9) a hyperbolic tangent profile for φ that goes from 1 to 0 over a finite distance λ and minimizing ΔF with respect to λ leads to:

$$\lambda \cong (12/\pi^4)^{1/3} \ (N^2 a^4/h)^{1/3} \cong (12/\pi^4)^{1/3} \ (Na\Sigma)^{1/3} \qquad (10)$$

and

$$\Delta F \cong 3/8 \ kT/\lambda a \qquad (11)$$

ΔF is the interfacial tension at the interface between a brush and a melt of same chemical nature! Note also that the interpenatration length λ scales as $N^{1/3}$ for a given density and can therefore, although small compared to h, be large compared to microscopic distances $\cong$ a.

The central result of this article is that **a droplet of long A' chains spreads only partially an A brush**. The spreading coefficient, neglecting $\gamma_{A'} - \gamma_{dA}$, is:

$$S \cong - \Delta F \cong - 3/8 \ kT/\lambda a < 0 \qquad (12)$$

Let us underline that $\delta\gamma = \gamma_{A'} - \gamma_{dA}$ is the difference of surface tension at a free interface between a dry brush and a melt of same chemical nature. In both cases air penetrates the polymer only on a short microscopic distance, say a'. The additionnal term due to the stretching of the brush chains can be estimated by a term similar to the second one of the right hand side of (9), evaluated for a φ profile varying from 1 to 0 over a distance $\cong a'$. We thus get:

$$\delta\gamma = \gamma_{A'} - \gamma_{dA} \cong - kTa'^2 h / N^2 a^5 \qquad (13)$$

This term can be neglected in (12) as $-\delta\gamma/\Delta F \cong (a^2/\Sigma N)^{2/3}$ for $a' \cong a$.

Equation (12) leads according to Young's relation [4] to a small contact angle θ for the equilibrium of the droplet:

$$\theta \cong 2 (\Delta F/\gamma)^{1/2} \qquad (14)$$

where $\gamma \cong \gamma_{A'} \cong \gamma_{dA}$ is a "chemical" A/air surface tension.

III- DISCUSSION

1- It is now natural to determine the typical values of P at which the cross-over between complete wetting and partial wetting occurs. This can be done by noting that we neglected in (9) a term describing the translational entropy of the P chains which in a Flory-like description reads:

$$\Delta F'a^3/kT \cong 1/P \int (1-\varphi)Ln(1-\varphi) \; dz \qquad (15)$$

This term of course favors the swelling of the brush, and scales as $-\lambda/P$ for $\varphi(z) \cong 1-th(z/\lambda)$. Comparing $\Delta F'$ to ΔF in (11), we find the cross-over value:

$$P^* \cong (\lambda/a)^2 \cong (N\Sigma/a^2)^{2/3} \qquad (16)$$

Complete wetting of the A brush by the A' melt is expected for $P<P^*$ [2].

2- In the case where the brush is due to the treatment of a surface by A-B block copolymers, the height h of the brush scales as $aN^{2/3}\chi^{1/6}$ where χ is the Flory parameter for the A/B system. As a result the contact angle for a drop of A'-melt of long chains $(P>P^*)$ decreases with N but increases

308

with the incompatibility of A and B. Semenov's theory of microphase separation in the strong segregation limit [5] enables one to calculate Σ and estimate the penetration length and interfacial tension ΔF:

$$\Delta F \, v/akT \cong 0.926 \, \chi^{1/18} N^{-4/9} \tag{17}$$

where v is the specific volume per monomer and N denotes the polymerization index of the copolymer (which has been assumed to be symmetric). Figure 3 shows the contact angle as a function of N for $\chi = 0.1$, $\gamma \cong 30$ dyn/cm, a $\cong 6.5$ A and $1/v \cong 6 \, 10^{21}$ cm^{-3}.

The cross–over length P* for the A' homopolymer chains can be estimated as $P^* \cong 0.16 \, N^{8/9} \chi^{-1/9}$, which leads for $N \cong 1000$ and $\chi \cong 0.1$ to $P^* \cong 100$.

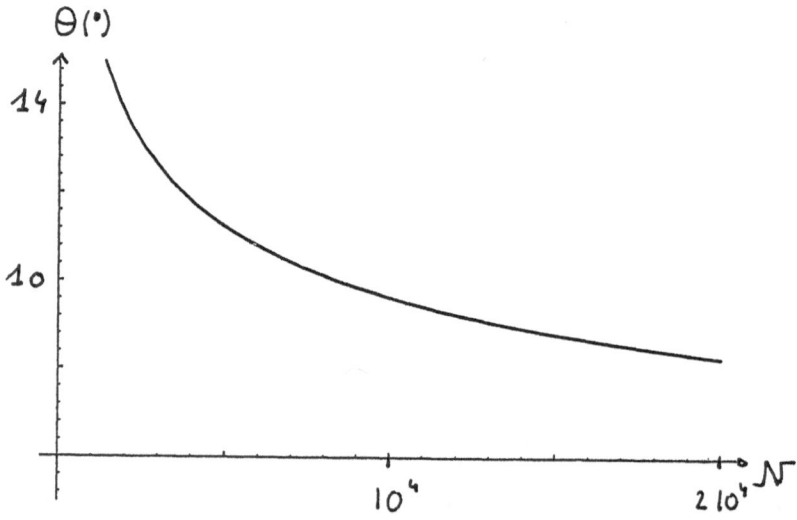

Figure 3: Variation of the contact angle θ with the length N of the copolymer. Parameters as defined in the text.

3– Preliminary experiments done in Institut Curie give some indication of partial wetting. The photograph presented here (Figure 4) has been obtained by atomic force microscopy, a droplet of polystyrene ($M_w \cong 91000$) having been laid on a 3L/2 structure of symmetric polystyrene/polymethylmethacrylate copolylmer ($M_w \cong 91000$) deposed on a silicon waffer. The observed angle (θ≅3°) is smaller than what formula (14) predicts (θ≅15°). We point out however, that the brushes are not very stretched in this system and that P is not much larger than P*, thus an increased penetration of the melt chains in the grafted layer is expected compared to the regime described in section II.

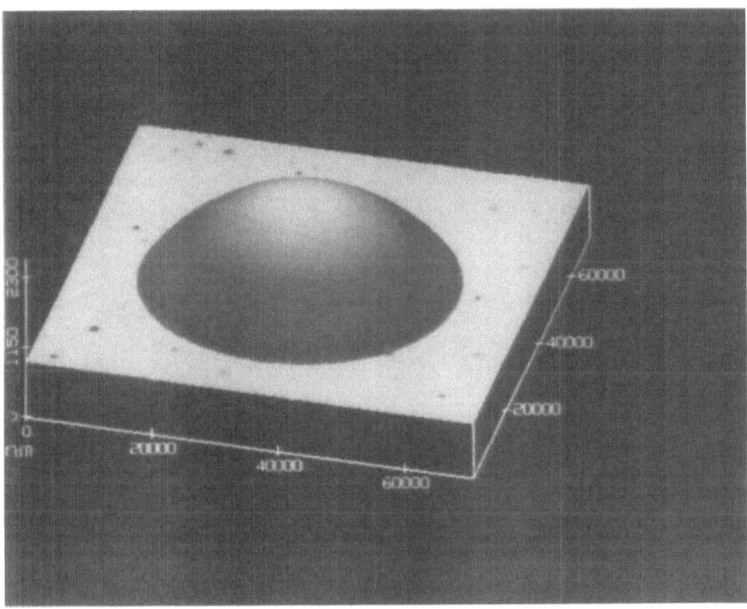

Figure 4: Equilibrium profile of a PS droplet on a PMMA/PS surface.

4- The surface tension ΔF can also be responsible for effective attraction between brushy copolymer aggregates [7,8,10,11] or polymer protected particles immersed in a melt of long chains. This effective attraction arises from the fact that bringing two brushes in contact replaces two brush/melt interfaces by a sole brush/brush interface. Note that in a melt of short chains one gets instead effective repulsion between the particles due to the osmotic repulsion of the brushes.

5- The precise shape of a polymer droplet over a polymer brush in the case P ≫ P* is rather subtle near the contact line, as a dry brush behaves as an elastic film of finite modulus [12]. It therefore deforms perpendicularly to the interface under the action of the pulling force $\gamma\sin\theta$ exerted on the contact line (Fig. 4). The profile of the brush surface on the two sides of the contact line should be different as the corresponding surface tensions are very different (γ on the free side, ΔF under the droplet). As surface tension forbids short wavelengths modes the free profile should relax on much larger distances ($\gamma \gg \Delta F$). These considerations are clearly of importance to extract the contact angle from an experimental measure of the overall polymer profile by Atomic Force Microscopy for example.

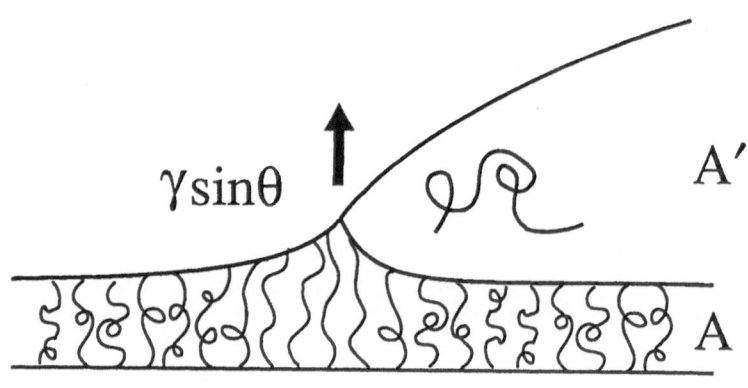

Figure 5: Schematic profile of the air/polymer interface close to the contact line.

6– The degree of interpenetration of melt and brush at contact is also of importance to analyze the dynamics of spreading of the droplet over the brush as it controls the amount of slippage that can occur at the interface. If the interpenetration length is somewhat smaller than a typical distance between entanglements in the brush, the relative friction of the two media can be modeled as that developping in a film of thickness λ and of Rouse–like viscosity $\eta \cong \eta_0(\lambda/a)^2$ [13]. On the other hand if entanglements exists between the two media, the friction can be significantly higher at low slippage velocities, although decreasing in a very non-linear way for higher velocities [14,15]. Of course, this is schematical as the notion of entanglement in a brush and the description of the corresponding relaxation processes constitute a very complicated problem [16]. Similar considerations hold to describe adhesion between brushy structures and melts.

Acknowledgments:
We are deeply indebted to Jacques Prost for his interest in this project and for very helpful discussions. We greatly appreciated Jim Harden's help in the preparation of the manuscript. L.L. thanks the organizers of the OUMS for giving him an opportunity to present this work and for their hospitality during his stay in Osaka.

REFERENCES

[1] S. Alexander, *J. Phys. France*, **38**, 983 (1977).

[2] P.G. de Gennes, *Macromolecules*, 13, 1069 (1980).

[3] A. Halperin, P.G. de Gennes, *J. Phys. France*, **47**, 1243 (1986).

[4] P.G. de Gennes, *Rev. Mod. Phys.*, 57, 827 (1985).

[5] A.N. Semenov, *Sov. Phys. J.E.T.P.*, 61, 733 (1985); *Zh. Eksp. Teor. Fiz.*, **88**, 1242 (1985).

[6] S.T. Milner, T.A. Witten, M.E. Cates, *Macromolecules*, **21**, 2610 (1988).

[7] A. Gast, L. Leibler, *J. Chem. Phys.*, 89, 3947 (1985).

[8] A. Gast, L. Leibler, *Macromolecules*, 19, 686 (1986).

[9] T. Witten, L. Leibler, P. Pincus, *Macromolecules*, 23, 824 (1990).

[10] A.N. Semenov, *Macromolecules*, 25, 4967 (1992).

[11] K.R. Shull, E.J. Kramer, *Macromolecules*, **23**, 4769 (1990).

[12] G.H. Fredrickson, A. Ajdari, L. Leibler, J.P. Carton, *Macromolecules*, **25**, 2882 (1992).

[13] F. Brochard, P.G. de Gennes, S. Troian, *C.R.Acad.Sci.Paris*, 310 II, 1169 (1990).

[14] F. Brochard, P.G. de Gennes, *Langmuir*, 8, 3033 (1992).

[15] A. Ajdari, F. Brochard-Wyart, P.G. de Gennes, L. Leibler, J.L. Viovy, M. Rubinstein, *Physica A*, submitted (1993).

[16] S. Obukhov, M. Rubinstein, *Macromolecules*, in press (1993).

Surface Segregation and Wetting from Polymer Mixtures

Ullrich Steiner, Erika Eiser, Andrzej Budkowski,
Lewis Fetters[a], Jacob Klein*,

Weizmann Institute of Science, Rehovot 76100, Israel
([a] Exxon Research and Engineering Co., Annandale, N.J. 08801, U.S.A)

Abstract: Coexisting binary polymer phases are characterised by very small interfacial energies even well below their critical solution temperature. By extension of Cahn's ideas concerning critical point wetting, one expects that such low energies should readily lead to the exclusion of one of the phases from any interface which favours the other; this phenomenon has implications for practical surface-related effects, ranging from welding to wear properties. Using nuclear reaction analysis, we have now observed such complete wetting behaviour from two different classes of binary polymer mixtures. These are mixtures of statistical olefinic copolymers of structure $-(C_2H_3(C_2H_5))_x((CH_2)_4)_{1-x}-$, with differing x values, and an isotopic pair of deuterated and protonated polystyrene. In the former case we have been able to follow the growth with time t of the wetting layer thickness l; our results indicate $l \sim \log t$.

1. INTRODUCTION

The formation of drops or thin liquid films on solid surfaces is not only commonplace, but also has considerable technological importance in a wide variety of applications. The transition between partial wetting of a solid by a liquid (which leads to the formation of drops for the case of thin spread films), and complete wetting which results in stable uniform films on the surface, was implicit already in the description by Young - nearly 200 years ago(1) - of the equilibrium contact angle θ of drops on a surface:

$$\cos(\theta) = (\gamma_{vs} - \gamma_{ls})/\gamma_{vl} \qquad (1)$$

Here subscripts v, l, and s refer to vapour, liquid and solid respectively, and γ_{xy} refers to the surface tension (or interfacial energy) between the x and y phases. The difference of $\gamma_{vs} - \gamma_{ls}$ determines the form of the liquid on the surface: if $\gamma_{ls} > \gamma_{vs}$, the liquid phase covers the substrate only partially and the contact angle θ is finite. If $\gamma_{vs} - \gamma_{ls} = \gamma_{vl}$, $\theta = 0$ and the substrate is covered by a continuous wetting film, which may attain macroscopic dimensions. Upon changing the balance of surface tensions - e.g. by changing the temperature - a transition from the droplet phase, called partial wetting to the phase featuring a continuous film - complete wetting - is observed. This is illustrated in fig. 1.

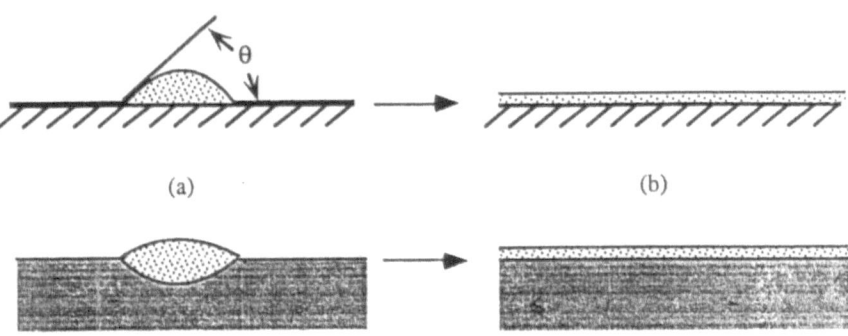

Figure 1: Showing the transition between (a) partial wetting and (b) complete wetting of a liquid phase in equilibrium with its vapour on a surface of a solid (top) or a liquid (bottom).

A. Teramoto, M. Kobayashi, T. Norisuje (Eds.)
Ordering in Macromolecular Systems
© Springer-Verlag Berlin Heidelberg 1994

It is instructive to consider a liquid in equilibrium with its vapour as a two-phase (one-component) system at coexistence. If such a system completely wets a wall with which it is in contact, a macroscopic liquid film forms, and the composition profile near the surface is as shown by the solid line in figure 2.

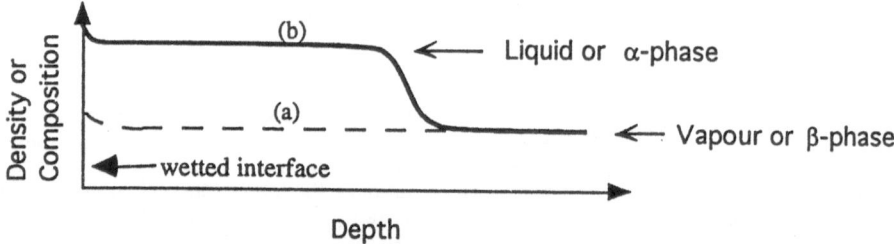

Figure 2: Schematic composition-depth profiles of the solid-favoured phase near the surface of a two-phase mixture. (a) partial wetting. (b) Complete wetting. The solid line may be viewed as the profile of a liquid at the surface in the case of complete wetting from a liquid-vapour system at coexistence.

The total surface energy associated with this wetting layer is simply the *sum* of the surface-liquid interfacial energy, γ_{ls}, and of the liquid-vapour interfacial energy γ_{vl}, since the liquid layer is sufficiently thick that these two contributions are unaffected by each other(2). The actual thick liquid film itself costs no energy (if we assume short ranged forces only are acting and ignore gravitational effects), because at coexistence the liquid and the vapour free energies are the same. At the wetting transition this sum must equal the surface energy γ_{vs} associated with completely excluding the vapour from the surface. Thus, as the condition for complete wetting, we may write

$$\gamma_{vl} + \gamma_{ls} \geq \gamma_{vs} \tag{2}$$

Inequality implies that there must be an energetically cheaper way to cover the surface than by a macroscopic wetting layer, namely by the vapour phase. Equality in equn.(2), on the other hand, implies that complete wetting by a liquid film takes place: it is exactly equivalent to the wetting implied by the disappearance of the dihedral droplet contact angle in the Young equation (1).

Droplets and liquid films are commonplace, but exactly the same type of partial-to-complete wetting transition can occur also when a solid is in contact with a mixture of two fluid components A and B one of which (say A) it prefers. When these components are only partially miscible two phases will coexist, an A-rich phase α and a B-rich phase β. In this case, as illustrated in fig. 2, the surface may be in contact with the β-phase (with a microscopically thin A-enriched layer) for the case of partial wetting (fig. 2(a)); or by a thick, macroscopic layer of the α phase in the case of complete wetting (fig. 2(b)). The discussion above for the coexisting liquid-vapour system may then be taken over exactly for the coexisting α/β system. We denote the interfacial energies γ_{ij}, where i,j refer to the α phase, the β phase or to the solid phase s. Then, rewriting equation (2), the following relation determines whether complete or partial wetting by the α phase will occur:

$$\gamma_{\alpha\beta} \geq \gamma_{s\beta} - \gamma_{s\alpha} \tag{2'}$$

In 1977 Cahn presented his seminal argument for the existence of critical point wetting(3). He assumed that the nature of the surface interactions was short ranged, and that the difference in the interaction energies $\gamma_{s\alpha}$ and $\gamma_{s\beta}$ originated in, and was proportional to the difference in compositions of the α and β phases, more specifically in the difference in content of the A component. Since the wall is not a 'critical component' this difference is simply the length of the tie-line (of the phase coexistence curve) at the appropriate temperature $T < T_c$, the critical temperature. Thus this difference scales as $(\gamma_{s\beta} - \gamma_{s\alpha}) \propto (T_c - T)^\beta$, where β is a surface spinodal exponent whose value is ca. 0.8. At the same time the interfacial energy scales as $\gamma_{\alpha\beta} \propto (T_c - T)^{2\nu}$, where 2ν is a bulk critical exponent whose value is ca. 1.3. Thus as T_c is approached the interfacial energy $\gamma_{\alpha\beta}$ will always decrease faster than the difference $(\gamma_{s\beta} - \gamma_{s\alpha})$, so that sufficiently close to the critical temperature the inequality (2') will become an equality and complete wetting will occur. This is Cahn's critical point wetting argument. It relies for its generality on the existence of short range surface forces alone, but it has been found to hold in most binary liquid mixtures examined to date, and has proved a compelling argument in the stimulation of much theoretical work(2, 4, 5).

When dealing with binary polymer mixtures, an interesting possibility has been pointed out by Binder and coworkers(6). This has to do with the fact that for polymer mixtures the role of translational entropy in promoting mixing is greatly reduced (by a factor of order N, the degree of polymerisation(7, 8)), so that the critical interaction parameter χ_c is very small, of order 1/N. This leads to the conclusion that even at temperatures far below T_c in the two phase region (for an upper critical solution temperature, UCST), one may have very low values of the segmental interaction parameter χ. I.e., even at $T \ll T_c$, one may have $\chi_c \ll \chi \ll 1$. At the same time, the interaction energies $\gamma_{s\beta}$ and $\gamma_{s\alpha}$ remain comparable to their values for the corresponding monomeric phases, as in the original Cahn assumption. Since $\gamma_{\alpha\beta}$ scales with χ, it follows that it can be very small even deep into the two-phase region of the phase diagram(6), while the difference $(\gamma_{s\beta} - \gamma_{s\alpha})$ remains comparable to its value for coexisting *monomeric* phases α and β. For this reason the inequality appearing in equn.(2) can - for the case of polymer mixtures - become an equality over a large range of the phase coexistence diagram (and not only in the vicinity of T_c as in the case of monomeric mixtures). Thus polymeric mixtures should be ideal candidates to exhibit complete wetting behaviour. Nonetheless, until very recently such behaviour has never been observed. Investigations of surface composition in binary polymer mixtures have shown only partial wetting, or surface enrichment, as in the case of isotopic blends(9) or of mixtures of compatible polymers with specific interactions(10).

In this lecture we describe a recent study of the surface compositions in samples of two polymeric mixtures: an isotopic mixture of microstructurally identical chains where the segmental interactions arise from the slight differences between deuterated and protonated species; and a binary mixture of random copolymers which differ slightly in their microstructure. In the former case we demonstrate for the first time the existence of a wetting layer in such an isotopic blend. In the second system we show unambiguously the growth with time - to macroscopic thickness - of a surface layer of one of the two phases in the mixture, from a coexisting composition of the other phase(11).

In the following section we describe the materials used, and the composition-depth profiling technique by which our samples are monitored. In section 3 we describe the determination of the phase coexistence behaviour for the binary polymer mixtures, and in sections 4 - 6 we discuss the manifestation of wetting in the different systems, and its dependence on time. In the final section we consider in more detail the question of the expected rate of growth of the wetting layer in our (finite-sized) system. We conclude with some remarks concerning the use of polymers in wetting studies.

2. MATERIALS AND EXPERIMENTAL

Two types of binary polymer systems were investigated. An isotopic mixture of perdeuterated polystyrene (dPS) and protonated polystyrene (hPS), of similar molecular weights M (or degrees of polymerisation N), obtained and characterised by Polymer Laboratories (UK); and statistical copolymers of ethyl-ethylene (EE) and ethylene (E) monomers, $(EE_{xi}E_{1-xi})_N$, where the monomers are arranged randomly along the polymer chain. The relative amount x_i of ethyl-ethylene monomers, EE = -(C_2H_3(C_2H_5))-, can be varied to any value $0 \leq x_i \leq 1$. In this sense the molecules (and thereby their interactions) may be viewed as continuously 'tunable' between essentially pure polyethylene and pure poly(ethyl ethylene). Some of these random copolymers were partially deuterated (to an extent e), a prerequisite of our analytic technique as described below. The miscibility of a mixture of such copolymers depends very sensitively on the difference in the x_i, as well as on the difference in deuteration(12, 13). The precise characteristics of the polymers used are indicated at the beginning of each section. It is noteworthy that mixtures of different x_i values (and extent of deuteration, e) can exhibit similar critical temperatures. As the material with the higher ethyl-ethylene content (i.e. higher x_i

value) is preferentially adsorbed at the free surface, this allows studies of systems with different effective surface interactions, while bulk miscibility is nearly unchanged.

Experimental

Samples were prepared by spin coating a film from toluene on a polished silicon wafer. When bilayer samples were required, a second film was spin coated onto mica and then float-mounted onto the first film. The samples were annealed in vacuum and subsequently quenched and stored at a temperature below the glass transition temperature (of 100°C for the PS samples, and ca. -40° C for the olefinic copolymers) until their composition-depth profiles could be analysed.

The composition-depth profile $\phi(z)$ of the samples following different annealing times was determined using $^2He(^3He,^4He)^1H$ nuclear reaction analysis (NRA). NRA is described in detail elsewhere(14, 15). In brief: a 900keV ^3He beam impinges on the polymer sample and the nuclear reaction $^3He + ^2H \rightarrow ^1H + ^4He + 18.35MeV$ takes place. The energy of the outgoing α-particles (^4He) reveals the deuterium distribution in the sample: the incident ^3He particles lose energy due to inelastic electronic processes as they penetrate the sample, so that the energy of the resulting α-particles depends on the depth in the sample at which the nuclear reaction occured. These α-particles lose additional energy on the way to the detector, and the overall energy spectrum can be related to the depth in which the reaction has taken place and to the local ^2H concentration at that depth. A schematic figure illustrating the geometry is shown in figure 3.

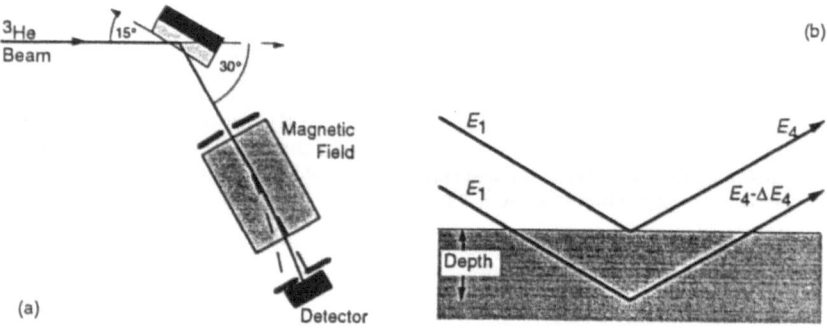

Figure 3: The geometry of the NRA used to profile the composition-depth characteristics of the deuterated polymer chains in these experiments. The magnetic field allows through only the required particles. (b) A magnified scale of the sample. The energy loss ΔE_4 reveals the depth at which the reaction occured.

3. DETERMINATION OF PHASE COEXISTENCE

In order to understand the nature of wetting from a binary mixture it is necessary to know the phase-coexistence characteristics. This was done for all the samples described in this study, using mainly the approach of reference (16). In this, a layer of one of the pure components (B say) is spin cast on the silicon wafer, and a similar layer of the other component (A) is laid on top. It is possible to ensure that surface enrichment or wetting effects do not perturb the final phase configuration; for example, by arranging for the surface preferred phase to be near the surface to start with. At temperatures above T_g the two components will interdiffuse, until the phases on either side of the original interface have reached their coexisting compositions ϕ_1 and ϕ_2 (of course, for $T > T_c$, *free* interdiffusion will take place). By determining these limiting coexisting compositions at different temperatures $T < T_c$, the phase coexistence diagram is determined. A typical profile for coexisting phases of the binary pair of olefinic copolymers designated d88 (x=0.88, e=0.37, N=1610)/h78 (x=0.78, e=0, N=1290) is shown in fig. 4. Also shown is the corresponding phase coexistence diagram, showing that for this pair the critical temperature is $T_c = 126\pm3°C$.

In addition to this method, it is also possible to determine the coexistence compositions ϕ_1 and ϕ_2 by making use of a 'single layer' approach. In this, a mixture of initial mean composition in the two phase region (for $T < T_c$) undergoes a phase separation where the each of the phases segregates to a different interface. The thickness of the respective phases is then regulated by the overall amount of the two components, and their compositions are simply the coexistence values for that temperature. Examples of this are shown in section 5.

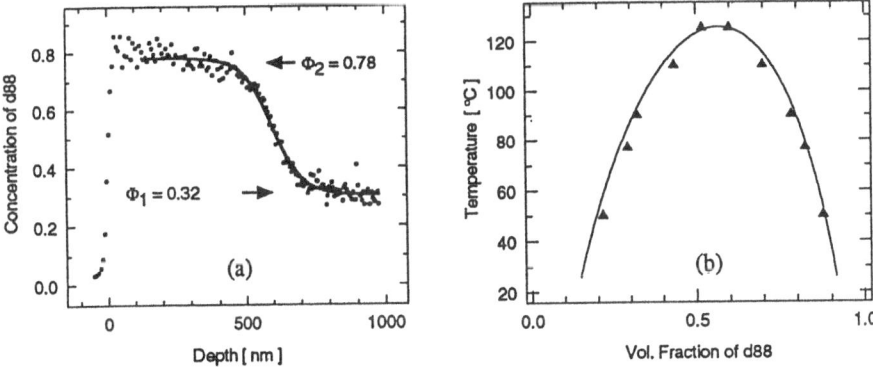

Figure 4: (a) Equilibrium composition-depth profile for the d88/h78 couple on silicon at 90°C, showing the coexisting compositions ϕ_1 and ϕ_2. (b) the corresponding phase-equilibrium diagram for this pair.

4. COMPLETE WETTING FROM THE ONE-PHASE REGION

We have studied the formation of a wetting layer of one polymeric phase in an equilibrium mixture from a coexisting composition of the other phase. To achieve this, we employed a "self-regulating" geometry: a film of the deuterated $EE_x E_{1-x}$ copolymer d88 is placed on the silicon substrate and covered with an h78 (x=0.78, e=0, N=1290) film. Initially, the d88-h78 interface broadens to a interfacial width comparable to the bulk correlation length; the materials partially interdiffuse until the two layers attain the composition of the coexisting phases ϕ_1 and ϕ_2, as described for determination of coexistence in the previous section. As d88 is preferentially adsorbed at the free surface, a d88 surface peak forms, as soon as there is a finite d88 concentration (ϕ) at the surface. As the d88 in the region adjacent to the surface approaches its value on the coexistence curve, the surface peak grows. For $\phi \approx \phi_1$, the surface layer attains a thickness which is macroscopic in the sense of being larger than the other length scales in the system (molecule size, correlation length). The layer-thickness continues to increase until the entire d88-rich phase is incorporated into the wetting layer, as expected in the case of complete wetting. This behaviour is reproduced in fig. 5, which shows the h78/d88 bilayer as prepared; after 30 minutes; and after 3 days at 110°C (16°C below the critical point of this mixture, $T_c=126°C$).

Complete phase inversion can be attained after long enough times. This shows unambiguously the complete wetting of the polymer-air interface by the phase rich in the d88 component as it grows from the phase rich in the h78 component. We are able to follow the growth of this layer in time - and we return to this point later in this lecture.

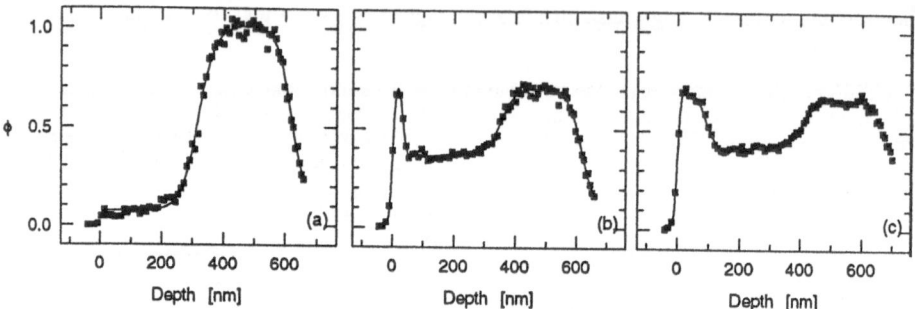

Figure 5. h78/d88 bilayers following annealing of t=0min (a), t=30min (b) and t=3days (c) at 110°C.

5. SURFACE SEGREGATION FROM THE TWO PHASE REGIME

In a second series of experiments thick monolayers of d86 (x=0.86, e=0.4, N=1520) and h75 (x=0.75, e=0, N=1267) mixtures were used with a d86 content varying from 5% to 40%. The phase coexistence diagram of this mixture shows Tc = 176±3°C. The depth profiles of samples of three initial compositions 20%, 30% and 40% and after annealing for 1day at 150°C are shown in Fig 2. For concentrations $\phi_\infty < \phi_1$, thermodynamic equilibrium is characterized by a peak with maximum d86 concentration $\phi_{max} < \phi_2$ and with a width smaller than the spatial extent of the polymer molecules. By analyzing the surface excess (the area under the peak) as a function of ϕ_∞, the surface interaction parameters for this mixture can be obtained(6).

For initial concentrations $\phi_{ini} > \phi_1$, the sample will undergo surface directed spinodal decomposition. Due to the surface activity of d86, the subsequent growth of the spinodal domains takes place in surface directed manner and at equilibrium the composition profile is characterized by a 'wetting layer' in contact with a "bulk" phase of composition ϕ_1. Similar experiments have been performed by Bruder and Brenn(17) on the mixture PS/brominated-PS.

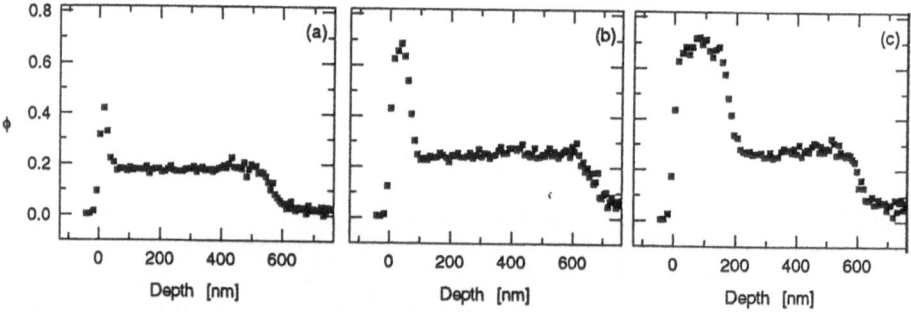

Figure 6. h75/d86 mixtures of initial d86 composition: (a) 20%, (b) 30% and (c) 40%, after 1 day at 150°C

It is important to note, however, that although the profiles in fig.6 (b) and (c) suggest that complete wetting has occured in a manner similar to the previous section, this is not necessarily the case. This is because the final configuration in these profiles is the one of lowest free energy (since the d86 component is the surface preferred one), irrespective of whether the system is in the complete wetting regime. This phenomenon is discussed in more detail elsewhere(18).

6. SURFACE SEGREGATION FROM ISOTOPIC MIXTURES

While wetting transitions in (small molecule) liquid mixtures have been reported for a number of different binary pairs close to their critical point, there are no reports of complete wetting in a case where the two components are chemically identical but differ only in their isotopic nature(19). In part this due to the fact that there are only very few non-polymeric isotopic mixtures which are immiscible at accessible temperatures. Nevertheless wetting from isotopic liquid mixtures presents a special interest: the bulk and surface interactions of isotopic mixtures can be modeled more easily than in systems where more complex chemical interactions are involved. The investigation of wetting from isotopic mixtures can thus be used to examine theoretical models of wetting transitions.

To study wetting from a binary isotopic mixture, we repeated the experiment outlined in the previous section, using a mixture of deuterated polystyrene dPS (N=17411) and its protonated analog hPS (N=17308). High molecular weight isotopic mixtures of polystyrene are known to phase separate(19), exhibiting an upper critical solution temperature. For this mixture, a critical temperature of ca. 221°C has been calculated. Fig. 7 shows single films of hPS/dPS mixtures, ca. 1.6 μm thick, containing 5%-40% of dPS, after annealing for 1 week at 218°C. While, for dPS concentrations < 30%, only a narrow layer enriched in dPS can be observed, there is a pronounced wetting-like layer at the free surface for samples containing more than 30% dPS.

It has to be pointed out, however, that this result is not obtained consistently every time the experiment is performed. This is most probably due, indirectly, to the low diffusion coefficient of this hPS/dPS system. While for the olefinic copolymers described in the previous sections D ~ 10^{-10} - 10^{-12} cm²/s, and the buildup of the wetting layer over ~100nm takes place on the time scale of 1 week, the diffusion coefficient for the hPS/dPS mixture is several orders of magnitude

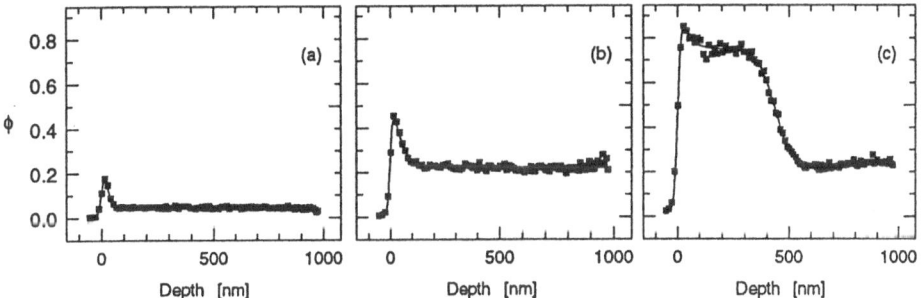

Figure 7 : hPS/dPS mixtures of initial dPS composition: (a) 20%, (b) 30%, (c) 40%, after 1week at 150°C

lower, D ≈ 10^{-19} cm²/s. For this reason, buildup of the wetting layer for the hPS/dPS mixture may in general take place over time scales which are beyond our experimental possibilities. However, NRA profiling of unannealed spin-cast films of this dPS/hPS mixture reveal non-homogeneous initial composition profiles, probably resulting from some variation in the spin-casting conditions. We believe that the reason why wetting-like layers, as shown in fig. 7, are not obtained reproducibly each time has to do with this. Films with a 'favorable' initial composition variation, that is, one where some of the surface- favoured (dPS) component is present at the surface already in the as-cast film, may reach equilibrium more quickly as compared to a homogeneous sample. Nonetheless, while this by no means constitutes a complete study of wetting from isotopic mixtures, our results suggest that wetting from isotopic mixtures may be obtained and studied.

7. DYNAMICS OF WETTING

In this section, we return to the results described in the section on wetting from the one phase region. As indicated in fig. 5, we are able to follow the build-up of the wetting layer with time on a time scale from 30min to 1 week. The width l of the wetting layer may be defined by the distance between the points of inflection of the profile at the air interface and at the interface where the wetting layer falls off. Experimentally, we find that the dependence of l on time t is best fitted by a logarithmic variation, $l \sim \log t$, as shown in figure 8(a).

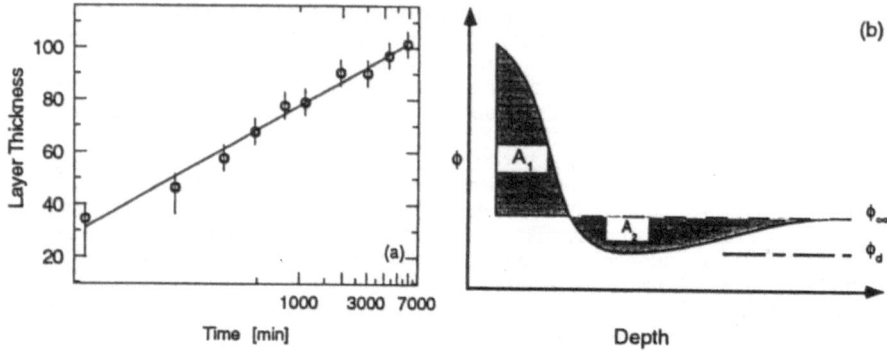

Figure 8 : (a) variation of layer thickness with time for the profiles of fig. 5. (b) schematic illustration of diffusion limited growth of a wetting layer.

While there is ample theoretical work on the equilibrium properties of wetting, there are few publications on the build-up of wetting layers. The discussion by Lipowsky and Huse[20], and by Jones[21], treats the case of a wetting layer of one phase growing at a surface by diffusion from a coexisting composition of the other. The following reproduces their argument, while the sketch in fig. 8b illustrates the geometry. The surface excess of the wetting layer $\Gamma = A_1$ is created by diffusion of of the surface active component from the region adjacent to the wetted surface. Conservation of material requires $A_1 = A_2$, the area of the depleted region from which the surface active component was incorporated into the wetting layer.

The surface excess Γ can also be expressed in terms of the thickness of the wetting layer $l \sim \Gamma$. The assumption here is that the wetting layer is in equilibrium with the concentration ϕ_d of the adjacent depleted zone. In that case the functional dependence of l on ϕ_d has been calculated by Cahn[3], for the case of $\phi_d \approx \phi_1$ ($\phi_d < \phi_1$):
$$l = -\text{const. } \ln(1-[\phi_d/\phi_1]) \qquad (3)$$
The prefactor in equn.(3) is of order of the correlation length in the mixture. Since the process is assumed diffusion controlled, the width of the depleted layer is of order $(Dt)^{1/2}$; thus the depleted area may be approximated as
$$A_2 \approx (Dt)^{1/2}(\phi_1 - \phi_d) \qquad (4)$$
Recalling $A_2 = A_1 = \Gamma \approx l$, we find, by eliminating ϕ_d from equns.(3) and (4), that at short times $l(t) \sim (Dt)^{1/2}$, while at long times
$$l(t) \sim \ln(t) \qquad (5)$$
While equn.(5) appears to be in agreement with our experimental result, it does not, unfortunately, accurately describe out experimental system. Because of the relatively high value of D and the finite thickness of the polymer films, the spatial extent of the depleted well $(Dt)^{1/2}$ very soon exceeds the width of the layer of composition ϕ_1 (fig. 1). This occurs already at the shortest times of measurement (ca. 30 minutes), so that the conservation conditions leading to equn.(5) are not valid in our system[22].

A more realistic description of the build-up of the wetting layer in our geometry has to take into account the finite size of our system, and in particular the "reservoir" (this is the phase of composition ϕ_2 next to the silicon substrate). It is likely that the growth-rate of the of the d88-rich wetting layer at the surface is limited by diffusion of material from this reservoir into the coexisting d88-poor phase adjacent to it (fig.5). A detailed description of wetting in such finite-sized systems has yet to come.

8. SUMMARY AND CONCLUSIONS

Because of the low interfacial energies between coexisting polymer phases even at temperatures far from the critical temperature, binary polymer mixtures should be good candidates to exhibit complete wetting behaviour, over a wide range of temperatures, at interfaces which favour one of the components. Using a high resolution composition-depth profiling method based on nuclear reaction analysis we have observed such wetting behaviour from mixtures of olefinic random copolymers, and also the formation of wetting-like layers from isotopic polymer mixtures. In particular, the large size and low mobility of polymeric chains allow convenient experimental studies of the spatial and time aspects of the wetting layers. Our results show that the growth with time t of such layers of thickness l from a finite size system obeys $l(t) \sim \ln t$, a slow variation which we believe is due to the diffusion-limited transport of surface active material in our system.

ACKNOWLEDGEMENTS

We thank the German Israel Foundation (GIF), the US-Israel Binational Science Foundation, the Ministry of Science and Technology (Israel) and the Commission of the European Communities for partial support of this work. This lecture was presented at the Osaka University Macromolecular Symposium on Ordering in Macromolecular Systems, and J.K. thanks the organisers for arranging a fruitful and very enjoyable symposium.

REFERENCES

1   Young T (1805) Phil. Trans. Roy. Soc. London 95:65
2   Schick M (1990) In: Charvolin J, Joanny JF, Zinn-Justin J (eds) Les Houches, session XLVIII, Liquids at Interfaces, pp. 419-497. North Holland, Amsterdam
3   Cahn JW (1977) J. Chem. Phys. 66:3667
4   de Gennes PG (1985) Rev. Mod. Phys. 57:827
5   Dietrich S (1988) In: Domb C, Lebowitz J (eds) Phase Transitions and Critical Phenomena. Academic Press, London
6   Schmidt I, Binder K (1985) J. Physique 46:1631
7   Flory PJ (1953) Principles of polymer chemistry, Cornell University Press, Ithaca
8   de Gennes PG (1979) Scaling concepts in Polymer Physics, Cornell University Press, London
9   Jones RALea (1989) Phys. Rev. Lett. 62:280
10   Bhatia QS, Pan DH, Koberstein J (1988) Macromolecules 21:2166
11   Steiner U, Klein J, Eiser E, Budkowski A, Fetters LJ (1992) Science 258:1126
12   Budkowski A, Klein J, Eiser E, Steiner U, Fetters LJ (1993) Macromolecules 26:3858
13   Graessley WW, Krishnamoorti R, Balsara NP, Fetters LJ, Lohse DJ, Schulz DN, Sissano JA (1193) Macromolecules 26:1137
14   Chaturvedi UK, Steiner U, Zak O, Krausch G, Schatz G, Klein J (1990) Appl. Phys. Lett. 56:1228
15   Klein J (1990) Science 250:640
16   Budkowski A, Steiner U, Klein J, Schatz G (1992) Europhysics Letts. 18:705
17   Bruder F, Brenn R (1992) Phys. Rev. Lett. 69:624
18   Steiner U, Klein J, Fetters LJ - to be published
19   Bates FS, Wignall GD (1986) Phys. Rev. Lett. 57:1429
20   Lipowsky R, Huse DA (1986) Phys. Rev. Lett. 57:353
21   Jones RAL - Personal communication.,
22   While there are no predictions for the case of wetting layer build-up for the precise situation corresponding to our geometry, there is a related theoretical study by Langer (Ann. Phys. (N.Y.), vol.65, p.53, 1971) which we recall in passing. This discusses how a series of one dimensional spinodal domains evolve with time. This situation bears some resemblance to our experiments in the following sense. Starting with an array of equidistant spinodal domains of the same width, every second domain grows while the domain adjacent to it shrinks, both with a logarithmic time dependence, while the

interdomain region at the coexisting composition retains a constant width. This is very reminiscent of the time-evolution of the adjacent domains in the profiles in fig. 5, though the driving force for the growth of the wetting layer differs from that of the spinodal decomposition.

(* ' - to whom correspondence should be addressed)

Interfacial Properties of Mixed Polymer Films Spread at the Air/Water Interface

M. Kawaguchi

Department of Chemistry for Materials, Faculty of Engineering, Mie University
1515 Kamihama, Tsu Mie 514 Japan

Abstract: Three binary mixtures of polymer films spread at the air/water interface have been investigated by surface pressure measurements and ellipsometry. Their compatibility and interfacial conformation are discussed as a function of the composition in the mixed polymer films.

INTRODUCTION

Monolayer films consisting of two or more film-forming compounds are also of much interest in many systems, such as biology, membranes, emulsions, and the food industry. The mixed films would be expected to provide much useful information regarding their orientation, packing, and interactions between the mixed compositions.

In most work on mixed films, the compatibility and incompatibility of the mixed compounds in the two-dimensional state is one of the most studies issues [1-2]. There are some criteria to determine either the compatibility or the incompatibility of the two compounds at the interface: (1) dependence of the mean area at a given surface pressure on molar fraction of one component in the mixture; (2) dependence of the collapse pressure on molar fraction; (3) dependence of the transition point on molar fraction.

Recently, with the aid of ellipsometry, second harmonic generation, fluorescence microscopy, and neutron reflection we would determine the structural properties of adsorbed polymer layers at the air/water interface [3,4]. Both fluorescence microscopy [5] and ellipsometry [3,6,7] are often used for studies of mixed monolayers. The former method is useful to investigate in situ the micro-morphology in the monolayer state and is able to reduce the molecular orientation in the monolayer state. On the other hand, the latter is one powerful technique to obtain the layer thickness and its refractive index of the binary mixed polymer spread at the air/water interface from two quantities of the phase difference, Δ and the azimuth, ψ of the amplitude ratio for light polarized parallel and normal to the plane of the incidence. In this paper, in order to investigate the interfacial properties of three binary mixed films, such as poly(ethylene oxide) (PEO)-poly(methyl methacrylate) (PMMA), poly(methyl acrylate) (PMA)-poly(vinyl acetate) (PVAc), and dipalmitoyl-phosphatidylcholine (DPPC)-poly-γ-methyl-L-glutamate (PMLG) mixtures, spread at the air/water interface, we used both techniques of surface pressure measurements and ellipsometry.

EXPERIMENTAL

Materials

One PEO sample with a narrow molecular weight distribution was purchased from Tosoh Co. Its molecular weight was determined to be 18 X 10^3 by light scattering. One fractionated PMMA samples, having a Mw = 280 x 10^3 as determined from the

A. Teramoto, M. Kobayashi, T. Norisuje (Eds.)
Ordering in Macromolecular Systems
© Springer-Verlag Berlin Heidelberg 1994

intrinsic viscosity measurement in benzene, was used. We employed a fractionated PMA with Mw = 589 x 10^3 and a fractionated PVAc sample with Mw = 300 x10^3. Their molecular weights were determined by intrinsic viscosity measurements in benzene for PMA at 25°C and in acetone for PVAc at 30°C. DPPC was purchased from Sigma Co. and it was used without further purification. Two PMLG samples were obtained by the polymerization of N-carboxy anhydride of L-gultamic acid γ-methyl ester. Their molecular weights were determined to be 10 x10^3 and 150 x 10^3 by intrinsic viscosity measurements in dichloroacetic acid at 25°C.

Spectrograde quality benzene and chloroform were used as the spreading solvent for PEO-PMMA and PMA-PVAc mixtures and for DPPC-PMLG mixtures, respectively. Deionized water, supplied from a Millipore Q-M system, was used in all experiments.

Surface Pressure Measurements

Surface pressures of the binary mixture films and their individual components spread at the air/water interface were determined by using the same instruments as previously used [3, 4, 6-9]. The polymer films were formed by the stepwise addition of the polymer solution using a microsyringe. The temperature of the water phase in a PTHF trough with a diameter of 15 cm placed in an aluminum water jacket was controlled to 25 ± 0.1°C by using circulating, thermostated water. The surface pressures were measured with a sandblasted platinum plate as a Wilhelmy plate hung on a Cahn 2000 electrobalance.

Ellipsometry

Our ellipsometer is based on conventional null-detection four zone methods, which have been reviewed in previous papers [8,9]. The precision in the readings of the goniometer mounted with the polarizer, the quater-wave plate, and the analyzer was 30 s. The experiments were preformed at a fixed incident angle of 70° by using monochromatic light with a wavelength λ = 546 nm. The changes in Δ and ψ between the clean water and the water surface covered with polymer films were expressed as $\delta\Delta = (\bar{\Delta} - \Delta)$ and $\delta\psi = (\bar{\psi} - \psi)$, respectively: $\bar{\Delta}$ and $\bar{\psi}$ are for the pure water, and Δ and ψ are for the water surface covered with the polymers.

RESULTS AND DISCUSSION

PEO-PMMA Mixtures

We performed the surface pressure measurements and ellipsometry of the binary mixtures of PEO and PMMA for six mixtures of 0.95/0.05, 0.90/0.10, 0.82/0.18 , 0.69/0.31, 0.36/0.64, and 0.22/0.78, of which these ratios of PEO and PMMA are expressed in monomer molar fraction. Figure 1 shows typical surface pressure-area (π -A) isotherms for the binary mixtures of PEO and PMMA, together with the isotherms of PEO and PMMA homopolymers. The isotherms at higher PEO contents have plateau regions around 10 mN/m, which corresponds to the collapse pressure of the PEO film. Below the plateau regions the surface pressure at the same area decreases with an increase in the PMMA composition. Above the plateau surface pressure, on the other hand, the composition dependence of the surface pressure is quite the reverse. According to the criteria for determination of the compatibility of the two compounds at the air/water interface, both the negative deviation from the additive line and dependence of the collapse pressure on the composition indicate that PEO-PMMA mixtures are compatible.

Ellipsometry of the mixed films shows that the $\delta\Delta$ values are much more sensitive to the spread amounts of both polymers than the $\delta\psi$ values: the $\delta\Delta$ values increase with

an increase in the spread amounts of polymer, whereas δψ values are less than 0.01°
over most surface concentrations for all compositions.   Thus, it is impossible to
calculate both the thickness and the refractive index of the adsorbed layer using the
Drude equation and the two values of Δ and ψ.

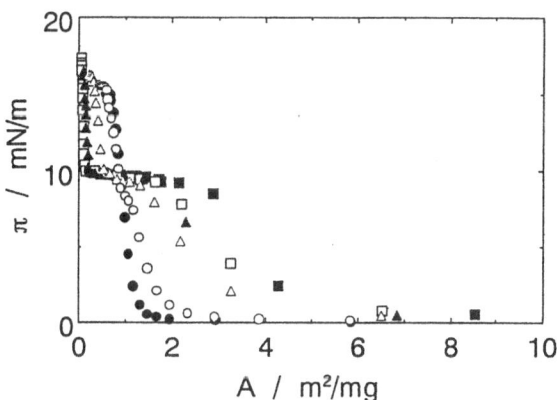

Fig.1.   π-A isotherms for mixtures of PEO and PMMA: PEO (■); PEO/PMMA =
0.95/0.05 (□); PEO/PMMA = 0.82/0.18 (▲); PEO/PMMA = 0.69/.031 (△); PEO/PMMA =
0.22/0.78 (○); PMMA (●) .

Typical plots of δΔ as a function of surface concentration of PMMA, Γ in the mixture
are given in Figure 2 for the PEO-PMMA mixtures of 0.95/0.05 , 0.69/0.31, and
0.22/0.78 , respectively, together with those of the PEO and PMMA homopolymers.
The δΔ values depend on the composition in the mixtures.   For the mixtures with
higher PEO contents , there is a kink around at Γ = 0.6-1.0 mg/m², where the collapse
surface pressure of PEO is observed.   For all the mixtures the δΔ values have a
plateau value above Γ = 1.7 mg/m² and the plateau δΔ is larger than that of the PMMA
homopolymer.

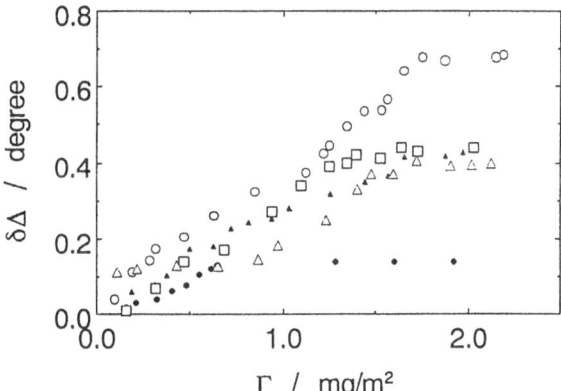

Fig. 2.   Plots of δΔ as a function of PMMA concentration in the PEO-PMMA mixtures
with molar ratio of 0.95/0.05 (△), 0.69/0.31 (○), and 0.22/0.78 (□) and concentration of
the PEO (●) and PMMA (▲) homopolymers.

For the mixtures with lower PEO contents, the $\delta\Delta$ values are in agreement with the simple addition of the $\delta\Delta$ for the individual polymers at a given concentration. This simple addition of $\delta\Delta$ is interpreted by a theoretical reason, that is, $\delta\Delta$ is sensitive to mass density at the interfaces and should be directly additive in the mixtures when both polymers are adsorbed at the air/water interface without desorption of the whole polymer chain. Since the concentration of PEO even at $\Gamma = 2$ mg/m^2 is lower than 0.7 mg/m^2, which corresponds to the onset of the full surface coverage of PEO homopolymer, PEO molecules also contribute to the changes in $\delta\Delta$ without its desorption from the water surface.

For the mixtures of the higher PEO contents, on the other hand, the values of $\delta\Delta$ at the higher Γ, where the PEO contents in the mixtures are above the threshold of full surface coverage of PEO homopolymer, decrease with the PEO contents, regardless of the high total amounts of PEO and PMMA. This result may be interpreted by the similar reason why the $\delta\Delta$ values for the PEO homopolymer do not change above the onset of the fully saturated PEO surface. In other words, PEO molecules do not significantly contribute to the changes in $\delta\Delta$ at the higher percent PEO mixtures due to the desorption of PEO from the air/water interface. Moreover, for the PEO-PMMA mixture of 0.95/0.05, the values of $\delta\Delta$ at the lower Γ are nearly consistent with the plateau value of the pure PEO.

PMA-PVAc Mixtures

Though PMA differs from PVAc only in the reversed position of the ester group, the PMA film had less expansion at the wide surface area than PVAc. This discrepancy was interpreted by the difference in the surface orientations, which are from the molecular model, at the air/water interface under the assumption of a flat conformation [10]. In order to prove that the less expanded-type PMA makes a more compact film layer on the water surface than PVAc, we used ellipsometry to yield the adsorbed layer thickness. Figure 3 shows the $\delta\Delta$ and $\delta\psi$ values for PVAc and PMA films as a function of concentration. The $\delta\Delta$ values for both polymers almost lineally increase with an increase in concentration. This concentration dependence is different from that for PEO and PMMA, whose $\delta\Delta$ values reach a plateau value above the concentrations where their surface pressures become constant. Since $\delta\Delta$ value

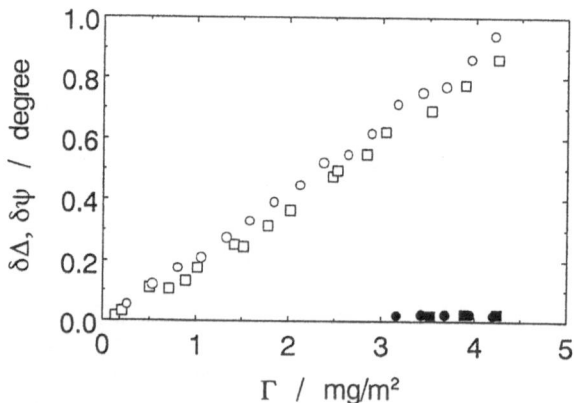

Fig. 3. Plots of $\delta\Delta$ and $\delta\psi$ as a function of concentration, Γ for PMA ($\delta\Delta$, $\bigcirc$; $\delta\psi$, ●) and PVAc ($\delta\Delta$, $\square$; $\delta\psi$, ■).

is a measure of mass density, which stems from the spread amount on the water surface, the continuous linear increase in $\delta\Delta$ with concentration means that both polymers are adsorbed at the air/water interface without any desorption of the whole polymer chain from the interface or any dissolution into the water phase.

On the other hand, the changes in $\delta\psi$ are less sensitive to those of the $\delta\Delta$. Above the concentration where both polymers are well in the plateau surface pressure, the value of $\delta\psi$ undoubtedly exceeds the precision of ψ for both polymers. Thus, one is able to calculate the thickness (t) and the refractive index (n_f) for both polymer films from Δ and ψ values under the assumption of a homogeneous layer. The thickness and refractive index for both polymers so calculated are 3.5±0.3 nm and 1.450±0.01, respectively, independent of the concentration. Unfortunately, it was found to be impossible to bring out the difference in the thickness between the films of PMA and PVAc spread at the air/water interface by ellipsometry.

Figure 4 shows π - A isotherms of the mixtures of PMA and PVAc for three mixtures of 1/4, 1/1, and 4/1, of which these ratios of PMA and PVAc are expressed in molar fraction, respectively. The mean surface areas at constant pressures almost fit on the additive line, which indicates that PMA and PVAc mixtures are ideally miscible or completely immiscible at the air/water interface from the criteria for determination of the compatibility of mixed films at the air/water interface. On the other hand, invariance of the collapse surface pressure on the compositions in the mixture was observed . Thus, from only the surface pressure measurements we can not lead to a conclusion for the compatibility of PMA-PVAc mixtures at the air/water interface. We will discuss again the compatibility of PMA-PVAc mixtures by taking into account the ellipsometric measurements described below.

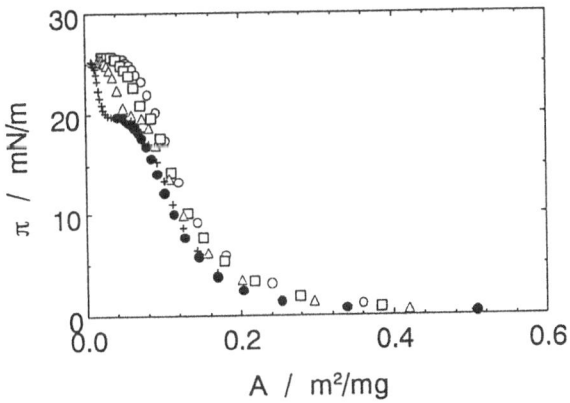

Fig. 4. π-A isotherms of PMA ($\bullet$), PVAc (o), 1/4 mixture of PMA-PVAc ($\square$), 1/1 mixture of PMA-PVAc ($\triangle$), and 4/1 mixture of PMA-PVAc (+).

Figure 5 displays the values of $\delta\Delta$ and $\delta\psi$ for the three mixtures as a function of PVAc concentration, Γ_{PVAc} in the mixture. The values of $\delta\Delta$ almost linearly increase with an increase in Γ_{PVAc} in the entire concentration range. Also the data points almost fit on the straight line calculated from the simple addition of $\delta\Delta$ for the individual polymers at a given surface concentration. Thus, the reason why the simple addition of $\delta\Delta$ stands up for the PMA and PVAc mixture films may result from their water-insoluble and strongly adsorbable character.

The δψ values for the three mixtures, which exceed the precision of ψ, increase with increasing Γ_{PVAc}. At the same Γ_{PVAc} the δψ value increases with an increase in total spread amount of PMA and PVAc, and this dependence is similar to the δΔ values. Thus, we can obtain the values of t and n_f for the mixtures and the results are illustrated as a function of total concentration in Table 1, in which errors in the t and n_f values are ca. 10 %. If both polymers are ideally compatible, we expect that

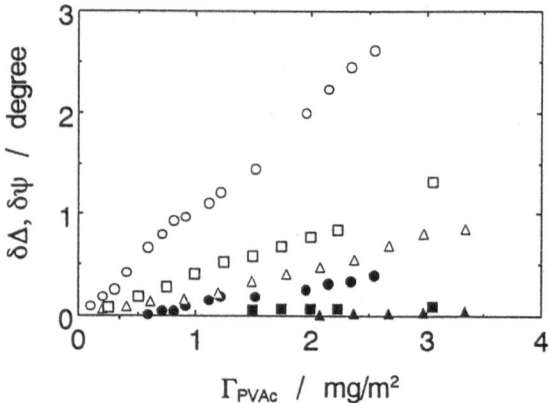

Fig. 5. Plots of δΔ and δψ for 1/4 mixture of PMA-PVAc (δΔ, △; δψ, ▲), 1/1 mixture of PMA-PVAc (δΔ, □; δψ, ■), and 4/1 mixture of PMA-PVAc (δΔ, ○; δψ, ●).

Table 1. Typical thickness (t) and refractive index (n_f) for PMA-PVAc mixtures as a function of total concentration Γ.

Samples (PMA/PVAc)	Γ mg/m²	t nm	n_f
1/4	2.96	3.8	1.417
	3.33	3.3	1.450
	3.70	5.5	1.417
	4.14	5.9	1.417
1/1	3.05	10.7	1.371
	3.56	12.5	1.371
	4.07	9.9	1.384
	4.57	11.0	1.376
	6.11	10.5	1.409
4/1	2.94	2.2	1.483
	3.44	5.5	1.417
	3.93	6.4	1.417
	4.42	12.6	1.384
	4.91	15.9	1.380
	5.89	23.1	1.371
	9.82	22.3	1.392
	10.70	24.2	1.394
	11.78	22.6	1.403

thickness and refractive index of a mixed polymer film should be independent of the mixture ratio at the same total amount of the two components in the mixture. However, a comparison of ellipsometric data for the mixtures shows that the thickness and refractive index of the mixed layer at the same total concentration depends on the mixed ratio. Furthermore, an increase in PMA in the mixtures further induces the repulsion forces between PMA and PVAc and thus it results in a thick and diluted film at the air/water interface. The repulsion forces may stem from the difference in the interfacial properties between PMA and PVAc; that is, PMA forms a more condensed film at the air/water interface than PVAc. Thus, these ellipsometry data lead to a conclusion that both polymers seem to be immiscible rather than ideally compatible.

Compatibility of PMA and PVAc at the air/water interface will be further discussed in the light of data from different preparation methods of the mixed films. We examined the effect of order of addition on the ellipsometric parameters: PVAc was first spread at concentrations of 0.5 and 1.0 mg/m^2 where PVAc was above and below the limiting area (1.75 m^2/mg) and then PMA was added to adjust the PMA-PVAc mixed ratios to 1/4, 1/1, and 4/1 and vice versa. The $\delta\Delta$ value for the separate spreading was quite different from that for the simultaneous spreading at the same concentration: irrespective of the order, however, the $\delta\Delta$ values were reversible, and at the PVAc concentration of 1.0 mg/m^2, the value of $\delta\Delta$ was almost equal to that for PVAc film. These results of the separately spread films seem to be sufficient evidence for the immiscibility of PMA and PVAc at the air/water interface.

DPPC-PMLG Mixtures

The plots of the mean areas at constant pressures as a function of molar fraction of PMLG show a negative deviation from the additive line and the collapse pressure and the transition point due to DPPC depend on the composition in the mixture. This means that the mixtures are compatible at the air/water interface according to the criteria for the miscibility rule of the mixed films. In fact, since naturally occurring cellular membranes mainly consisting of proteins and phospholipids are thermodynamically miscible [2], it is expected that artificial membrane-like mixtures of lipids and polypeptides would be compatible.

The $\delta\Delta$ values for the DPPC and PMLG increase with an increase of Γ and attain the plateau value above a surface concentration where the surface pressure is well in the collapse pressure, respectively. Similar results for DPPC were observed by Ducharme et al. [11]. The $\delta\Delta$ values of two PMLG samples are almost independent of the molecular weight within the experimental errors. Since the $\delta\Delta$ value is a measure of mass density and should be proportional to the amount adsorbed at the interface, the fact that the $\delta\Delta$ values for both samples remain constant for a further increase in concentration indicates desorption of some portions of the spread DPPC and PMLG from the air/water interface.

Figure 6 shows $\delta\Delta$ as a function of the concentration of DPPC in the mixtures with 4/1, 1/1, and 1/4 molar ratios of DPPC and PMLG, which are prepared by spreading chloroform solutions of the DPPC and PMLG mixtures (simultaneous spreading). Data points almost linearly increase with the surface concentration and then they gradually level off. A surface concentration where the $\delta\Delta$ values reach a plateau, deceases with an increase in the component of PMLG in the mixture and the plateau value of $\delta\Delta$ increases.

Dashed lines drawn in Figure 6 correspond to $\delta\Delta$ values calculated from the weighed sum of $\delta\Delta$ for DPPC and PMLG at a given concentration by assuming an

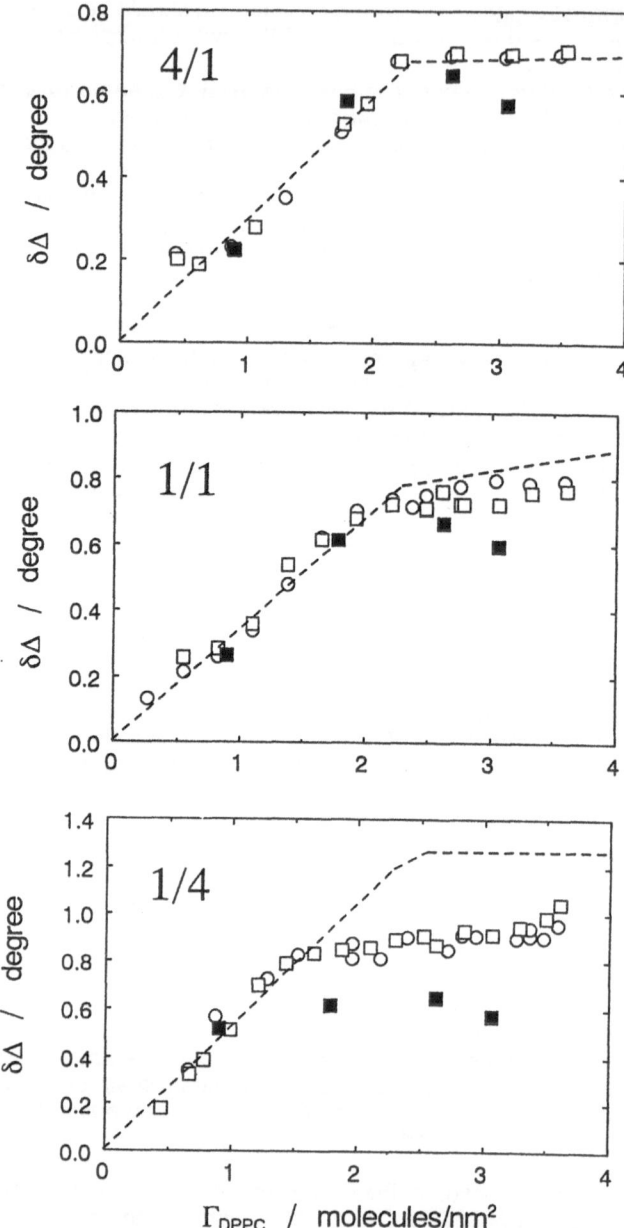

Fig. 6. Plots of δΔ the 4/1, 1/1, and 1/4 mixtures of DPPC and PMLG as a function of concentration of DPPC in the mixtures: molecular weight of PMLG, Mw = 10 x 10³ (○); Mw = 150 x 10³ (□). The filled squares indicate the δΔ for the separate spreading. The dashed line corresponds to the weighed sum of δΔ.

approximately linear increase in the δΔ values of each component with its concentration up to the plateau value. Some portions of the calculated line are in agreement with the observed δΔ values for all mixtures and this means that all spread amounts are really adsorbed at the air/water interface without desorption because of the miscibility of DPPC and PMLG in the monolayer state reduced from the surface pressure measurements. For the 4/1 mixture of DPPC and PMLG, the observed δΔ value is coincident with the calculated one in the entire concentration range. However, a negative deviation of δΔ from the calculated line is observed for the 1/1 and 1/4 mixtures of DPPC and PMLG and the difference in δΔ between the observed and calculated ones increases with an increase in PMLG component in the mixture. Such a negative deviation indicates that some portions of the really spread amounts are desorbed from the air/water interface.

We also examined the effect of order of addition (separate spreading) on the ellipsometric parameters and compared them with the results for the simultaneous spreading. DPPC was first spread at a given concentration (DPPC concentrations of 0.89, 1.78, 2.62, and 3.06 molecules/nm^2) and then PMLG was added to adjust the mixed ratios of 4/1, 1/1, and 1/4 of DPPC and PMLG. The δΔ values for the separate spreading are displayed as a filled square in Figure 6 and we notice interesting features in the figures. At a DPPC concentration of 0.89 molecules/nm^2 the δΔ values for the respective mixtures are almost coincident with that for the simultaneous spreading, indicating that PMLG and DPPC are adsorbed at the air/water interface without desorption. With an increase in the first spreading concentration of DPPC the negative deviation of δΔ from that for the simultaneous spreading is observed and a DPPC concentration where the negative deviation is first observed, is lower with an increase in PMLG component in the mixtures. Except for the DPPC concentration = 0.89 molecules/nm^2 the δΔ value reaches $0.6 \pm 0.02°$ for irrespective of the mixture.

The other ellipsometric parameter δψ was less sensitive to the changes in the addition of spread amounts than the δΔ value and the reliable data were obtained above a concentration where the δΔ attains the plateau value. A layer thickness of 2.5 ± 0.3 nm and a refractive index of 1.475 ± 0.005 were obtained for DPPC at the plateau. The thickness is not far from a thickness of 2.2 nm for DPPC in the solid state monolayer [11]. For PMLG films a layer thickness of 2.0 ± 0.3 nm and a refractive index of 1.475 ± 0.005 were given at the plateau. Such a thin layer thickness indicates that PMLG chains have α-helical conformation with its helical axis laid on the water surface. The thickness so calculated is almost coincident with a thickness of 2.2 and 1.9 nm determined from an electron microscope [12] and ellipsometry [13] for the LB films of PMLG, respectively.

On the other hand, a thickness of the DPPC/PMLG mixtures prepared by the simultaneous spreading was obtained in the range of 4.0 - 5.0 nm with the refractive index range of 1.475 ± 0.005 at the plateau for the respective mixtures. The thickness is larger than that for each component and roughly equal to the sum of the thicknesses of DPPC and PMLG. The thicknesses of the mixtures prepared for the separate spreading at higher DPPC concentrations than 1.78 molecules/nm^2 were obtained 2-3 nm with the refractive index range of 1.475 ± 0.005. Thus, from such layer thickness and refractive index, it is difficult to determine whether DPPC or PMLG is preferentially adsorbed at the air/water interface.

CONCLUSIONS

We have demonstrated the interfacial properties of three binary mixed polymer films spread at the air/water interface by use of surface pressure measurements and ellipsometry. From the surface pressure measurements, the PEO-PMMA and the DPPC-PMLG mixtures were found to be compatible, whereas for the PMA-PVAc mixtures we could not reach a conclusive determination of their compatibility. However, ellipsometry led to the PMA-PVAc mixtures incompatible in terms of a composition dependence of the thickness and a comparison of the separate and simultaneous spreadings. For the PEO-PMMA mixtures, the thickness and refractive index could not be extracted from ellipsometry. For the PMA-PVAc mixtures, at the same total adsorbed amounts the thickness increases and the refractive index decreases with an increase in PMA composition in the mixtures. For the DPPC-PMLG mixtures, the thickness was ca. 4.5 nm, independent of the mixed ratio and PMLG mass.

REFERENCES

1 Gains JrGL (1966) Insoluble monolayers at liquid-gas interface, Plenum, New York
2 Birdi KS (1989) Liquid and biopolymer monolayers at liquid interfaces, Plenum, New York
3 Swalen JD, Allara DL, Andrade JD, Ahandross EA, Garoff S, Israelachvili J, McCarthy TJ, Murray R, Pease RF, Rabolt JF, Wynne KJ, Yu H (1987) Langmuir 3:392
4 Kawaguchi M (1993) Prog Polym Sci 18: in press
5 Ahlers M, Muller W, Reichert A, Ringsdorf H, Verzmer J (1990) Angew Chem Int Ed Engl 29: 1269
6 Nagata K, Kawaguchi M (1990) Macromolecules 29:3957
7 Kawaguchi M, Nagata K (1991) Langmuir 7:1478
8 Kawaguchi M, Tohyama M, Mutoh Y, Takahashi A (1988) Langmuir 4:407
9 Kawaguchi M, Tohyama M, Takahashi A (1988) Langmuir 4:411
10 Crisp DJ (1949) J Colloid Sci 1:49
11 Ducharme D, Max J-J, Salesse C, Leblanc RM (1990) J Phys Chem 94:1925
12 Takeda F, Matsumoto M, Takenaka T, Fujiyoshi Y (1981) J Colloid Interface Sci 84:220
13 Baglioni P, Dei L, Gabrielli G (1983) J Colloid Interface Sci 93:402

Polymer Monolayer Dynamics

Hyuk Yu

Department of Chemistry, University of Wisconsin
Madison, Wisconsin 53706 U.S.A.

Abstract: Polymer chain dynamics at interfaces of vapor/liquid and liquid/liquid have been probed by confining macromolecules as monolayers or thin films and examining their field induced and spontaneous capillary waves in addition to their lateral diffusion and in-plane steady shear viscosity, all in a Langmuir trough in conjunction with simultaneous surface pressure measurements. Various experimental methods have been used to examine the chain dynamics and this paper deals with two of such, namely surface quasielastic light scattering and surface canal viscometry.

Focus has been place on thin film viscoelastic properties of monolayers as examined by retardation effects of the surface films on the propagation and damping characteristics of the capillary waves. With the aid of a dispersion equation, two viscoelastic parameters, i.e., the surface dilation elasticity ε and the corresponding viscosity κ, are deduced from the observed spatial wave length or wave propagation rate and the wave damping constant. By virtue of ε and κ representing short-ranged, local packing states and the attending dynamics of polymeric monolayers, they are independent of molar mass. Thus we have chosen to examine structural parameters of some vinyl polymers together with polyethers and poly(dimethyl siloxane).

Parallel with this effort to understand the monolayer dynamics in terms of structural parameters, we present a very recent endeavor in determining steady shear viscosity of monolayers by observing the surface flow rate of monolayers through a narrow canal on a Langmuir trough, rather analogous to capillary viscometry under a constant driving head. Here the aim is to confirm the hydrodynamic coupling model of Harkin and Kirkwood, and probe the molar mass dependence of the in-plane steady shear viscosity η_S of a vinyl polymer, poly(t-butyl methacrylate)

INTRODUCTION

Polymers at interfaces are a new field of research in polymer physics. A good deal of the activity is directed toward ordered molecular films(1, 2), which are in a thickness range from a monomolecular depth to several hundred nanometers, in part because of their technological potential and of new scientific questions they elicit. Electronic and optical devices presently incorporate structures that are in this thickness range or approaching it, and organic thin films have been proposed to replace both passive and active components traditionally fabricated with

A. Teramoto. M. Kobayashi, T. Norisuje (Eds.)
Ordering in Macromolecular Systems
© Springer-Verlag Berlin Heidelberg 1994

other materials. Basic research into molecular interactions in thin film structures leading to an understanding of the collective properties of ordered arrays have only recently been possible with organic films, characterized in more details by a number of new surface science techniques. In most of these molecular film studies, uses of polymeric films as structurally robust alternatives to simple amphiphiles seems to be rather commonplace(1).

The area of Langmuir-Blodgett (L-B) films, monolayer structures transferred from the air-water interface (henceforth written A/W for short) to a solid substrate, has now passed the initial stage of enthusiasm world wide. On the other hand, its fundamental appeal has not diminished while the earlier vision of prompt fabrication of technologically important devices has not materialized. The appeal is the facile manner with which a single monolayer or multilayers can be deposited onto substrate surfaces. There still appears to be a long range optimism in the attempts to fashion these films toward the uses in electron beam-, x-ray- and photo-resists(3), insulation and passivation layers, high density optical recording devices, non-linear optical components and permselective membranes(4). Particularly noteworthy are novel classes of topotactic polymerizations and chemical reactions on A/W and L-B films(5-8).

In the context of L-B films, our activity has been centered around the problems of monomolecular films in situ at A/W, prior to their transfer to solid substrates. We have been focusing on the chain dynamics in monolayer state on A/W and the interface of oil-water(O/W). Aside from the impacting technologies attending these activities, there emerge intriguing new scientific issues. These have to do with the elucidation of the two-dimensional state of the chain configuration via the scaling laws, introduced to the field by the French school of des Cloiseaux(9), Daoud and Jannink(10), and de Gennes(11). Until recently, the theoretical tools were limited(12) to interpret the chain conformations in the "two-dimension" that can be realized by adsorbing surface active polymers on A/W as monolayers. More recently a new generation of experimental studies, mainly static surface pressure measurements, have appeared to test positively the scaling laws in the two-dimension(13). Thus the time seems ripe in focusing on the two-dimensional character of polymer chains in order to provide the basic science support for crucial development of emerging technologies with these molecular films. The pivotal issue we have confronted so far is; what is meant by the two-dimensional character of polymeric configurations at A/W and O/W, and how we can go about probing the chain dynamics in such a state as well as how to detect any departure from such a state?

Monolayer Dynamics

Many common vinyl polymers, mostly insoluble in water, are amphiphilic in character by virtue of their backbone structure being hydrophobic while the side chains being hydrophilic. Alternatively, the backbone structure itself is amphiphilic as in polyethers such as poly(ethylene oxide) and poly(tetrahydrofuran). These polyethers and poly(vinyl esters) such as poly(vinyl acetate), and polyacrylates and polymethacrylates all form stable monolayers on A/W. With a considerable literature accumulated on the static surface pressure studies on polymer monolayers for the past four decades or more, its resurgence in the past ten years in the context of the 2D scaling models has been most remarkable indeed. Parallel with this development, studies of dynamic properties of these monolayers by means of the surface quasielastic light scattering (SLS) have begun at about the same time, initiated by Langevin and coworkers[14], and its application to various monolayers of the conventional amphiphiles such as fatty acids and phospholipids was started in earnest in 1977[15] although the inception of the method with pure liquids surfaces could be traced to Katyl and Ingard as early as 1967[16]. The method is still macroscopic in character in deducing viscoelastic parameters of monolayers on A/W and O/W with a spatial wave length range of 10-100 µm, and it relies on the mode coupling characteristics[17] of the spontaneously induced, transverse capillary waves to the longitudinal compressional waves. Thus, the parameters are the dynamic longitudinal elasticity ε_d and the corresponding viscosity κ. The method have been implemented in our laboratory since 1984[18], and initially we have begun with the studies of simple amphiphiles[19-20] to make contact with the studies performed earlier by others; the references to them can be found in Langevin's review[14]. While fully quantitative interpretations of results to date have been elusive, we have been able to identify in some cases the dynamic elasticity ε_d to the static compressional modulus deduced from the surface pressure-area isotherms. Thus the long wavelength limit of the light scattering experiment is firmly established, hence the dilational viscosity κ must also be regarded as a macroscopic viscoelastic parameter.

In analogy with the viscoelasticity in bulk state, the above referred dilational elasticity and viscosity are expected to represent the local packing dynamics, hence they would not depend on the molar mass of polymers forming the monolayers. Thus, a more surgical probe for the conformational dynamics should be chosen to correlate chain dimensions to dynamic parameters. For this purpose, we have chosen the in-plane steady shear viscometry technique, which has been around for some time but hitherto been fully implemented for polymeric monolayers.

TECHNIQUES

Surface Quasielastic Light Scattering

The starting literature should begin with a book by Levich[21], titled "Physicochemical Hydrodynamics", which gives the hydrodynamic basis for the thermal capillary waves. Essential elements of the governing hydrodynamics are the linearlized Navier-Stokes equation with surface boundary conditions. In addition to the transverse capillary wave mode, there are two other surface wave modes, i.e., lateral dilational-compressive mode and lateral shear mode.

Basic physics of the technique parallels that of the sound wave propagation in bulk liquids due to spontaneous density fluctuations, giving rise to the Brillouin doublet in addition to the Rayleigh central peaks in the scattered power spectrum[22,23]. Arising from the same mechanism, there exist spontaneous sound waves propagating on the surfaces of simple liquid/vapor interfaces. The transverse waves relative to the surface plane are called thermal capillary waves or ripplons. Hence, from such surfaces, the light scattering is predominantly contributed by the capillary waves because of its dielectric permittivity fluctuations being the largest. Hence, the light scattering spectrum from the thermal capillary waves on simple liquid interfaces gives two observable parameters, Doppler shifted peak with a frequency f_S, which signifies the propagation rate of a given capillary wave, and the spectral width of the peak Δf, which stands for the temporal damping rate of the capillary wave. From the first, f_S, one deduces the surface tension of the liquid since it corresponds to the square root of the ratio of surface tension to subphase liquid density ρ, and from the second, Δf, the shear viscosity of subphase liquid since it corresponds to the kinematic viscosity η/ρ. Analogy with the Brillouin spectrum of bulk liquids is now complete. Choice of a particular capillary wave with the spatial wave number k is specified by the scattering angle. Under certain conditions where k is large enough such that the gravity wave contribution can be neglected, f_S goes as $k^{3/2}$ and Δf as k^2. These conditions are called the Kelvin and Stokes limits, respectively:

$$f_S = (\sigma/\rho)^{1/2} \cdot k^{3/2} \tag{1}$$

$$\Delta f = (2\eta/\rho\pi) \cdot k^2 \tag{2}$$

Once the liquid surface is covered with monomolecular layers or thin films, the propagation characteristics of the capillary waves are now influenced by other surface waves that can couple efficiently with the capillary waves. The situation then becomes more interesting since the light scattering spectrum could be used to probe the surface viscoelastic properties of these layers or films. In particular, the capillary wave mode couples efficiently with the dilational-compressive wave mode on the liquid/vapor interfaces, whereby we can deduce the elastic modulus and the corresponding viscosity of the wave mode under certain simplifying assumptions. We illustrate the situation schematically in Fig. 1.

Air/Water: The dispersion relation given below describes the capillary wave motions at A/W (24) and is used to extract film viscoelastic properties from the SLS spectra. It is represented as

$$[\eta(k+m) - \varepsilon^* k^2/i\omega] \, [\eta m(k+m)/k - \sigma^* k^2/i\omega] = \eta^2 (k-m)^2 \tag{3}$$

where η = shear viscosity of bulk water, $k = (2\pi/\lambda)\sin\phi\cdot\cos\theta$, λ = laser wavelength, ϕ = scattering angle, θ = incident angle measured from the surface normal, $m = [k^2 - i\omega\rho/\eta]^{1/2}$, $\omega = 2\pi i \, (\Delta f_{s,c}/2)$, f_s = frequency shift (or peak frequency), $\Delta f_{s,c}$ = instrument width corrected frequency full width at half-height, ρ = density of water, $\varepsilon^* = \varepsilon - i\omega \kappa$, ε = surface dilational elasticity, which was referred to earlier as the dynamic elasticity with a subscript d but we will henceforth write without the subscript for brevity, κ = surface dilational viscosity, $\sigma^* = \sigma - i\omega\mu$,

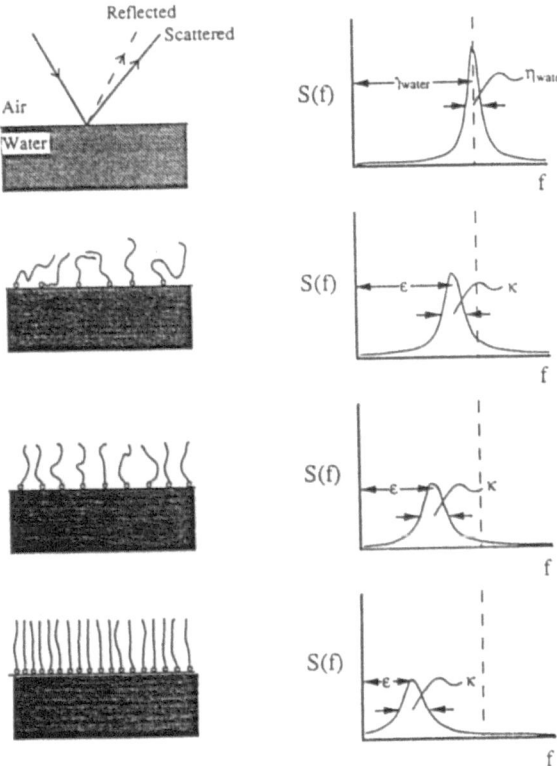

<u>Figure 1:</u> A schematic representation of surface light scattering power spectra, from which one can deduce surface elastic or viscoelastic properties, depending on whether the surface is monolayer covered or not.

σ = surface tension, and μ = surface transverse viscosity. The difference between the left-hand side and the right-hand side of Equation (3) is set equal to zero to solve the equation for the parameters of interest. For simple liquids without any surface films, this equation is simplified because $\epsilon^* = 0$ and $\mu = 0$. Hence, the two observable quantities of the power spectrum, f_s and $\Delta f_{s,c}$, are directly related to σ/ρ and η/ρ. For film-covered surfaces, the situation becomes more interesting as alluded to earlier because the longitudinal(dilational-compressive) waves on the surface couple with the transverse capillary waves. Although the longitudinal waves make a negligible contribution to the light scattering intensity, they can affect the propagation characteristics of the transverse capillary waves. The longitudinal waves are controlled by the film parameters ϵ and κ, both of which have shear components that cannot be separated from the pure dilational components by SLS. A frequency-dependent term in σ^* has been defined as the surface transverse viscosity μ. Goodrich (25) has earlier provided a simple physical picture of μ.

<u>Oil/Water:</u> A SLS study on an oil/water interface was first performed in our laboratory with spread films of a phospholipid(26). It was found that the light scattering spectra were completely different from those obtained at A/W relative to the damping behavior whereas the film elastic

properties were rather comparable at both interfaces. Some years ago Lucassen-Reynders and Lucassen(29) have already seen a large difference between A/W and O/W with the induced capillary wave motions for adsorbed films. The governing dispersion relation at O/W (24) parallels that of A/W and is given below using the notation of Löfgren et al.(27)

$$[\eta_w(k + m_w) + \eta_d(k + m_d) - \varepsilon^* k^2/i\omega] \times$$
$$[\eta_w m_w(k + m_w)/k + \eta_d m_d(k + m_d)/k - \sigma^* k^2/i\omega] =$$
$$[\eta_w(k - m_w) - \eta_d(k - m_d)]^2 \tag{4}$$

where subscript w refers to water, subscript d refers to the oil phase, and the other variables are as defined after Equation (3). Solving Equation (4) with the input parameters of η_w, η_d, ρ_w, ρ_d, k, and σ(here it stands for the static interfacial tension), one obtains the dependence of f_s and $\Delta f_{s,c}$ on ε at different values of κ and μ. Because the densities and viscosities of the two phases are comparable at O/W, there exists negligible coupling between the longitudinal dilational waves and transverse capillary waves. In the most sensitive case, i.e., $\kappa = 0$, $\Delta f_{s,c}$ increases only by about 8% as the coupling becomes progressively more efficient from the bare interface ($\varepsilon/\sigma = 0$) to the maximal extent when the reduced elasticity reaches a value of 0.2. The corresponding change on A/W comes to almost 200%.

Instrument and Calibration: The instrument setup is illustrated in a block diagram as shown in Fig. 2A. A Langmuir trough made of Teflon with the inside dimensions of 28 x 11 x 1.25 cm^3 was placed horizontally inside a Plexiglas box. The glass tubing coil is placed at the bottom to control temperature to within 0.1°C by circulating a heat exchanging liquid through the coil. Both liquid and vapor temperature and humidity inside the box were monitored during the experiment. A platinum Wilhelmy plate, 2.5 cm long, 1.0 cm high, and 0.1 mm thick attached to a Cahn 2000 microbalance, is used to determine simultaneously static surface tension as well as any variations of the surface tension due to contamination or evaporation of the liquid during a scattering experiment. It is also a sensitive tool for testing the purity of the sample.

The optical configuration of the heterodyne light scattering is illustrated in Figure 2B. The light source, a single-mode 7-mW He-Ne laser is so oriented that the polarization of the electric field lay perpendicular to the plane of incidence. The beam was guided by (1/10)λ mirrors to a position incident on the liquid surface about 5 cm from the edges of the trough. If an incident spot on the liquid is not perfectly flat, a serious error in the measurement of the central frequency may arise. The incident angle is fixed to $\theta = 64.36 \pm 0.07°$ and the beam diameter on the surface was ≈0.4 mm.The pinhole in front of PMT was adjusted large enough to pass the entire portion of a chosen diffraction order. The whole apparatus is assembled on a concrete table weighing 2 tons, which in turn is supported by large coil springs from a steel frame. This configuration has been shown to filter out the noises in acoustic frequency range relevant to our experiment (28). A typical run accumulates about 1000 counts and has the ratio of peak intensity to the background intensity of 2 to 6 depending on the diffraction order. All data accumulations are controlled by microcomputers, and the final spectral analysis is performed on a VAX computer.

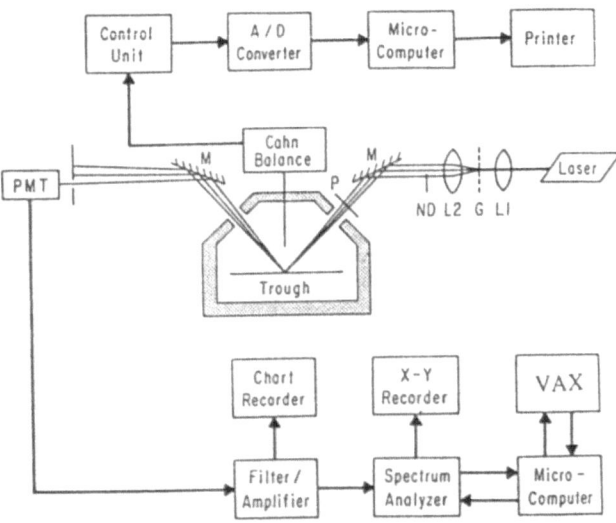

Figure 2: (A) Block diagram of the SLS instrument with the Wilhelmy plate film balance.

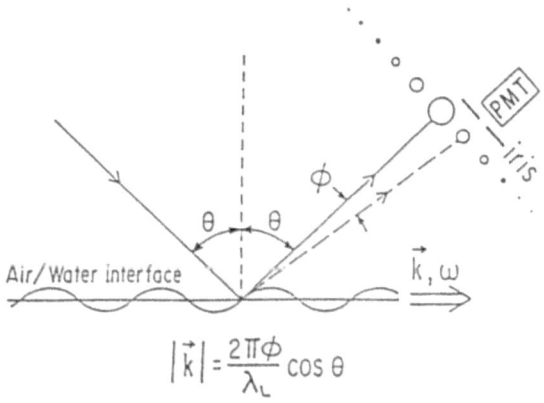

$$\left|\vec{k}\right| = \frac{2\pi\phi}{\lambda_L}\cos\theta$$

Figure 2: (B) Optical configuration of the scattering geometry for the SLS instrument with an optical grating giving arise to a diffraction pattern at the plane of PMT; θ is the incident angle normal to the liquid plane, φ the scattering angle, λ_L the incident laser wavelength, and ω and |k| (=k) are the capillary wave frequency and the magnitude of scattering wave vector, respectively.

For the calibration standards, we chose water, anisole and ethanol. Since f_S depends strongly on the scattering wave vector k, it was crucial to determine the value of k to within 0.1%. We have calculated k ≐ 260.6, 323.5, and 386.1 cm^{-1} for the 4th, 5th, and 6th diffraction order of the optical grating, respectively. The calibration results are detailed elsewhere(30), so that we show here only the limiting behaviors of pure liquid surfaces. In Fig. 3, we plot f_S and $\Delta f_{S,c}$ as dictated by the Kelvin and Stokes limits, Equations (1) and (2), in order to show how the two variables scale with k. The standard deviations of σ and ν are < 0.1% and 5%, respectively, which are associated with the measurements of f_S and f_{exp}.

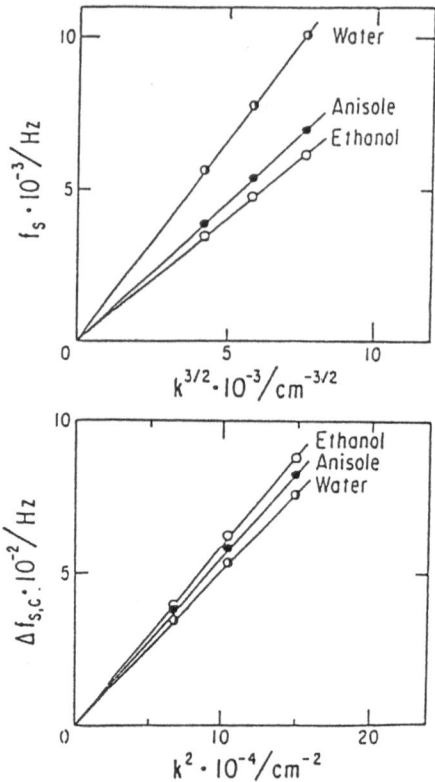

Figure 3: Dispersion relation for the three standard liquids, showing that Kelvin and Stokes limits are followed to good approximation.

Surface Steady Shear Viscosity

Interfacial systems, like bulk matter, undergo deformation and flow in response to an applied stress and possess the basic rheological properties such as viscosity and elasticity that have been thoroughly characterized for bulk phases. One surface rheological parameter that can be directly measured for monolayer films at the air/water interface is the in-plane steady shear viscosity (η_s). This quantity has been measured by many different techniques in the past, and the whole subject of surface rheology including experimental, instrumental and theoretical developments has been reviewed(29-31). Briefly, we describe the construction of a canal surface viscometer based on the original design of earlier workers(32-35) but modified to overcome significant artifacts and to provide consistently reproducible results. The technique involves forcing a monolayer to flow through a canal under a constant surface pressure head and measuring the area flow rate. The instrument is analogous to a constant driving head capillary viscometer used for the steady shear viscosities of bulk liquids.

Some details of our canal surface viscometer and validation of the working instrument with representative monolayers will now be discussed(36). In the course of testing this instrument, we have compared two distinct designs of a glass-paraffin canal and a Teflon canal. The former design was among the earliest used(35) and is the basis of a few later studies using this technique(37-39). We found that the glass-paraffin canal design can introduce artifacts into surface viscosity determinations, and that the resulting anomalous behavior is no longer seen with the new Teflon canal. From this study it appears that the problems associated with the glass-paraffin canal, in large part, account for the discrepancy in η_s magnitudes obtained with the two canal designs.

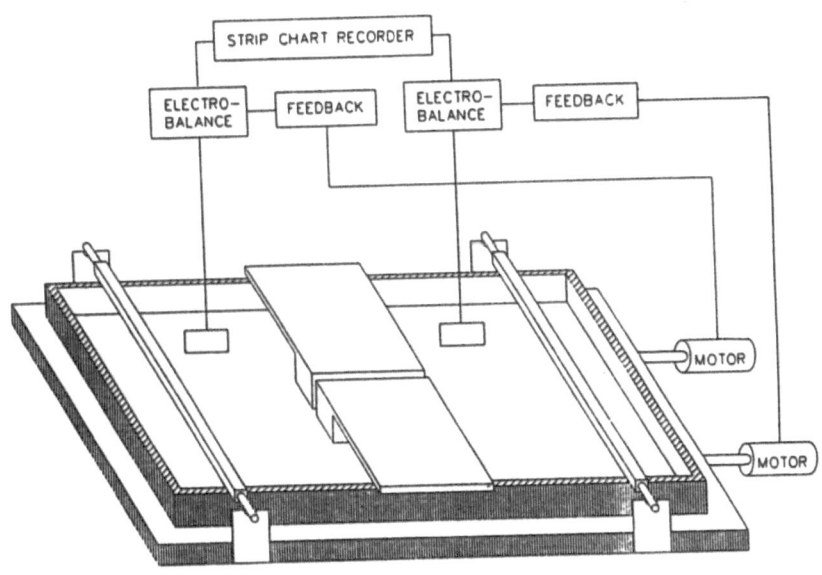

Figure 4: A block diagram of the surface canal viscometer, whose construction and calibration details will appear in Rev. Sci. Instrum., 1993

Instrument and Calibration: A schematic representation of the instrument is shown in Fig 4. The mechanical apparatus consists of a Langmuir trough made of Teflon with the inner dimensions of 61.6cm (length) x 11.9cm (width) x 1.8 cm (height), mounted in an aluminum container; on the front of which is attached a linear scale graduated in millimeters (not shown).The trough and container are mounted on an aluminum base plate, underneath which are two worm screws that produce rectilinear motion of the two Teflon barriers.These barriers serve as pistons to confine the monolayer to a particular area. In the center of the trough is the canal. For the Teflon canal, a step is machined into the canal wall to produce a sharp edge and promote a 90° contact angle when the aqueous subphase is filled up to the top surface; without this step, the air/water interface was found to form a curved, concave-down meniscus for sufficiently small canal widths, which is undesirable. The canal width was made to be variable for a range of 1.00 mm to 2.00 mm(± 0.05 mm) in order to check for shear rate dependence of the area flow rate.

A feedback electronic circuit was used to maintain the surface pressure constant to within 0.1 dyn/cm on each side of the canal during monolayer flow. As the monolayer flows in the direction depicted, the surface pressure (Π) on the right side will decrease. One can maintain Π constant by moving the right barrier in the indicated direction. A given surface pressure corresponds to a corresponding output voltage from the electrobalance which in turn serves as the input to the feedback system to maintain the voltage constant. The feedback system accomplishes this operation by sending the appropriate current to the armature of a DC motor, generating a torque which turns the worm screw and hence moves the barrier. In addition, as monolayer flows into the left compartment, there is an increase in Π on that side which necessitates a second feedback system if the left compartment Π is to be kept constant during flow.

For the parallel plane wall geometry of the canal viscometer design, Harkins and Kirkwood(41) developed a hydrodynamic theory to relate the in-plane steady shear viscosity (η_s) to observable quantities. The result of the derivation in an approximate form valid for η_s greater than 10^{-4} surface poise (sP = 1 g/s and millisurface poise, msP = 10^{-3} g/s), in the limit of a canal depth much greater than the width, is

$$\eta_s \approx \frac{\Delta\Pi w^3}{12LQ} - \frac{w\eta}{\pi} \tag{5}$$

where $8/\pi^4$ is approximated as $1/12$, $\Delta\Pi$ is the (positive) surface pressure gradient across a canal, w the canal width, L the canal length, Q the area flow rate and η the subphase steady shear viscosity. The subphase viscosity in the second term of Eq.5 appears since as the monolayer acquires momentum there is momentum transfer down into the subphase. Thus, the subphase exerts a drag stress on the monolayer which, for a given $\Delta\Pi$, L and w, slows down the area flow rate Q. Accounting of this hydrodynamic coupling of a monolayer with its subphase is achieved by solving the linearized Navier-Stokes equations for the subphase and assuming the boundary conditions of no-slip at the walls and floor of the canal with a stress balance (under steady-state conditions) at the monolayer covered air/water interface (41-43). We have recently provided(44) an experimental confirmation of the hydrodynamic coupling by examining the two limits of Eq.5, i.e.,the large surface viscosity limit with the second(coupling) term still finite and small surface viscosity limit when the first term in the right hand side is equal to the second, thus Q is now independent of η_s but inversely proportional to η. While there has been an absence of experimental test of Eq. 5 until our efforts, we are now prepared to accept its validity including the hydrodyanmic coupling term.

The primary experimental quantity here is the surface area flow rate Q, which is determined by monitoring the linear displacement of the pressing barrier as a function of time and multiplying the barrier displacement by the width of the trough (11.9 cm). Typically, a fixed distance for the barrier travel is chosen, and the time elapsed in moving that distance is monitored in successive intervals during the run to assess the steadiness of the flow. Normally, 10 to 30 time measurements can be made in a single run and the average time. To estimate the uncertainty in the steady state area flow, we have taken one standard deviation of the average time during the run, which ranged from 5 to 15%.

RESULTS AND DISCUSSION

Surface Light Scattering

We now turn to a set of selected examples of polymers together with the relevant discussion. The chosen set is a series of six homopolymers at A/W (45) in the order of increasing hydrophobicity, poly(ethylene oxide) (PEO), polytetrahydrofuran (PTHF), poly(vinylacetate)(PVAc),poly(methyl acrylate) (PMA), poly(methylmethacrylate)(PMMA) and poly(t-butyl methacrylate) (PtBMA). The second part deals with a poly(ethylene oxide)-polystyrene block copolymer (PEO-PS) at an oil/water interface(46).

Homopolymer Monolayers at A/W. The first four polymers form liquid-like monolayers which are ideally suited for SLS because of their intermediate elasticities. We present the results in the following order. First, we will show the static surface pressure profiles with respect to area per unit mass (Π-A) for four polymers and present how the two types of monolayers, the expanded and condensed, differ in the profiles. Next, we present the spectral shift f_s and width $\Delta f_{s,c}$ profiles with respect to the surface concentration Γ for all six polymers and distinguish them relative to the type of monolayers. For all three dependent variables, the qualitative features relative to A, surface area per unit mass, or Γ are delineated here. For more quantitative analyses of the deduced viscoelastic parameters, we will defer to a later part of this chapter.

The surface pressure Π is plotted as a function of A (Π-A isotherm) for PTHF, PMA, PMMA, and PtBMA monolayers in Figure 5. PTHF and PMA monolayers are of the expanded type where a finite value of Π is observed over a wide range of A. In general, Π gradually increases with decreasing A and finally shows a relatively steep increase before attaining a plateau value. This type of Π-A isotherm is similarly observed with PVAc and PEO monolayers, which have been reported as early as in 1970 by Shuler and Zisman(47).

Contrary to those of PEO, PTHF, PVAc, and PMA, the Π-A isotherms for PMMA and PtBMA monolayers have sharper increases in Π as A decreases compared to the expanded monolayers. This is commonly ascribed to the presence of strong lateral interchain interactions. Starting at large areas, there is almost no increase in surface pressure until the abrupt increase in Π begins, whereby such a Π-A isotherm is referred to as a condensed isotherm. The limiting areas extrapolated to $\Pi = 0$ through the linear portion of the Π-A isotherms are smaller for these two polymers than those of expanded monolayers.

Without dwelling on details, we just present the final results of the dilational elasticity ε and coresponding viscosity κ. Avoiding for the moment the Γ dependences of ε and κ for the 6 polymers, the results are summarized in Table 1 where A_0, the limiting area per unit mass, (with precision of ± 0.01 m^2/mg), is contrasted to $A_{0,m}$, the limiting surface area per monomer unit

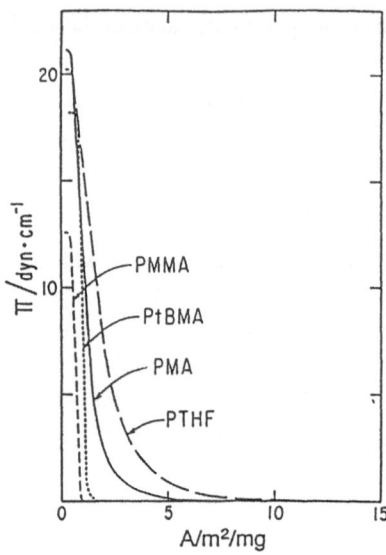

Figure 5: Surface pressure-area(Π-A) isotherms for two polymer monolayers of the condensed type, PMMA and PtBMA, and another two of the expanded type, PTHF and PMA, all at 25°C.

Table 1: Summary of monolayer parameters at 25°C.

Polymer	A_0 (m^2/mg)	$A_{0,m}$[a] $(\mathring{A}^2/mon)$	δ[b] $(cal/cm^3)^{1/2}$	$\Gamma(\varepsilon_{s,max})$[c] $(\times 10^5 \ mg/cm^2)$	$\Gamma(\varepsilon_{max})$[d] $(\times 10^5 \ mg/cm^2)$	$\Gamma(\Delta f_{max})$[e] $(\times 10^5 \ mg/cm^2)$		$\Gamma(\kappa_{max})$[f] $(\times 10^5 \ mg/cm^2)$
						1	2	
PEO	4.16	30.4	17.7	4	4	4	—	—
PTHF	2.90	34.7	16.5	5	5	3	10	—
PVAc	1.73	24.7	15.8	9.5	9.5	5	16	11
PMA	1.67	23.8	16.0	9.0	9.0	5	13	10
PMMA	0.95	15.8	15.5	13	—	6	—	—
PtBMA	1.14	26.9	14.6	10	—	8	—	10

[a]Limiting area per monomer unit, calculated from A_0

[b]Mean solubility parameters of a polymer segment on the water surface are calculated from the geometric mean $(\delta_{water}\delta_{polymer})^{1/2}$, where δ_{water} and $\delta_{polymer}$ are the solubilities of water and the polymer segment, respectively.

[c]Surface concentration of the polymer when its static Gibbs elasticity is a maximum.

[d]Surface concentration of the polymer when its static longitudinal elasticity is a maximum.

[e]Surface concentration of the polymer when its corrected spectral wideth $\Delta f^{s,c}$ reaches a peak; columns 1 and 2 list the Γ-values of the first and second (where present) peaks.

[f]Surface concentration of the polymer when its longitudinal viscosity is a maximum.

(with precision of ± 0.4 Å^2/monomer), by listing them side by side. Simultaneously, we indicate the progressive decrease of the geometric mean of solubility parameter relative to that of water, a measure of hydrophilicity. In the last four columns, we list the observed surface concentrations when one of the dependent variables maximizes.

Block Copolymer at O/W. Finally we present a study of PEO-PS block copolymer at the heptane/water interface as an example of O/W(46). The sample was oligomeric diblock copolymer synthesized anionically with the degrees of polymerization of blocks to be on the average 34 and 25, respectively, for PEO and PS segments. for the sake of brevity, we do not show the experimental results here. For comparison sake, we have also performed SLS measurements at A/W though we do not present them here but refer readers to the original paper(46). We summarize the findings as follows. Unlike those at A/W, f_s at O/W has a very small range of invariance with Γ at low range and in fact f_s decays immediately and monotonically with Γ, which is an indication of an enhanced surface activity of the copolymer at O/W. A second point of the contrast has to do with only a slight hint of the wave mode coupling between the longitudinal and capillary waves. Most importantly, the absence of the coupling indicates that ε and κ have alomst no influence on the the capillary wave propagation, thus ε and κ cannot be determined at the expense of making possible precise determinations of the dynamic surface tension σ and the transverse viscosity μ (without making any assumptions about them). The values of σ so determined turn out to be close to the static values by the Wilhelmy plate method and the transverse viscosity μ is found to be zero within an experimental error of $\pm 5 \times 10^{-6}$ g/s(surface stoke) over the the whole range of Γ. Thus this in part lends support to our starting assumption of μ being zero for A/W although we must offer a caveat that this invites an unproven extension of what is found in O/W to A/W, however plausible it may be.

Surface Canal Viscosity

First of all we present the Π-A isotherms for the various molecular weight samples of PtBMA in Fig. 6. Note that at 25.0 °C, when $\Pi > 1$ dyn/cm, the film is in the semi-dilute state as per $\Pi \propto A^{-y}$ independent of molecular weight, in analogy with the osmotic pressure of a bulk semi-dilute solution scales with polymer concentration, independent of molecular weight. As has been known earlier(44), PtBMA monolayers on A/W at 25°C can be regarded as in close to a theta condition. We determine the y value using all of the molecular weight data to be 15.0 ± 1.3 (std. dev.).

Next, we consider the quantitative relation between η_s and M for PtBMA for the data shown in Fig. 7. Note that each line in this graph is an isobar in that we compare the different molecular weight data at the same surface pressure, which amounts to essentially the same surface monomer density. One notes here that in almost 3 decades of molecular weight, the power law behavior of η_s relative to molecular weight is generally well followed. There may be some curvature for the lower isobars. In any case, for $\Pi \geq 4.5$ dyn/cm the power law is amply established for the entire molecular weight range, $10 < M(\text{kg/mol}) < 3400$; $\eta_s \propto M^\beta$, with the exponent β collected in Table 2, where it may be seen that a small dependence of β on Π exists.

346

The 95% confidence limits, however, are too large to attribute the small increase in β to be statistically significant.

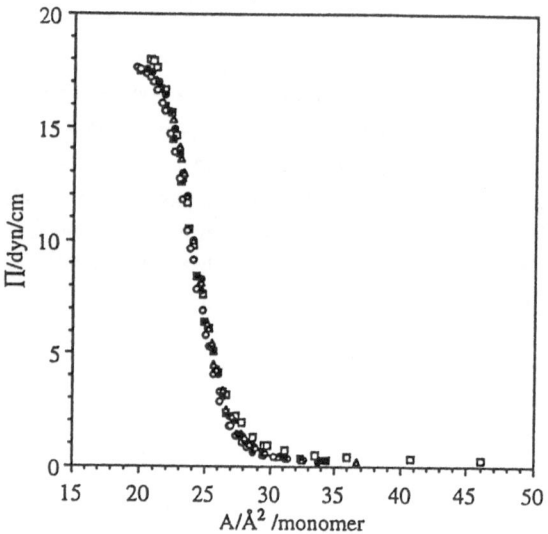

Figure 6: Surface pressure-area isotherms of PtBMA at 25°C at different molecular weights, 10 - 3.4 x10³ kg/mol

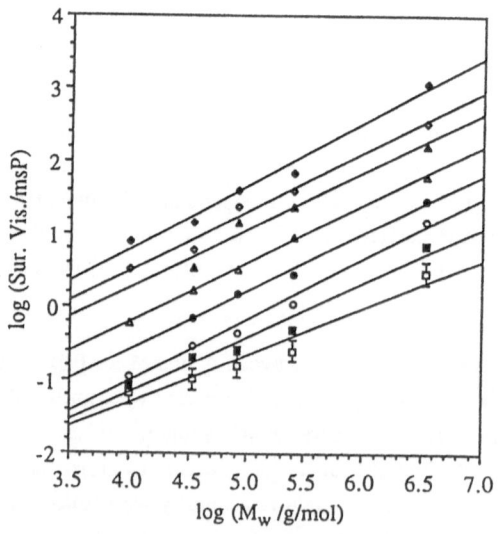

Figure 7: ηs vs. M at different P

Table 2: Molecular weight exponent β for the surface steady shear viscosity, $\eta_s \propto M^\beta$ at different surface pressures.

$\Pi/(dyn/cm)$	β [a)]
0.5	0.60 ± 0.16
1	0.69 ± 0.07
2	0.74 ± 0.16
4.5	0.71 ± 0.07
6	0.76 ± 0.03
8	0.79 ± 0.07
10	0.80 ± 0.05
12	0.84 ± 0.06

[a)] All $\pm$ values are 95% confidence intervals for $n = 4$ or 5.

The results in Fig. 7 encompass the surface pressure range $0.5 < \Pi(dyn/cm) < 12$, which is to say that the surface viscosity results are for the most concentrated 2-dimensional polymer solution we can form before the film collapses. In monolayers, we see the exponent β to be in the range $0.60 < \beta < 0.84$ all in the semi-dilute regime, which close in magnitude to the value 1 observed in Rouse-type dynamics of concentrated polymer solutions in 3 dimensions. Therefore, the rheological results for PtBMA when confined to 2 dimensions suggest that the molecules, at the various Π values, do not exhibit any entanglement effects, but appear to remain segregated even though $M > M_e$ and the solution is concentrated.

ACKNOWLEDGMENT

This is a brief summary of the work contributed by many of my former students and current collaborators including Prof. George Zografi of the School of Pharmacy of the University of Wisconsin, Prof. Masami Kawaguchi of Mie University, Drs. Masahito Sano, Bryan B. Sauer, Mark Sacchetti, Yen-Lane Chen, Mehran Yazdanian, Randy J. Skarlupka, and Taihyun Chang. Support for these studies has been generously provided by the U.S. National Science Foundation, Eastman Kodak Company and Procter and Gamble Company. Partial support of the Donors of the Petroleum Research Fund of the American Chemical Society is also acknowledged.

348

References:

1 Swalen JD, Allara DL, Andrade JD, Chandross EA, Garoff S, Isrealachvili J, McCarthy TJ, Murray R, Pease RF, Rabolt JF, Wynne KJ, Yu H (1987) Langmuir 3:932
2 Adv Mater Special Issue: Organic Thin Films 3
3 McCord MA, Pease RFW (1986) J Vac Sci Technol B 4:86
4 Ulman A (1991) An Introduction to Ultrathin Organic Films, Academic Press, New York
5 Day DR, Ringsdorf H (1978) J Polym Sci, Polym Lett Ed 16:205
6 Day DR, Lando JB (1980) Macromolecules 13:1478
7 Dorn K, Klingbiel RT, Specht DP, Tyminski PN, Ringsdorf H,O'Brien DF (1984) J Am Chem Soc 106:1627
8 O'Brien K, Long J, Lando JB (1985) Langmuir 1:514
9 des Cloiseaux J (1975) J Phys (Leo Ulis Fr) 36:1199
10 Daoud M, Jannink G (1976) J Phys (Leo Ulis Fr) 37:973
11 de Gennes P-G (1979) Scaling Concept in Polymer Physics, Cornell University Press, Ithaca NY; (1981) Macromolecules 14:1637
12 Gaines, Jr GL (1966) Insoluble Monolayers at Liquid-Gas Interfaces, Interscience Publishers, New York NY
13 Douglas JF, Cherayil BJ, Freed KF (1985) Macromolecules 18:2455
14 Langevin D (1981) Colloid Interface Sci 80:412
15 Hård S, Löfgren H (1977) J Colloid Interface Sci 60:529
16 Katyl RH, Ingard U (1968) Phys Rev Lett 19:64; (1967) ibid 20:248
17 Lucassen-Reynders EH, Lucassen J (1969) Adv Colloid Interface Sci 2:347
18 Sano M, Kawaguchi M, Chen Y-L, Skarlupka RJ, Chang T, Zografi G, Yu H (1986) Rev Sci Instrum 57:1158
19 Chen Y-L, Sano M, Kawaguchi M, Yu H, Zografi G (1986) Langmuir 2:349
20 Sauer BB, Chen Y-L, Zografi G, Yu H (1988) Langmuir 4:111
21 Levich VG (1962) Physicochemical Hydrodynamics, Prentice Hall, New York NY
22 Stanley HE (1971) Introduction to Phase Transitions and Critical Phenomena, Oxford University Press, Chapter 13
23 Kohler F (1972)The Liquid State, Verlag Chemie, Düsseldorf, Chapter 9
24 Kramer L (1971) J Chem Phys 55:2097
25 Goodrich FC (1962) J Phys Chem 66:1858
26 Sauer BB, Chen Y-L, Zografi G, Yu H (1986) Langmuir 2:683
27 Löfgren H, Neumann RD, Scriven LE, Davies HJ (1984) J Colloid Interface Sc, 98:175
28 Shaya SA, Han CC, Yu H (1974) Rev Sci Instrum 45:280
29 Joly M (1972) In: Matijevic E (ed) Surface and Colloid Science. Interscience Vol 5, p 1, New York
30 Goodrich, FC (1973) In: J F Danielli, Rosenberg MD, Cadenhead DA (eds) Progress in Surface and Membrane Science. Vol 7, p 151. Wiley, New York
31 Edwards DA, Brenner H, Wasan DT (1991) Interfacial Transport Processes and Rheology, Butterworth-Heinemann, Boston
32 Joly M (1937) J Phys Radium 8:471
33 Joly M (1939) Kolloid-Z 89:26
34 Joly M (1938) J Phys Radium 9:345
35 Myers RJ, Harkins WD (1937) J Chem Phys 5:601
36 Sacchetti M, Yu H, Zografi G (in press) Rev Sci Instrum
37 Nutting GC, Harkins WD (1940) J Am Chem Soc 62:3155
38 Jarvis NL (1965) J Phys Chem 69:1789
39 Adamson AW (1990) Physical Chemistry of Surfaces, Wiley, New York
40 Gaines GL (1966) Insoluble Monolayers at Liquid-Gas Interfaces, Wiley, New York
41 Harkins WD, Kirkwood J G (1938) J Chem Phys 6:53
42 Hansen RS (1959) J Phys Chem 63:637
43 Sacchetti M (1990) MS Thesis, University of Wisconsin-Madison, Madison
44 Sacchetti M, Yu H, Zografi G (in press) J Chem Phys
45 Kawaguchi M, Sauer BB, Yu H (1989) Macromolecules 22:1735
46 Sauer B B, Yu, H, Tien C-f,Hager DF (1987) Macromolecules 20:393
47 Shuler RL, Zisman W A (1970) J Phys Chem 74:1523

Author Index

Subject Index

Springer-Verlag
and the Environment

We at Springer-Verlag firmly believe that an international science publisher has a special obligation to the environment, and our corporate policies consistently reflect this conviction.

We also expect our business partners – paper mills, printers, packaging manufacturers, etc. – to commit themselves to using environmentally friendly materials and production processes.

The paper in this book is made from low- or no-chlorine pulp and is acid free, in conformance with international standards for paper permanency.